PROCEEDINGS OF THE FIFTH CONFERENCE ON MECHANICAL BEHAVIOR
OF SALT, MECASALT V / BUCHAREST, ROMANIA / 9-11 AUGUST 1999

Basic and Applied Salt Mechanics

Edited by

N.D.Cristescu
Department of Mechanics and Engineering Science, University of Florida, USA

H.R.Hardy, Jr.
Department of Energy and Geo-Environmental Engineering, The Pennsylvania State University, USA

R.O.Simionescu
Department of Mechanics, Bucharest University, Romania

A.A. BALKEMA PUBLISHERS LISSE/ ABINGDON/ EXTON (pa)/ TOKYO

Cover Photograph

Three cubic specimens of salt deformed to failure at different stress geometries in the "true triaxial rig" at the BGR. Original size: 53 mm edge length.

Photograph provided by Udo Hunsche, Bundesanstalt fur Geowissenschaften und Rohstoffe
(BGR), Hannover, Germany.

ISBN 90 5809 383 2

Published by: A.A. Balkema, a member of Swets & Zeitlinger Publishers
www.balkema.nl and www.szp.swets.nl

Printed in the Netherlands

Basic and Applied Salt Mechanics, Cristescu, Hardy, Jr & Simionescu (eds)
© 2002 Swets & Zeitlinger, Lisse, ISBN 90 5809 383 2

Conference organization

Organizing Committee
N.D. Cristescu (Chairman) – University of Florida
H.R. Hardy, Jr. (Co-Chairman) – Pennsylvania State University
R.O. Simionescu (Co-Chairman) – University of Bucharest
C. Cristescu (Secretary) – University of Florida

Local Organizing Committee
University of Bucharest
R.O. Simionescu (Chairman)
I. Mihailescu, Rector
D. Massier
S. Roatesi
IPROMIN, Bucharest
E. Medves (Secretary)
SALROM, Bucharest
P. Reisz, Director
C. Stoicescu

International Advisory Committee
M. Aubertin (Canada)
M. Ghoreychi (France)
F.D. Hansen (USA)
U. Hunsche (Germany)

Basic and Applied Salt Mechanics, Cristescu, Hardy, Jr & Simionescu (eds)
© 2002 Swets & Zeitlinger, Lisse, ISBN 90 5809 383 2

Foreword

The international Conferences on the Mechanical Behavior of Salt have a long tradition. This conference series was initiated by Professors Reginald Hardy and Michael Langer, the first conference being held in 1981 at Pennsylvania State University, USA. The second Conference was organized in 1984 in Hannover, Germany, the Third in Palaiseau, France in 1993, and fourth in Montreal, Canada in 1996. The present conference, the fifth of the series, was held in Bucharest, Romania, August 9-11, 1999.

The 5th Conference was organized by The University of Bucharest, Romania, and The University of Florida, USA, with the support of Sandia National Laboratories USA, and SALROM (Autonomous Administration of Salt), Romania, and with the collaboration of The Pennsylvania State University, USA, the Romanian Academy, the Romanian Association for Engineering Geology, and the Romanian National Agency of Science, Technology and Innovation (ANSTI).

The conference brought together mining engineers, researchers, and university professors interested in the mechanical behavior of salt. An important aspect of this Conference, as compared with the previous ones, was a strong participation by Central and East European scientists and engineers. The conference, held in the Aula Magma hall of the University of Bucharest, was officially opened on August 9 by a series of brief presentations from the following:

N.D. Cristescu, Chairman MECASALT5, Professor, University of Florida, USA
Ioan Mihailescu, Rector of the University of Bucharest, Romania
Viorel Barbu, Vice President of the Romanian Academy, Romania
W.M. Phillips, Dean of the Graduate School, University of Florida, USA
H.R. Hardy, Jr., Co-Chairman MECASALT5, Professor, The Pennsylvania State University, USA
R. Reisz, Director SALROM (Autonomous Administration of Salt), Romania.

The technical presentations continued through till noon on August 11. At that time, the local organizing committee provided special eyeglasses and an optimum location for viewing the total solar eclipse which occurred in the Bucharest area.

On August 12, a special two-day technical/tourist field trip for conference participants and their families, fully supported by SALROM (Autonomous Administration of Salt), left Bucharest by bus. On the first day there was a visit to the Slanic Prahova Salt Mine, and after crossing the Carpatian Mountains, visits to the towns of Brasov and Sighisoara. Supper and accommodations were provided in the town of Targu Mures. The next day, the group visited the salt lakes at Sovata, and the salt mine at Praid. On the way back to Bucharest, there was an opportunity to view the plains of

Transilvania and Bran Castle, where according to tradition, "Dracula" had spent some time. The tour arrived in Bucharest late on the evening of August 13.

The conference included ten technical sessions. A total of 94 participants from 12 countries (Bulgaria, Canada, France, Germany, Netherlands, Norway, Romania, Spain, Sweden, USA, Ukraine, and United Kingdom) contributed to the conference activities. Due to the ongoing military activities in Yugoslavia, a number of participants were unable to attend in person. The present proceedings include 44 full papers and summaries that clearly describe the main findings and ongoing studies of the different research groups attending the Bucharest conference.

We take the opportunity to thank all those who helped to organize and provide support to this highly successful event, namely:

- The international advisory committee.
- The local organizing committee, and in particular Professor D. Massier and Dr. Medvés, for a high level of assistance.
- P. Resiz and C. Stoicescu directors of SALROM.
- Dr. F.D. Hansen – For the support of the Conference by SANDIA.
- Professor M. Mihailescu, Rector of the University of Bucharest.
- Professor E. Simion, President of the Romanian Academy.
- Professor W. Phillips, Dean of the Graduate School, University of Florida.

Special thanks are due to Dr. Cornelia Cristescu, the secretary of the organizing committee, for her dedication and diligence in sending letters and E-mails to the authors, checking and retyping some of the short and extended summaries and assisting in preparing the papers for the Conference proceedings. Final preparation and assembly of the conference proceedings was undertaken at The Pennsylvania State University. The assistance of Mrs. Donna Baney in this phase of the project is gratefully acknowledged.

Tentatively the next salt conference, MECASALT6, will take place in Carlsbad, New Mexico, at the site of WIPP project. At the conclusion of the conference, Dr. F.D. Hansen, Sandia National Laboratories, described the Carlsbad area and the WIPP site, and invited participants to take part in the MECASALT6 conference.

Nicolaie D. Cristescu
University of Florida
Proceedings Editor and
Conference Chairman

H. Reginald Hardy, Jr.
Pennsylvania State University
Proceedings Editor and
Conference Co-Chairman

R. Olivian Simionescu
University of Bucharest
Proceedings Editor and
Conference Co-Chairman

Basic and Applied Salt Mechanics, Cristescu, Hardy, Jr & Simionescu (eds)
© 2002 Swets & Zeitlinger, Lisse, ISBN 90 5809 383 2

Dedication to Wolfgang Dreyer

Professor Wolfgang Dreyer 1920-2000

The international community of scientists in the field of salt mechanics lost a respected and well-known member of the fraternity with the unexpected death of Professor Dreyer in November 2000 – only a few months after a scientific honorary colloquium was held on the occasion of his 80th birthday. It was a great pleasure for all the participants of this colloquium, associated with both science and industry, to see the former university lecturer, the enthusiastic and rousing scientist and the professional consultant of numerous salt-mining and salt cavern construction projects in such great physical and mental shape. His free spoken closing words of thanks, including a specific interpretation of the famous $E = mc2$ formula in his inimitable manner, somewhere between scientific seriousness and hidden roguishness, will long be remembered by all of the participants. Although no one in the audience realized it, these words were to be his farewell to friends and colleagues.

Following his study of mineralogy, gaining his doctorate in 1955 as well as his habilitation in 1966 with the venia legendi in petrography and petromechanics, Dr. Dreyer, in 1971, became a professor at the Clausthal University of Technology. In 1975, he habilitated in the field of rock mechanics, and became a university professor in soil and rock mechanics in 1978. He retired in 1985.

During his 20 years as a university teacher Dr. Dreyer gained a considerable international reputation as an expert in rock mechanics. After years of preparing numerous students for their exams, particularly in mathematics, physics and mechanics, many successful mining engineers from Clausthal still agree that their professional career would probably have been completely different without Wolfgang Dreyer.

After his first scientific studies in the field of petrophysics, the young scientist increasingly dedicated himself to salt rocks and mechanics in the early 60's. Subsequently, his interests turned to salt mining and the construction of salt caverns for the storage of liquid and gaseous energy sources such as crude oil and natural gas, a new development in Germany towards the end of the 60's. In these areas Professor Dreyer found the basis for his life's work. Here, following a mining tradition he developed model mechanics for salt rocks and consequently a dimensioning concept for salt caverns with respect to different operational phases. At that time the mechanical specifics of repository rocks had to be recorded and integrated into the dimensioning of the supporting structure without the help of highly sophisticated computer-aided calculation processes. Since the material characteristics of salt are characterized by a distinctive non-linearity it was virtually impossible to analyze such problems using simple theoretical models. Due to his understanding of the mechanical characteristics of salt rocks, gained in laboratories and underground, it was Professor Dreyer, who in Germany rapidly increased the volume of typical gas storage caverns by an order of magnitude, from some 10,000 m3 to several 100,000 m3. Thanks to his design concepts he obtained the full support of the licensing authority for the increase of storage volume capacity into previously unknown engineering territory.

Dr. Dreyer published a number of books, papers and reports. For example; "Die Festigkeitseigenschaften natürlicher Gesteine insbesondere der Salz- und Karbongesteine: ein Beitrag zur Gesteinsmechanik" (The strength characteristics of natural rocks, particularly of salt and carbon rocks: a contribution to rock mechanics), "Gebirgsmechanik im Salz" (Rock mechanics salt mass), as well as his textbook, which has also been published in English, entitled "Underground storage of oil and gas in salt deposits and other non-hard rocks". These all testify to his creative engineering activities. Later, with the increasing use of computer modeling in the field of rock mechanics, it was Professor Dreyer, who supervised the first dissertation in Germany dealing with the numerically supported dimensioning of supporting structures for salt caverns – a dissertation written in Canada and among the last to be supervised by him. It was also Professor Dreyer who recognized the innovative potential of computer-aided planning and who recommended it to the responsible authorities, asking them to consider this new process and to use the greatly extended possibilities regarding detailed proofs of safety and optimized efficiency of cavern system utilization.

A special part of Professor Dreyer's scientific work was his cooperation with East German scientists during the cold war. Initially, the procedural setting was provided by the "Internationales Büro für Gebirgsmechanik", the International Bureau for Rock Mechanics, during the 1960's. Subsequently, the East German government terminated this means of scientific exchange, and there were years of silence and mutual observation of new developments only through publications. The exchange taking place within the setting of a study group called "Salzmechanik" (salt mechanics), now also open to East German experts. Professor Dreyer was a member of this group from its foundation in the early 80's. For many years he contributed to this group and until his death he participated in their meetings, enriching the sessions over the last years by means of historical reminiscences, generous kindness and the attitude of a grand seigneur.

One of the East German scientists and fellow travelers from the earlier days described Professor Dreyer at his 80th birthday with the following words: "With his active scientific appearances in Leipzig and Freiberg (mining academy) Professor Dreyer tried to counter political confrontation with scientific cooperation. He performed his scientific work with great enthusiasm, and was always open to the problems of the people in East Germany."

Besides his wide range of activities in Germany, Professor Dreyer was also internationally well known in the field of salt mining and salt cavern construction. For example, for many years he provided rock-mechanics supervision of a cathedral in the salt rocks of Columbia and contributed to maintaining its functionality for the benefit of the local people.

Many personal friends and associates from the scientific and engineering communities will long remember Professor Dreyer for his prominent and charming personality as a miner, scientist, engineer, and university teacher. The Proceedings of the Fifth Conference on the Mechanical Behaviour of Salt are dedicated to his memory.

K.H. Lux
Technical University of Clausthal

Basic and Applied Salt Mechanics, Cristescu, Hardy, Jr & Simionescu (eds)
© 2002 Swets & Zeitlinger, Lisse, ISBN 90 5809 383 2

Table of contents

Part III – Creep/damage and dilatancy

Part IV – Constitutive modeling

Part V – Crushed salt behavior

Part VI – Numerical modeling

Part VII – Storage and disposal projects

Part VIII – Mining application

Part IX – Case studies

Part X – Salt pillars and cavities

Part I

Laboratory and in-situ testing

Basic and Applied Salt Mechanics, Cristescu, Hardy, Jr & Simionescu (eds)
© 2002 Swets & Zeitlinger, Lisse, ISBN 90 5809 383 2

Acoustic emission in salt during incremental creep tests

H.Reginald Hardy, Jr.
Rock Mechanics Laboratory, Department of Energy & Geo-Environmental Engineering,
The Pennsylvania State University, University Park, USA

ABSTRACT

Acoustic emission (AE) techniques provide a convenient tool for the study of deformation and failure of geologic materials. These range from small scale phenomena such as dislocation motion, dissolution, grain growth, and microfracture generation and propagation; to larger scale phenomena such as inter- and intra-granular displacement, compaction, and macroscopic fracture. Unfortunately these techniques have not been extensively applied in the study of mechanical behavior of salt.

The current paper will briefly describe the basic AE concepts and experimental techniques, and review the literature relative to the application of AE techniques to the study of the mechanical behavior of salt. The major thrust of the paper will be the application of the AE technique to the study of the incremental creep behavior of a number of different salt types. In these studies both short term (1×10^3s) and long term (100×10^3s) incremental loading experiments have been carried out on intact specimens, and AE and longitudinal and transverse strains have been measured. The time-dependent variation of longitudinal strain, and acoustic emission is considered along with the correlation of creep strain and AE.

INTRODUCTION

Experimental studies on a variety of materials indicate the most solids, including geologic materials emit low-level seismic events, or acoustic emissions (AE), when they are stressed or deformed (Hardy, 1981). This AE activity provides an indirect means of monitoring the internal deformation and failure characteristics of a material. A wide range of studies on "hard" rocks, during the last 20 years, indicates that AE techniques provide a convenient tool for the study of deformation and failure of geologic materials. These range from small scale phenomena such as dislocation motion, dissolution, grain growth, and microfracture generation and propagation; to larger scale phenomena such as inter- and intra-granular displacement, compaction, and macroscopic fracture. Unfortunately these techniques have not been extensively applied in the study of mechanical behavior of salt.

The current paper will consider the relationship of AE and time-dependent (creep) strain in geologic materials. Earlier, related studies on hard rocks and artificial (recompacted) salt will be briefly discussed. The major thrust of the paper will be the analysis of experimental data obtained from incremental creep tests on salt specimens from Louisiana and Texas salt domes. During these tests, axial and transverse strains were monitored by foil-type, SR-4 strain gages. AE was detected by an attached high frequency AE-transducer and total counts continuously recorded. Loading programs were established to investigate strain and AE behavior in the elastic and inelastic region, at the yield point, in the dilatancy region, and during low level incremental creep.

EARLIER STUDIES

1. Hard Rocks

In-house studies have clearly indicated that geologic materials exhibit well-defined acoustic emission (AE) activity during creep deformation. Studies reported by Hardy et al., (1970) indicate that there is a linear relationships between AE and axial creep strain for a number of different hard rocks, namely: Tennessee sandstone, Indiana limestone, and Barre granite. The associated experiments were carried out using a lever-type loading facility; strains were monitored using attached SR-4 strain gages; and AE data was monitored by an attached accelerometer, and an associated signal conditioning system and analog tape recorder. The frequency response of the overall AE monitoring system was 300 Hz to 15,000 Hz. Further details on the experimental techniques are given elsewhere (Kim, 1971).

The loading pattern and typical results for Tennessee sandstone are shown in Figure 1. The accumulated AE activity versus axial creep strain data for all three rock types were found to fit the general equation,

$$N = A + Be \qquad \text{(Eq. 1)}$$

where N is the number of AE events, e is the creep strain in μs, and A and B are constants which depend on the rock type and stress level. Table 1 lists the associated data for the three hard rock types studied.

2. Artificial Salt

At Penn State, Roberts (1981) investigated creep and AE activity in artificial salt, recompacted ground salt used for agricultural purposes. Details of the composition of this material (denoted here as salt type S15) are given in the appendix. Further details regarding various mechanical properties of this material are presented in Hardy (1982).

4

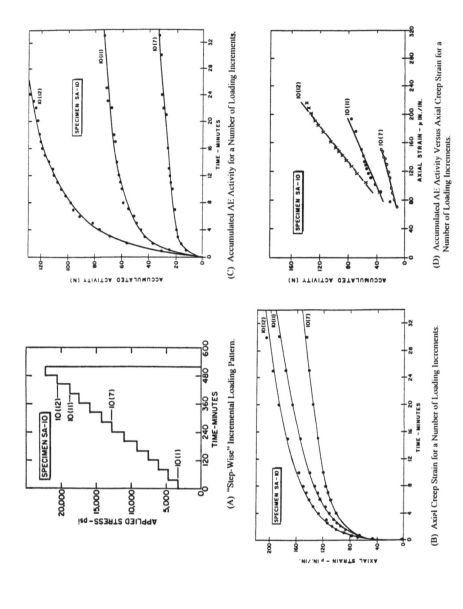

Figure 1 - Data from Incremental Creep Tests on Tennessee Sandstone (after Hardy et al., 1970). [Stress levels for load increments 10(7), 10(11) & 10(12) are given in Table 1.]

5

Table 1

Results of Least-Square Fit of Data Relating Axial Creep Strain and
Accumulated AE Activity for Three Types of "Hard" Rock
(After Hardy et al., 1970)

Rock Type	Specimen Number	INC No.	$\Delta\sigma$ psi	σ_f psi	σ_f/Co	A (1) Events	B (1) Events/μs
Sandstone	SA-10	10(7)	1,716	12,725	0.53	-4.9	0.25
		10(11)	1,292	19,001	0.79	-2.4	0.43
		10(12)	1,661	20,662	0.86	-26.9	0.80
Indiana Limestone	LI-5	5(5)	782	7,174	0.75	-12.2	1.91
		5(6)	781	7,955	0.83	- 9.6	1.65
		5(7)	782	8,737	0.91	- 2.3	1.30
Barre Granite	GR-13	13(7)	1,829	13,394	0.43	-20.0	2.63
		13(10)	1,619	18,414	0.59	-11.33	1.95
		13(14)	1,599	24,956	0.79	-6.2	1.49
		13(15)	1,718	26,674	0.85	+11.6	0.99

(1) $N = A+Be$, where N is total AE events for a specific stress increment, e is the associated creep strain in μs. A and B are coefficients, where in particular B is the number of AE events per μs of creep/strain.

(2)   Co is the ultimate compressive strength of the various rock types.

Studies were carried out using the same lever-type loading facility used by Kim (1971) for studies on hard rocks; and strains were monitored by attached SR-4 gages and recorded on strip chart recorders. A major difference in Roberts (1981) experimental facilities was associated with the AE monitoring system. Here AE data was detected using a high-frequency resonant-type AE-transducer (D-140B) and a commercial AE monitoring system (3000 series) both manufactured by Dunegan/Endevco. The frequency response of the overall AE monitoring system was 100 KHz to 300 KHz. With the 3000 series a single output pulse was generated for each AE event processed or for each signal threshold crossing. Figure 2 presented later clearly indicates the difference in the two types of output. Individual event data were generated using a built-in digital envelop processor (DEP). A reset clock was also incorporated in the system to provide an analog output signal equivalent to AE rate. Figure 2 illustrates AE signals observed during creep of artificial salt. Based on the included time scales, the high frequency character of the signals is evident.

The loading pattern and typical result for creep tests on a specimen (S-14) of artificial salt are presented in Figure 3. The AE activity rate versus axial creep strain rate for a number of loading increments is plotted in log-log form in Figure 3C and seen to be

Event 1   Event 2

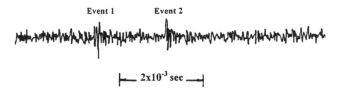

|— 2x10⁻³ sec —|

(A) AE Signals Observed During Creep.

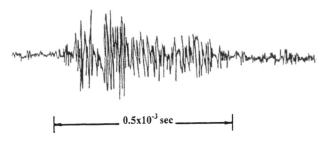

|————— 0.5x10⁻³ sec ————|

(B) Expanded Form of AE Event 2 Shown Above.

Figure 2 - AE Signals Observed During Creep of Artificial Salt (after Roberts, 1981).

statistically linear in character. Based on tests of eight specimens the data was found to fit the general equation,

$$\dot{N} = C(\dot{e})^m \qquad \text{(Eq. 2)}$$

where $\dot{N}$ is the rate of AE activity in number/sec, $\dot{e}$ is the creep strain rate in μs/sec and C and m are constants which vary from specimen to specimen. Results are presented in Table 2. Here the range of creep strain rates and AE activity rates are presented along, with the computed values of C and m. The listed correlation coefficients indicate that the quality of data fit to Equation 2 is excellent. The major difference in results for specimens where AE events (DEP active) and AE counts (DEP inactive) were involved is clearly evident.

RECENT STUDIES

1. Sample Material

Studies have been underway on specimens of two types of salt, S5 and S12, obtained from salt domes in Texas and Louisiana, respectively. Bulk specific gravity for both salts were nominally 2.13. The average grain size for both was approximately 0.2 in, however, the S5 salt appeared to have a wider range in grain size, i.e. 0.1-0.5 in. The sample materials utilized in the current study were in the form of core, ≈2.12 in. in diameter prepared earlier from large blocks obtained in the field. Future details on the two salt types are given in the appendix.

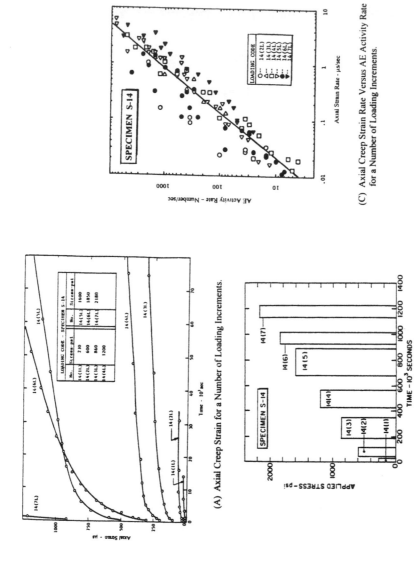

(C) Axial Creep Strain Rate Versus AE Activity Rate for a Number of Loading Increments.

(A) Axial Creep Strain for a Number of Loading Increments.

(B) "Zero-Based" Incremental Loading Pattern.

Figure 3 - Data from Incremental Creep Tests on Artificial Salt (after Roberts 1981). [Stress levels for load increments 14(1L) to 14(7L) are given in (A).]

Table 2

Results of Least-Squares Fit of Data Relating Rate of AE Activity and
Axial Creep Strain Rate for a Number of Artificial Salt Specimens
(after Roberts 1981)

Specimen Number	$\dot{e}$ Range μs/sec	$\dot{N}$ Range events/sec	C (1)	m (1)	Correlation Coefficient
S-15(2)	.01-9	0.2-50	7.5	0.889	0.987
S-17(2)	.01-5	0.1-90	35.2	1.16	0.876
S-18(2)	.01-9	0.3-40	17.0	0.845	0.993
S-21(2)	.01-5	0.5-60	25.5	0.859	0.967
S-8(3)	.01-10	0.9-7000	287.1	1.401	0.931
S-12(3)	.01-10	0.1-9000	365.5	1.406	0.907
S-14(3)	.01-8	4-8000	1186.4	1.258	0.938
S-16(3)	.01-8	5-6000	542.9	1.410	0.887

(1) $\dot{N} = C(\dot{e})^{m}$ where N is the rate of AE activity in counts/sec, $\dot{e}$ is the creep rate in μs/sec and C & m are constants.

(2) DEP-active, (3) DEP-inactive

2. Experimental Procedures

Test specimens were nominally 2.12 in. in diameter and 4.35 in. in length, and were prepared following procedures described earlier by Hardy (1982). Specimens were instrumented with SR-4 resistance strain gages, type E-06-10CBE-120, attached using M-Bond AE-10 cement. Both the gages and cement were supplied by Micro-Measurements, Inc. Two sets of three strain gages, located at 120 degree intervals around the enter of the specimens, were used to monitor axial and transverse strain. During testing, the specimens were located between special stress equalizer loading heads. These loading heads contained hard rubber inserts which also provided a degree of acoustic noise isolation to increase the signal-to-noise ratio (SNR) during AE measurements. The specimen was loaded uniaxially using an MTS programmable testing facility, and the applied load was monitored using an integral load cell. Strains and applied load were recorded using a computer-based data acquisition system. AE activity was detected by a Dunegan type D-140B transducer, coupled to the specimen with Dunegan AC-WS couplant and held in place with masking tape. AE signals were amplified by a Dunegan model 1801-190B preamplifier (40dLB) and then fed to a single channel Dunegan model 3000 system. The resulting AE channel involved a total system gain of 90dB. The accumulated AE ring-down counts were recorded by the same computer-based data acquisition system used for strain and load recording. Figure 4 shows a block diagram of the experimental arrangement.

3.   Loading Program

The experimental results presented in this paper are associated with uniaxial compressive tests on specimens S5-1 and S12-7. The "zero-based" incremental loading program used employed twenty stages of loading, load-hold, unloading and zero load rest. The loading and unloading rates were 5 psi/sec, and both the load-hold and zero load rest periods were 15 min. in duration. Each subsequent load-hold level was increased by approximately 100 psi over the preceding one, resulting in load-hold stresses of approximately 100, 200, 300, —, 1900, 2000 psi. As a result, during each loading stage the current load-hold stress level was 100 psi greater than the previous maximum stress level. In the current paper, creep and AE behavior were investigated for a number of loading increments early in the loading program.   For example, in the case of specimen S12-7, creep strain and AE activity was investigated for stress increments 1-8 as illustrated in Figure 5.

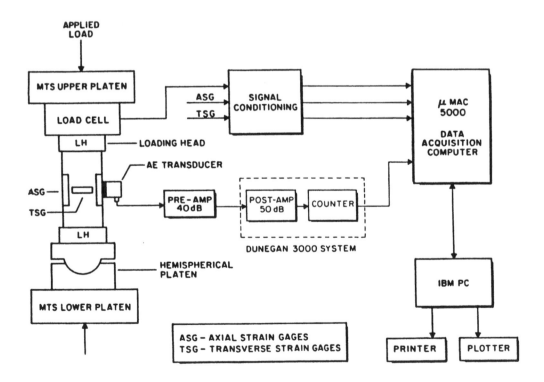

Figure 4 - Block Diagram of Experimental Arrangement Used for Acoustic Emission
          Studies on Salt.

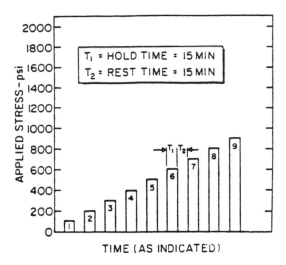

Figure 5 - Section of Overall Incremental Loading Program Over Which the
Relationship of Creep Strain and AE Activity was Investigated for
Specimen S12-7. (Data was analyzed for stress increments 1-8).

4. General Experimental Data

Each of the 20 incremental tests carried out on specimens S5-1 and S12-7 required
a total of approximately 12 hours. During this period all data was recorded on a computer-
based data acquisition system. Post-test analysis involved processing and graphical
presentation of load, strain (axial and transverse) and AE total count data for each of the
twenty load increments. For example, Figure 6 shows the graphical data for load
increments 5 & 6, specimen S12-7. Although some differences in the experimental data
were observed for stress level and salt type, the data shown in Figure 6 is generally typical.
During loading, an elastic strain component was observed at low stresses, however, it was
difficult to separate it from the inelastic component. Over the range 0-800 psi, earlier
studies indicate that the Young's modulus during unloading varies from 0.6 to 1.0 x 10^6
psi. See Hardy and Shen, 1998 for more detailed data relative to the analysis of tests on
specimen S12-7; and Hardy 1993a, 1993b for similar analysis on specimen S5-WS/1 test
data.

5. Creep and AE Results for S12-Type Salt

Detailed analysis was carried out on the creep and AE data observed during tests
on specimen S12-7 (Hardy & Shen, 1998). Graphical presentations of axial creep strain
and accumulated AE activity as a function of time are presented in Figures 7 and 8. In
each case, data is presented for five different stress increments ($\Delta\sigma$) 408, 510, 615, 720 &
810 psi. The creep strain data in Figure 7 indicates a nearly transient type behavior during
the first 100-300 seconds, followed by steady state behavior during the remaining time of

the test. It is interesting to note that the AE activity versus time data for the same stress increments, Figure 8, mimics the strain versus time data. Similar behavior has been noted for tests on hard rocks and "artificial" salt illustrated earlier (Figure 1 & 3). To investigate the apparent relationship between AE activity and axial creep strain the associated data were plotted against each other. The results are presented in Figure 9.

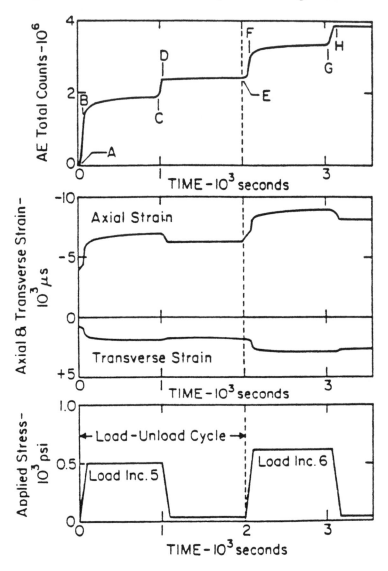

Figure 6 - Typical Graphical Data from Tests on Salt Specimen S12-7 (after Hardy & Shen, 1998). [Load increments 5 & 6.]

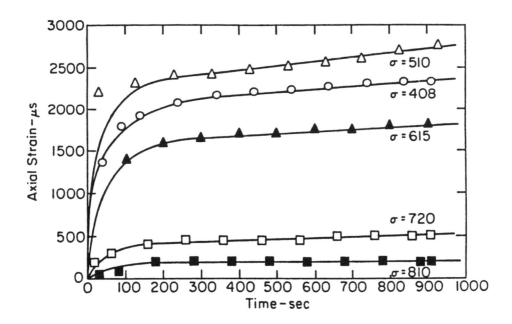

Figure 7 - Axial Creep Strain for a Number of Load Increments, Specimen S12-7.

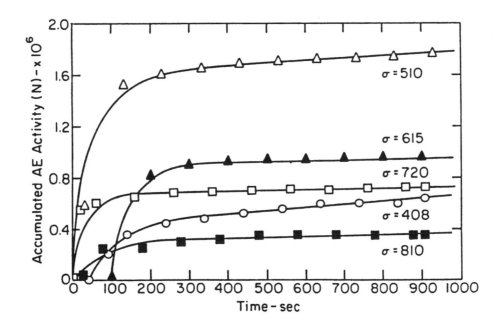

Figure 8 - Accumulated AE Activity for a Number of Load Increments, Specimen S12-7.

13

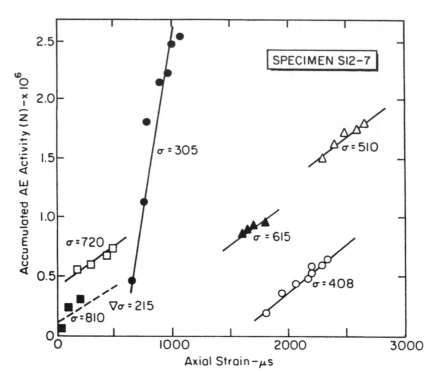

Figure 9 - Accumulated AE Activity Versus Axial Creep Strain for a Number of Stress
Increments, Specimen S12-7. [The B-factors shown in Table 3, are the
slopes of the straight line plots.]

Table 3

Results of Graphical Analysis of Data Relating Accumulated AE Activity and
Axial Creep Strain for Salt Specimen S12-7

INC No.	$\Delta\sigma$ psi	σ_f psi	σ_f/Co —	A (1) 10^6 counts	B (1) 10^3 counts/µs
3	305	305	0.16	-3.60	6.10
4	408	408	0.21	-1.30	0.80
5	510	510	0.17	-0.23	0.75
6	615	615	0.32	-0.35	0.75
7	720	720	0.38	+0.41	0.70
8	810	810	0.43	+0.11	0.63

(1) $N = A+Be$ where N is the total AE counts for a specific stress increment, e is the associated
strain in µs. A and B are coefficients, in particular B is the number of AE counts per µs of
creep strain.

(2) Co, the ultimate compressive strength, was approximately 1900 psi.

14

In general there is a clear linear correlation between accumulated AE activity and axial creep strain for tests on the S12-7 salt specimen. The results of the graphical analysis are presented earlier in Table 3. A similar linear relationship was noted for tests on hard rock specimens (Hardy et al., 1970) although the values of the B-factor were considerably smaller due to the lower sensitivity and limited frequency response of the AE monitoring system utilized.

6. **Creep and AE Results for S5-Type Salt**

Detailed analysis was carried out on creep and AE data observed during earlier tests on specimen S5-WS/1 (Hardy, 1993, 1996). These tests were carried out following procedures similar to those used for tests on specimen S12-7. Typical data are shown in Figure 10. As noted earlier for specimen S12-7, the variation of AE activity and creep strain with time for specimen S5-WS/1 are very similar. To investigate the apparent correlation further, the AE data were plotted against the creep strain data for a number of loading increments. The results are presented in Figure 11 and indicate an excellent linear relationship. Statistical analysis was carried out on the data and the results are presented in Table 4. The high level of the correlation coefficients obtained, $0.9025 \leq r \leq 0.9924$, indicates the validity of the assumed linear relationship between accumulated AE activity and creep strain.

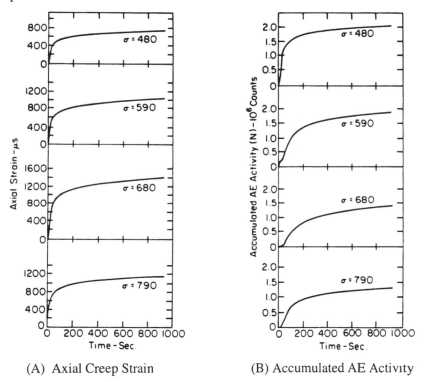

(A) Axial Creep Strain   (B) Accumulated AE Activity

Figure 10 -   Axial Creep Strain and Accumulated AE Activity for a Number of Load Increments, Specimen S5-WS/1.

15

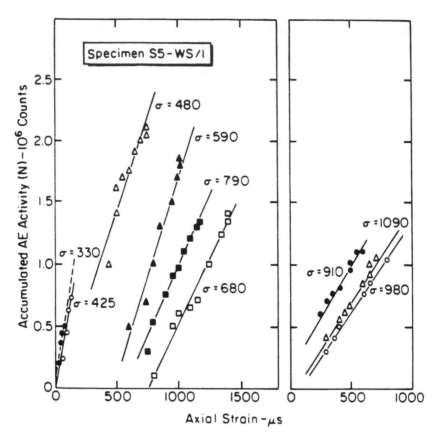

Figure 11 - Accumulated AE Activity Versus Axial Creep Strain for a Number of Stress
Increments, Specimen S5-WS/1. [The B-factors shown in Table 4 are the
slopes of the straight line plots.]

7. Comparison of Results for S5 & S12 Salt Types

Generally the two salt types behave similarly during incremental creep tests. Over
the stress range (200-1000 psi) and loading duration (1000 sec.) presented in this paper,
both salts exhibit transient creep at lower stresses changing to transient-steady state creep
at higher stress levels. In these short-term studies the observed steady state creep will be
referred to as "pseudo steady-state creep." Both salt types exhibit AE activity during
loading, constant load (creep) and unloading phases of the test program.

Figure 12 shows the pseudo-steady state creep rate versus applied stress data for
a number of load increments for the S5 and S12 salts. Also included is the stress at the so-
called 2% yield point (σ_y & σ_y'), and the stress (σ_t') at which dilation, i.e., specimen
volume increases beyond original no-stress volume, begins to occur.

Table 4

Results of Statistical Analysis of Data Relating Accumulated AE Activity and
Axial Creep Strain for Salt Specimen S5-WS/1.

INC No.	$\Delta\sigma$ psi	σ_f psi	r (2)	A (1) 10^6 counts	B (1) 10^3 counts/μs
2	210	210	0.9025	-0.25	11.58
3	330	330	0.9286	+0.06	5.99
4	425	425	0.9814	+0.00	5.69
5	480	480	0.9268	-0.03	2.90
6	590	590	0.9777	-1.52	3.24
7	680	680	0.9869	-1.51	2.04
8	790	790	0.9924	-1.38	2.50
9	910	910	0.9821	+0.15	1.66
10	980	980	0.9844	-0.10	1.62
11	1090	1090	0.9825	-0.13	1.53
12	1180	1180	0.9719	+0.19	1.52

(1)   N = A+Be, where N is the total AE counts for a specific stress increment, e is the
       associated strain in μs. A and B are coefficients, in particular B is the number of
       AE counts per μs of creep strain.
(2)   Correlation coefficient

Figure 13 illustrates the variation of the B-factors as a function of stress level
(see Tables 3 and 4), for specimens S12-7 and S5-WS/1. In the later case (Figure 13B)
the variation of an estimated B'-factor, computed from the ratio of the total number of
AE events divided by the total creep strain during each increment, is also shown.
Generally there is a reasonable correlation of B- and B'-factor values. For both salt
types, below a stress of approximately 400 psi, the values of B rise rapidly with
decreasing applied stress (6.1 - 11.6 x 10^3 counts/μs). This is a transitional region
involving both elastic and time-dependent inelastic behavior, specimen crack closure,
etc., which may be responsible for the high B values.

Above an applied stress of approximately 400 psi the values of the B-factor
become more consistent and show a decreasing level of B-factor, with linear
dependence on applied stress level. Table 5 indicates the associated data. Clearly, salt
type S5 exhibits a higher average level of AE activity than type S12 ($\overline{B}$ = 2.126 and $\overline{B}$
= 0.726 x 10^3 counts/μs respectively). Furthermore, the rate of change of B with stress
level is largest for specimen S5-WS/1 (D = 0.0024) compared with specimen S12-7
(D = - 0.0004).

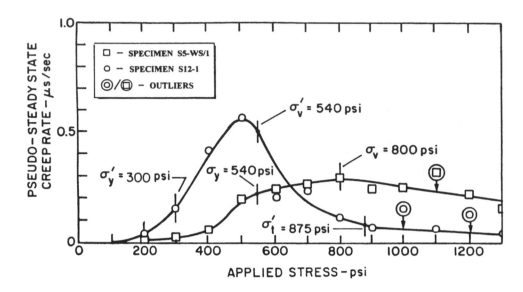

Figure 12 - Pseudo-Steady State Creep Rate Versus Applied Stress for a Number of
Load Increments for Two Different Salt Types. [Factors σ_y, σ_y', σ_v, σ_v' and
σ_t' are defined in the text.]

Table 5

Statistical Analysis of B-Value Data for Two Different Salt Types.

Specimen	Stress Range psi	C(1)	D(1)	r(2)	$\overline{B}$ (3) 10^3 counts/μs
S5 - WS/1	480 - 1180	4.15	- 0.0024	-0.8798	2.126
S12 - 7	400 - 800	1.00	-0.0004	-0.9529	0.726

(1) $B = C + De$ where B is the B-factor, the number of AE counts per μs of creep
     strain. C and D are coefficients, in particular D is the number of AE counts/μs
     per psi stress level.
(2) Correlation Coefficient
(3) Average B-value over specified stress range

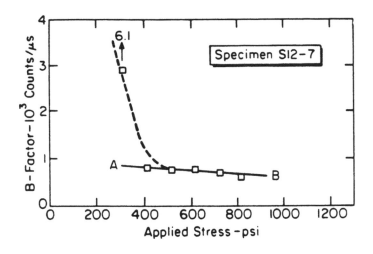

(A) Specimen S12 - 7

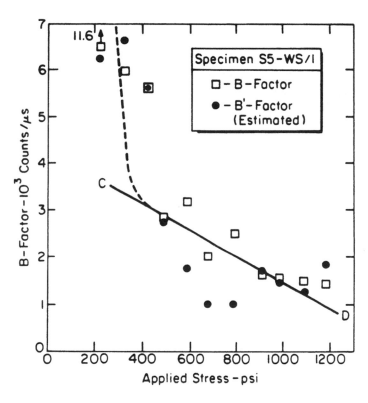

(B) Specimen S5 - WS/1

Figure 13 - Variation of B-Factor with Applied Stress for Two Different Salt Types.

DISCUSSION

The results of tests on two different salt types (S5 and S12) relative to the variation of AE activity and strain during short-period (1000 sec) incremental creep tests are presented in the previous sections. A review of the more important aspects of these tests are as follows:

(1)   Well defined AE activity occurs during all phases of the loading, constant load, and unloading sequence.

(2)   There is a well defined, correlation between strain and AE activity during the constant load (creep) phase.

(3)   The AE versus creep strain relationship is accurately described by Eq. 1, $N = A + Be$, where N is the total AE counts for a specific stress increment, e is the associated strain in µs, and A and B are coefficients. In particular B is the number of AE counts per µs of creep strain. A similar relationship has been found earlier for creep tests on "artificial salt" and on a number of types of hard rock (granite, limestone and sandstone).

(4)   Over an applied stress range of approximately 400 - 1200 psi, the B-values appear to be linearly dependent on the applied load level. Average values in this range for S5 and S12 type salts are 2.126 and 0.725×10^3 counts/µs, respectively.

(5)   Both the AE versus axial strain, and the pseudo-steady state creep rate versus applied stress data appears to be unique to the salt type. Additional experiments are necessary to validate this observation. Furthermore, future experiments are planned to investigate the variation of strain and AE activity during much longer creep tests (hours through days).

REFERENCES

Hardy, H. R., Jr., 1981, "Applications of Acoustic Emission Techniques to Rock and Rock Structures: A State-of-the-Art Review," Proceedings ASTM Symposium on Acoustic Emission in Geotechnical Engineering, Detroit, June 1981, ASTM STP 750, pp. 4-92.

Hardy, H. R., Jr., 1982, "Theoretical and Laboratory Studies Relative to the Design of Salt Caverns for the Storage of Natural Gas," Monograph published by The American Gas Association, Arlington, Virginia, A.G.A. Cat. No. L-51411, 709 pp.

Hardy, H. R., Jr. 1993, "Evaluation of In-Situ Stresses in Salt Using Acoustic Emission Techniques," Proceedings Seventh Symposium on Salt, Kyoto, Japan, April 1992, Elsevier Science Publishers, Amsterdam, Vol. I, pp. 49-58.

Hardy, H. R., Jr., 1996, "Application of the Kaiser Effect for the Evaluation of In-Situ Stresses in Salt," Proceedings 3rd Conference on the Mechanical Behavior of Salt, Ecole Polytechnique, Palaiseau, France, September 1993, Trans Tech Publications, Clausthal-Zellerfeld, Germany, pp. 85-100.

Hardy, H. R., Jr., and H. W. Shen, 1998, "Acoustic Emission in Salt During Elastic and Inelastic Deformation," Proceedings 6th Conference on Acoustic Emission/Microseismic Activity in Geologic Structures and Materials, Penn State University, June 1996, Trans Tech Publications, Clausthal, Germany, pp. 15-28, 1998.

Hardy, H. R., Jr., et al., 1970, "Creep and Microseismic Activity in Geologic Materials," Rock Mechanics - Theory and Practice, Proceedings 11th Rock Mechanics Symposium (Berkeley, June 1969), pp. 377-413, published by AIME.

Kim, R. Y., 1971, "An Experimental Investigation of Creep and Microseismic Phenomena in Geologic Materials," Ph.D. Thesis, September 1971, Department of Mineral Engineering, The Pennsylvania State University.

Roberts, D. A., 1981, "An Experimental Study of Creep and Microseismic Behavior in Salt," M.S. Thesis, March 1981, Department of Mineral Engineering, The Pennsylvania State University.

APPENDIX-Sample Material

Salt Type S5-This material was obtained from a Texas salt dome in the form of two blocks (7 x 7 x 24 in.) of approximate weight 100 pounds each. The material was massive, white, and fine grained. There was no obvious macroscopic grain orientation or major discontinuities. Grains were transparent and nearly colorless. Grain size varied from 0.1-0.5 in. with an average grain size of approximately 0.2 in. The results of limited laboratory tests indicated that the material had a uniaxial compressive strength of 2,166 psi, a bulk specific gravity of 2.13, and a porosity of 1.3%.

Salt Type S12-This material was obtained from a Louisiana salt dome in the form of two large field cores, 16 in. in diameter and approximately two ft in length. Total weight of the two cores was approximately 700 pounds. The material was massive, white, and fine grained. The grain size varied from 0.2-0.4 in. with an average grain size of approximately 0.2 in. Bulk specific gravity varied from 2.12-2.14 with a mean value of 2.12.

Salt Type S15-This material is a low density polycrystalline "artificial salt." Available in 50 pound blocks, it is normally used to provide livestock with a source of sodium chloride for their diet. The composition of the material is 99.9 percent sodium chloride with only trace amounts of impurities. This salt was a very white, fine grained (1/32 inch), polycrystalline mass with no apparent preferential grain orientation. Visual observations indicated that the fabric consisted of fuse inter-granular contacts, containing a fine salt matrix. The bulk specific gravity of this salt was found to vary between 1.92 to

2.00 with a mean value of 1.95. The porosities determined using these specific gravity data and the known specific gravity of salt single crystals was found to be in the range of 7.6-11.8 percent.

Tennessee Sandstone-This rock is light brown in color with pigment streaks running in random directions. It is fine grained, the average grain size being 6.3 x 10^{-3} in. It exhibits no apparent bedding planes. The uniaxial tensile strengths of air-dried and wet specimens were in the range of 850-900 psi and 400-500 psi, respectively. Values of uniaxial compressive strength ranged from 18,000-20,000 psi for the air-dried state. The average values of the elastic moduli, Young's modulus and Poisson's ratio, in uniaxial tension were found to be 4.09 x 10^6 psi and 0.088, respectively. The apparent porosity and permeability were found to be 6.7% and 3.7 x 10^{-4}, Darcy respectively. The moisture content in the rock in the air-dried state was found to be 0.50%. The average mineralogical composition for this rock was as follows: Quartz, 83.66%; Matrix, 9.33%; Opaques, 6.33% and Others, 0.68%.

Basic and Applied Salt Mechanics, Cristescu, Hardy, Jr & Simionescu (eds)
© 2002 Swets & Zeitlinger, Lisse, ISBN 90 5809 383 2

The acoustic emission of the salt massifs

Mariana Ionita & Ervin Robert Medvés
S.C. IPROMIN S.A. Bucharest

ABSTRACT

For determining the stress status variation in the salt massifs, in the paper we use the method which studies the acoustic emissions phenomenon of theses rocks.

To this end was realized the correlation of the acoustic activity of the salt with the mechanical micro processes which appear due to the modification of the local stress status by means of studding the indexes which are characterizing globally this activity: the amplitude of the acoustic events, the energy transported by the acoustic signals, the speed of the acoustic emission, the number of the registered events as well as of the parameters which are characteristic to the singular signals from the salt acoustic spectrum: the frequency of the phenomenon characteristics, the calculation of the energetically traces and of the alternation indexes.

This data sets are completed with those corresponding to the force and deformation measurements. Finally are presented the laboratory research results made on salt samples obtained on the basis of the used methodology.

1. INTRODUCTION

The deformation-fissuring processes that appear in the rock salt massif are accompanied by generation of certain acoustic vibrations when its mechanical resistance's are exceeded. The registration and appraisal of the specific parameters of these phenomena's allows for the quantification of the stress state, as wen as for the follow up, monitoring signalling in time of certain potential phenomena of geomechanical instability.

2. RESEARCH METHODOLOGY

In order the clarity the acoustic emission phenomena of the salt rocks and to correlate it with the evolution in time of the mechanical microprocessors which occur in the rock as a result of the local stress state alteration it was necessary to develop a research methodology of the acoustic behaviour of these types of rocks.

These researches were conduced in laboratory, on various salt samples, especially under monoaxial stress as well as in situ. The detection and registration of the acoustic and mechanical data was performed by means of an automatic system whose components are the following:
- acoustic emission transducers;
- tensometric force and deformation devices;
- A/D converter;
- PC computer;
- data acquisition and detection programs.

The methodology of analysis of the acoustic emission phenomena in salt rocks implies:

a. Global characterisation of the acoustic activity by registering and monitoring the following parameters:
- frequency of the signals;
- variation of the relative amplitude of the acoustic events;
- variation of the energy conveyed by the acoustic signals;
- number of events registered in a certain period of time;
- cumulated level of the acoustic emission during a given interval;
- velocity of the acoustic emission.

b. Examination of the indicators that characterise the singular events of the acoustic spectrum of the salt rock:
- characteristic frequencies of the phenomenon;
- attenuation indicators;
- form of the acoustic signals;
- spectral characteristics of the signal;
- rate of the acoustic emission in a certain domain of frequencies.

c. Location of the acoustic emission hipocenters.

3. RESULTS OBTAINED IN LABORATORY AND IN SITU

The above mentioned methodology and apparatus were used to correlate the acoustic activity with the stress condition in the salt massive of Slanic Prahova. In this respect samples were taken and acoustic and mechanical tests were conducted, were the axial loading speed applied upon the samples is high (quick creeping), as well as long-term tests.

Fig.1 presents the evolution in the of the acoustic signal amplitude registered on the rock salt samples.The essential changes of the longitudinal/states deformations (variations of 45%) determine alterations of the value of the acoustic signal amplitude, reaching 65%.

Fig.2 and 3 present the time variation diagrams of the acoustic emission velocity and of the cumulated number of events. These acoustic data outline the fact that the intensity of the acoustic level depends on the character of the deformation process. Thus, if on the average the cumulated number of events is of 1500-2000, at an acoustic emission velocity of ~9 events/sec during the moments that precede a major change of the structure of the sample (fracturing or intensification of the microfracturing process), these parameters Therese suddenly to 7,000 and 45 events/sec, respectively. The correlation between these acoustic parameters and the mechanical ones can be quantified by a relationship of the type: increase of the strain by 25% of the compressive strength R_c inducing the amplification of the acoustic values by at least 3 times.

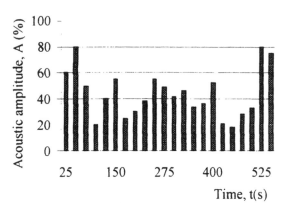

Fig.1 Evolution of the acoustic amplitude signal

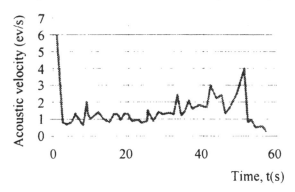

Fig.2 Variation of the acoustic emission velocity

Fig.4 presents the results obtained by the spectral analysis of the registered signals. The characteristic frequencies of the acoustic signals in rock salt ranges around 1500-6000 Hz. Characteristic to the process are the frequencies ranging between 2250-4000 Hz.

Major fractures occur when the deformations and stresses increase by 25% and the dominant frequencies are ~6000 Hz.

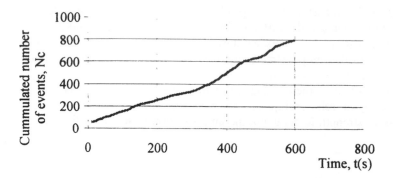

Fig.3 Variation of the cummulated numbe of events

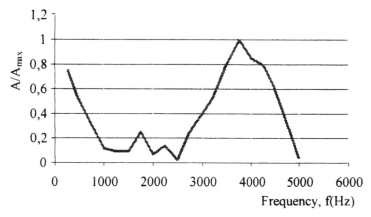

Fig.4 Spectral analysis of signal

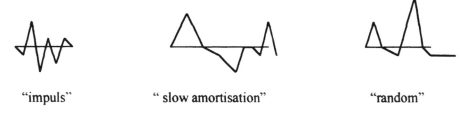

"impuls"   " slow amortisation"   "random"

Fig.5 Type of emisson signal acoustic

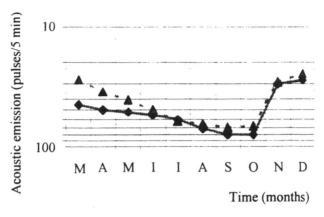

Fig.6 Acoustic emisson curves in Slanic Salt Mine

On the tested samples 3 types of acoustic signals were identified: "impulse"type, "slow attenuation" type and "random" type signals (fig.5). According to a statistical analysis of the mechanical and acoustical data, in the interval corresponding to a strain that does not exceed 40% of the R_c compressive strength, the "impulse" type signals represent 73% as compared to the period when the strain applied on the sample ranges between 60-90% of R_c and when the "slow attenuation" or "random" type signals represent 80%.

In that concern the attenuation factor, it is characterised by values varying by ~30% from the moment when a strain is exceted upon the sample and until the moment of its fracturing.

The acoustic activity was periodically measured at the Slanic Prahova mine from March until November 1992, on the walls and on the pillars on the extraction chambers [1]. The mining works lead to the occurrence of some phenomena of instability, to the deformation of the roof and floor, to intense fracturing etc.

The subsidence measurements varied between 3 mm/month in 1985 and 14 mm/day in October 1992.

The acoustic measurements in some active areas of the salt mine in 1992 showed an increasement of the acoustic activity (fig.6), correlated with the increment of the instability by the intensification of the roof fracturing and an increasement of the rock deplacement.

The correlation of the in situ data with the laboratory owes allowed for the prognosis of the dangerous increase of the stress-deformation state, registered during the next period and the closure of the activity in December 1992.

3. CONCLUSIONS

The results of the acoustic emission measurements in rock salt, correlated with the state of deformation strain confirm the use of the acoustic emission method in the appraisal of the deformation and in the prognosis of certain instabilities in the mining work conduced in rock salt layers.

REFERENCES

[1] C. Marunteanu & V. Niculescu – Acoustic emission in salt-work (Slanic Salt Mine) – Seventh International Congress International Association of Engineering Geology, 5-9 sept./Lisbon/Portugal.

Basic and Applied Salt Mechanics, Cristescu, Hardy, Jr & Simionescu (eds)
© 2002 Swets & Zeitlinger, Lisse, ISBN 90 5809 383 2

Dependence between loading speed and fracture strength

Ervin Medvés & Sorin Berchimis
S.C. IPROMIN S.A., Bucharest

Liviu Draganescu
Salina Slanic. Prahova, Romania

ABSTRACT

It was established that values of the mechanical strength obtained in laboratory on the salt samples depend besides the mineralogic-petrografic quality of the material, the degree of the sample relaxation, etc., also on the loading speed with which we are acting on the tested sample. For characterizing the dependence between the leading speed and the pursued mechanical parameter we are elaborated a series of experiments, specially loading by compression on salt samples from Slanic Prahova deposit. In the paper are described the tests made and are presented the obtained results, stressing specially the aspects tied to the dependencies already mentioned. The following steps of our research will enlarge the investigation range of this type and monoaxial compression test in steady creeping and loading – unloading, and the results will be mathematically modeled.

1. INTRODUCTION

With a view to a qualitative and especially quantitative assessment of geomechanical phenomena generated by rock salt workings, it is necessary a precise determination of mechanical parameters of the salt. These determinations, normally, are made in laboratory observing prescriptions of International Office for Rocks Mechanics and Standards, also the Romanian ones and others international regulations. In many cases, there have been recorded incongruences between determined mechanical parameters and the results, obtained by their usage at the assessment of geomechanical phenomena, the various values appeared when the dynamics of phenomena was different. Analyzing the factors that lead to these situations we have reached the conclusion that, at the carrying on of the lab tests, loading rates of the specimens must be adjusted to dynamics of in situ phenomena. Of course, this is very difficult to accomplish, especially when geomechanical phenomena, that are the object of the study, are not so well removed. To break this deadlock we have adopted, as work methodology in the lab, tests with various loading rates. Thus, for each salt samples, at least three loading rates are used in tests, followed by a results processing and drawing of typical curves of rule of variation of mechanical parameter followed in connection with the value of loading rate.

2. DETERMINATION OF VARIATION RULE BETWEEN LOADING RATE AND MECHANICAL RESISTANCE TO RUPTURE OF ROCK SALT

The way of operation for such tests will be exemplified on a series of determinations carried out at Slanic Prahova salt mine, where we would study the way specimen loading rate variation affects resistance to rupture at compression.

2.1 Geological characteristics of Slanic Prahova salt deposit

Section which includes the salt Slanic Prahova massifs belongs to coverlet sheets made of sediment formations (mostly of flish type and subordinated of molasa type) overthrust masses outwards over the platforms from the front of Carpathian Mountains. The deposits are belonging to the stratigraphical period of Palaeogene and Quaternary (L. Drăgănescu [1998]). Eocenous deposits are represented by Plopu strata (clay-sabulous sandy flish with globigerine marl intercalations) and strata with scribbling in Colti facies. Inferior Oligo-Miocenous deposits of Slanic syncline are located upon the inferior Oligocenes of Tarcau sheet, are made of two litofacieses, one of Pucioasa Fusaru with stratigraphic sequence: Pucioasa marnes with Fusaru sandstones, Vinetisu formation and sequence of Superior Disodiles with menilites intercalations); the other one bituminous with Kliwa sandstone (with units: Inferior Melinites, Inferior Disodiles, Inferior Kliwa sandstone, Podul Morii formation, Superior Kliwa sandstone and Superior Menilites). Inferior Miocene deposits from Slanic syncline are: Inferior evaporitic formation, Cornu formation and Doftana molasa). Medium miocene deposits are represented by Badenian deposits predominantly spreading Slanic tuff (zeolitiferous tuffs sequence and globigerine marls), Superior saliferous formation, radiolan shales and marls with Spirialis in Superior saliferous formation were delimited two lithostratigraphical sequences: one detritico-"evaporitic" and one breccious. In first of them can be individualized two lithological facies: a) one evaporitico-bituminous (at which must be added at basis the horizon with gypsum under the salt) and b) detritic-lutitic facies.

The Superior part of evaporitic-bituminous sequence is represented by salt accumulation known as Slanic salt massive with lenticular shape with a development on three directions with lengths of 2.7 km, width of 8-900 m and thickness of 45-500 m. Superior Miocene deposits are represented by Sarmatians deposits (marls, sands and limy sandstones).

Pliocene deposits are sporadic and represented by yellow greysh sands with limy sandstone intercalations: quaternary deposits, alluvial cones, soil and landslides.

2.2 Macroscopic characteristics of rock – salt samples

Salt samples were taken from Fs3 drilling, that crosses the salt massive (see section of figure 1).

From the petrographical point of view, the salt is an aggregate of crystals, forming a compact mass within which very often occur liquid and gaseous hydrocarbons inclusions with specific odor. Generally, Slanic Prahova salt has high level of purity. When it is clean has a colorless and transparent aspect (99% NaCl), but there are also

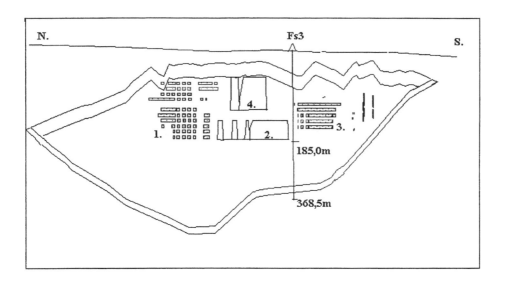

Fig. 1. N-S Section through Slanic Prahova salt massive: 1. Victoria salt mine , 2. Unirea salt mine, 3. Cantacuzino salt mine, 4. Carol (United Principalities) salt mine, Fs3 – research drilling for salt rheological and physic-mechanical characteristics on 185 m 365.5 m interval (from the application "Geologist" of Slanic salt mine).

impure sequence with clay-marlous material, that gives well defined grayish black colors to the salt (97-99%).

Rock salt has a few complex physico-mechanical characteristics, differentiated on geological structure, deposit age, depth, etc.

With a view to establish salt rheological and physico-mechanical characteristics from Slanic Prahova deposit, was taken into consideration a great diversity from the petrographical point of view.

2.3 Testing and caring out of experiment

Sample drawing was performed on the whole length of Fs3 (368.5 m) drilling, distinguishing 17 samples.

For monoaxial compression tests were made cylindrical specimens with 100 mm length and 50 mm diameter. The loading was exercised within a hydraulic press EDZ – 100 type, that allows different loading rates. In our tests were used three loading rates: 2.5 MPa/min , 1.25 MPa/min and respectively 0.8 MPa/min. For each sample and rate were accomplished 3-5 tests, thus 204 trials in all. Processing the recorded data were calculated the compressive stress (σ_c) and strains (ε_l-longitudinal, ε_t-transversal and ε_V-of volume) diagrams σ - ε being drawn. Then, in order to characterise the influence of loading rate, upon the determined values of mechanical resistance of rock salt, a statistical processing of tests data was made.

3. TESTS RESULTS

To make a more complex characterization as possible of the interdependence of loading rate and compression resistance of Slanic Prahova salt, in a first stage was drawn the bunch of medium curves σ - ε for the three loading rates applied.

In figure 2 it could be seen how with the diminution of loading rates lowers resistance to rupture at compression and rises the distortion value for the similar loading levels.

Secondly, was plotted curve that characterizes the interdependency rule σ_c – v (resistance to rupture – loading rate) computing its regression function. In figure 3 you can see this polynomial type curve with the following equation:

$$\sigma_{cr} = 0.6069 \ v^4 + 5.8659 \ v^3 - 21.893 \ v^2 + 38.594 \ v + 0.1867$$

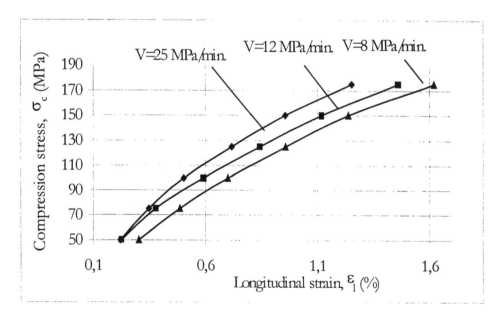

Fig.2 Relation $\sigma - \varepsilon$ for some rate of loading

4. CONCLUSIONS

Following the tests was proved without no doubt the existence of an influence of the way the tests were performed (various rates) upon the evolution of the stresses (σ_e) and $(\varepsilon\%)$ generated within the specimen mass. These interdependencies characterization

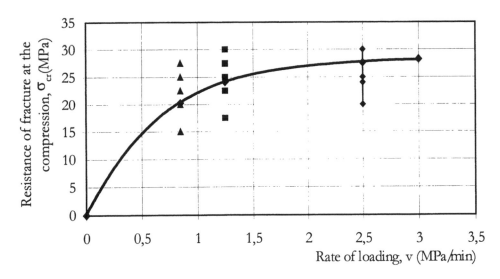

Fig.3 Relation between $\sigma_{cr} - \varepsilon$

and results transposition using relations typical to each salt massive. This partly makes possible to analyze and/or foresee geomechanical phenomena by mathematical and physical simulation, resorting to typical relation of similitude.

REFERENCES

L. Drăgănescu , [1998] "Geological Study of the salt deposit of the Slanic Prahova and the impact of the exploitation on the Environmental", University of Bucharest.

Basic and Applied Salt Mechanics, Cristescu, Hardy, Jr & Simionescu (eds)
© 2002 Swets & Zeitlinger, Lisse, ISBN 90 5809 383 2

Salt extrusion-rates in the Zagros

C.J.Talbot
Department of Earth Sciences, Uppsala University, Sweden

Introduction

Back-stripping of sediments deposited in salt - withdrawal basins indicate that natural salt structures develop very slowly. Deep conformable pillows of salt can be inferred to have risen at 0.01 mm a^{-1} for a hundred millions of years (Edgell, 1996) and intrusive diapirs typically rise at rates of 0.02 to 2 mm a^{-1} for 20 to 2 Ma (Jackson & Talbot, 1985). However, these rise rates are for subsurface (blind) diapirs buried beneath other rocks. Rate of salt flow can increase dramatically when the salt breaches the surface and is constrained only by air or water. Diapiric salt is then subject to maximum buoyancy forces which may be supplemented by lateral tectonic forces. Huge (160 X 90 X 9 km) nappes of salt have been shown to have advanced at average rates up to 275 mm a^{-1} from Middle Miocene to Plio-Pleistocene over the floor of the Gulf of Mexico where the Mississippi delta is sliding on salt into the open basin (Wu et al., 1990).

Monitoring salt extrusion in Iran

The 130 or so diapirs of Hormuz (Neoproterozoic-Cambrian) salt currently at the surface in the desert scenery of the Zagros mountain range in Iran provide a unique field laboratory for studying salt mechanics. All stages in the development of salt structure are present in three generations of diapirs (pre-, syn and post Zagros thrusting and folding). These range from deep pillows through blind diapirs to all stages of extrusion (Talbot and Alavi, 1996). Diapiric salt typically breaches the surface in ring dikes, which rapidly rise to bun-shaped domes that soon reach c. 1 km above their orifice which can be gravel plains near see level or mountain massifs already c. 1 km above sea level. The steep flanks of the salt domes then flow downslope over the surrounding scenery to generate sheets of allochthonous salt (namakiers) flanking a summit dome. These salt mountains have the characteristic profiles of viscous fountains. Rapid extrusion soon exhausts and closes the local deep salt source layer. No longer supplied by new salt from below, the summit domes subsides into a salt plateau with the distinctive profile of a viscous droplet. The salt profile then degrades beneath increasing accumulations of soils of insoluble Hormuz sequence which fill a hollow above a breccia chimney (Talbot, 1998).

The long-term *Zagros Halokinetics* research programme is run by geologists from the Hans Ramberg Tectonic Laboratory of the Department of Earth Sciences at Uppsala University, and the Institute for Earth Sciences of the Geological Survey of Iran (GSI) under the auspices of GSI's general director. Since 1993, teams of geologists have spent about a month a year visiting most of the diapirs emergent in the Zagros and choosing ten for monitoring. By end 1998, the positions of about 270 markers on the salt had been measured up to five times from 15 theodolite stations installed in different years, all but one on bedrock beside the salt. Our measurements to date suggest that five of the ten monitored mountains are dissolving faster than they flow and that the other five are flowing faster than they dissolve.

Hormuz rock salt (usually >90 vol % halite) typically extrudes from an orifice c. 2 km across and rises c. 1000 m above the orifice before it spreads over the surrounding country rocks (for up to c. 10 km thick) in the salt that emerges in the summit domes, subdued color banding ($\leq$ c. 10 m thick) is taken to represent the original bedding in salt that has flowed at least 5 km along the deep source layer and 3 to 12 km up the diapir before being exposed. Most of this salt has a grain size of 1-3 cm with a slight gneissose foliation but pegmatite of halite (grain size c. 10 cm) a few dm thick parallel the foliation in places and provide useful strain makers.

On scales > 100 m, the color banding and foliation generally parallels the nearest boundary to the following salt whether this is free to air or non-slip against country rocks or underlying grovels. On smaller scales the foliation is truncated by steep erosion surfaces in an outer zone of dilated elastic salt characterized by brittle fractures (Talbot 1998). The foliation is axial planar to generally recumbent folds that are gentle in the flanks of the summit dome but increase in amplitude to isoclinal nappes downslope. The grain size of the salt generally decreases downslope along foliation-parallel shear zones that probably root to irregularities in the bedrock channel or inflections in internal dirt bands. Minor recumbent folds with mylonites of halite only c. 10 cm thick in their lower limbs (Talbot, 1998) emphasize that the flow profiles of allochthonous salt sheets can be very irregular (Talbot, 1979). Laboratory experiments (Cristescu and Hunsche, 1993) indicate that fine grained halitic mylonites are likely to be orders of magnitude weaker than adjacent layers of course-grained gneissose salt.

Salt kinematics

Theodolite readings from baselines on stable country rock are used to measure vertical and horizontal angles to natural markers on the salt. Apparent displacements are calculated by multiplying the sine of the angular difference between successive theodolite readings by the distance between the theodolite and the marker measured on 1:50,000 maps. Taking our errors in lengths of site-lines to reach 5%, and in theodolite readings to be $\pm$ 0.005 radians, suggests that the apparent displacements we calculate, may be in error by as much as 10%. The presentation will illustrate apparent displacements of 43 markers dispersed over one of the largest and most active salt fountains over 4.5 years (June 1994 to November 1998).

Salt velocities vary yearly but are generally fastest on the lip of the summit dome(between c.1 and 10 m a^{-1}). The salt slows rapidly down the 8 km length of the namakier to the snout which comparison between topographic maps dated 1952 and 1977 indicates can locally advance at rates that average c. 8 m a^{-1}. In one 18 month period, a bulge c. 1 m high backed up behind a country rock buttress on one flank of this mountain but migrated over 4 km downslope and largely dissipated over the following 2 years.

The response times of this salt extrusion and its 3 km diapiric source to tectonic forces (such as nearby faulting in the country rocks) are likely exceed a hundred years. The significant changes in shape and kinematics we have measured over 4.5 years are therefore unlikely to reflect changes in the supply rate of salt from depth. It is anticipated that salt kinematics will correlate with annual rainfall figures (for rainfall can be expected to accelerate salt flow as well as salt dissolution: Talbot and Rogers, 1980).

The rainfall in the region is 200-400 mm a^{-1} and, if all this drained off the salt fully saturated, it could potentially dissolve a vertical thickness of salt of 3 to 7 cm a^{-1}. However, most of the precipitation evaporates on the salt or the cloaking soils and the runs-off is rarely fully salt-saturated. The range above is therefore judged to be on the high side of reality and, even then, small compared to vertical and horizontal movements of markers that commonly exceed 1 m a^{-1}. The flow of salt on this mountain therefore greatly outpaces salt dissolution.

Modelling

The dimensions of a viscous fountain being dissolved over its top surface is a measure of the vigor of its extrusion and the time it has been extruding. A simple 3D numerical model of the mountain mentioned above assumes a viscous fluid to be extruded from a circular orifice with a diameter that is a specific ratio of the diameter of the extrusion (Medvedev, 1998). The viscous salt spreads under gravity and dissolves over all its top surface at a constant vertical rate. This model greatly simplifies nature by assuming steady dissolution and flow of the salt in a consistent pattern over a smooth horizontal surface. A very thin layer of salt has to be assumed along the top of the country rocks for the emerging salt to spread over them. Profiles in time simulate well the young hemispherical domes of salt in the Zagros but not the concave-upward slopes around the summit dome of salt fountains. There must be some direct relationship between the diameter of the summit dome, the vigor of salt rise, and the dimension of the orifice; but as this is not yet known; the 1.7 km width of the orifice, a sensitive parameter, is merely "slightly smaller than the summit dome" in the current simple model.

The numerical model is non-dimensional but the profiles it generates have been tuned to the width and height, and the maximum horizontal velocity, on a north-south centered profile along the main stream of the mountain referred to above. The maximum velocity is chosen to fit the measured values (between 4 and 6 m a^{-1} at the lip of the summit dome) but the other velocities predicted by the model compare well with nature, both qualitatively and quantitatively).

Dimensionalisation of the model allows us to constrain unknown quantities. These are:

- the (constant) vertical dissolution rate of the salt surface is between 2 and 3 cm a^{-1} ;
- the salt has a viscosity in the range 10^{16} to 10^{17} Pa s.
- the vertical extrusion rate of salt out of the orifice is between 2 and 3 m a^{-1} ;
- the shape of the mountain is close to equilibrium and has taken something like 55,000 years to reach its present size. As growth slows towards equilibrium, the inferred age of the mountain is imprecise ($\pm$ 20,000 years).

All these modeled values appear to be acceptable.

Implications

Our observations in Iran emphasize that the salt liberated into the sky or sea from a diapir that breaches the restraint of its cover rocks, can extrude 3 or 4 orders of magnitude faster than blind diapirs intrude. Salt extrusion at rates of meters a year is only likely as long as salt is supplied from depth. At an extrusion rate of 2 to 3 m a^{-1}, our best known mountain extrudes salt at c. 5×10^6 m^3 a^{-1}. In its c. 55,000 year lifetime, it has therefore already expelled approximately 200 km^3 of salt from the local geological record. This is likely to have closed its source (c. 1 km thick over an area of radius c. 8 km) to a primary weld. We therefore infer that episodes of extrusion are comparatively short-lived, significantly short-lived, significantly less than a million years. Brief temporary episodes of extrusion which end when they starve their source layer account for how sufficient Hormuz salt remains at depth to feed the current third generation of extrusions in the Zagros.

Backstripping of salt structures cannot constrain their flow rates when extruding salt was recycled without deposition constantly infilling the withdrawal basin. Backstripping can therefore give a misleading impression of natural flow rates of diapiric salt, particularly if the geological control requires averaging past salt velocities over intervals longer than 10^5 years. The maximum salt velocity of salt reported in a particular German diapir, of 0.125 mm a^{-1} during Cenomanian-Late Paleocene times (Zirngast, 1996), was averaged over 39×10^6 years. This average could still be correct even if it include a brief cryptic interval (of say 10^5 years) when the salt flowed 4 orders of magnitude faster.

Structural observations of salt extrusions in Iran and measurements of their flow rates are also relevant to how salt flow might affect wells drilled to hydrocarbons overpressured beneath sheets of allochthonous salt. Diggs et al. [1997] inferred that wells developed through a particular salt sheet buried beneath the Gulf of Mexico would be safe from rupture over 11 years (Talbot [1993]). However, they "calculated" likely strain rates from the stresses they inferred from halite subgrain sizes in isolated sidewall cores of salt. Diggs et al. [1997] then interpolated their inferred translations of <1 cm a^{-1} to a smooth profile through their salt sheet. This smooth contour contrasts strongly with the irregular flow profiles seen in layers of gneissose salt tens of meters thick spreading over salt mylonites often <10 cm thick in several salt extrusions in southern Iran. Development wells through sheets of allochthonous salt might still be moving deserve 100% coring through all salt sections.

References

Diggs, T.N., Urai, J.L. and Cartyer, N.L. [1997]. "Rates of salt flow in salt sheets, Gulf of Mexico: Quantifying the risk of casing damage in subsalt plays." *Terra Nova,* **9**, Abstract 11/4B29.

Cristescu, N. and Hunsche, U. [1993]. "A constitutive equation for salt." *Proceedings 7th International Congress Rock Mechanics, Aachen, Sept. 1991.* Balkema, Rotterdam **3**, 1821-1830.

Edgell, H.S. [1996]. "Salt tectonism in the Persian Gulf basin." *Geological Society of London Special Publication* **100**, 129-151.

Jackson, M.P.A. and Talbot, C.J. [1985]. "External shapes, strain rates and dynamics of salt structures." *Bulletin of the Geological Society of America,* **97**, 305-328.

Medvedev, S. [1998]. "Thin sheet approximations for geodynamic applications." *Acta Universitatis Upsaliensis.* Comprehensive summaries of Uppsala PhD dissertations from the faculty of Science and Technology, **368**.

Talbot, C.J. [1979]. "Fold trains in a glacier of salt in southern Iran." *Journal of Structural Geology,* **1**, 5-18.

Talbot, C.J. [1993]. "Spreading of salt structures in the Gulf of Mexico." *Tectonophysics,* **228**, 151-166

Talbot, C.J. [1998]. "Extrusions of Hormuz salt in Iran." In: Blumdell, D.J. and Scott, A.C. (eds), L*yell: the Past is the Key to the Present.* Geological Society of London, Special Publications, **143**, 315-334.

Talbot, C.J. and Alavi, M. [1996]. "The past of a future syntaxis across the Zagros." *Geological Society Of London Special Publication* **100**, 89-109.

Talbot, C.J. and Rogers, E.A. 1980. Seasonal movements in a salt glacier in Iran. *Science, Washington.* **208,** 395-397.

Wu, S., Bally, A.W., and Cramez, C., [1990]. "Alochthonous salt, structure and stratigraphy of the North-eastern Gulf of Mexico." Part 11: Structures: *Marine & petroleum Geology,* 7, 334-370.

Zirngast, M. [1996]. "The development of the Gorleben salt dome (northwest Germany) based on Qualitative analysis of peripheral sinks." *Geological Society of London Special Publication* **100**, 203-226.

Basic and Applied Salt Mechanics, Cristescu, Hardy, Jr & Simionescu (eds)
© 2002 Swets & Zeitlinger, Lisse, ISBN 90 5809 383 2

Geodynamic research on the mirovo salt deposit near Provadia, NE Bulgaria

Georgi Valev
University of Architecture, Sofia, Bulgaria

Peshka Stoeva
University of Mines and Geology, Sofia, Bulgaria

Georgi Janev
Director GEOSOL, Provadia, Bulgaria

The Mirovo salt deposit is located approximately 5 km south-east of the town of Provadija, in the Varna province, Bulgaria. That is a unique natural phenomenon with its origin, composition, location and geophysical characteristics. The Mirovo salt diapir has a depth of approximately 4000 m and its top (the "salt mirror") is 15 to 20 m deep. The diapir has a base area of the order of 200 km^2 and generally has the shape of a truncated cone. Approximately 73% of the diapir is sodium Chlorid, the remainders comprising insoluble elements and other evaporites, irregularly distributed through the salt mass. It was excavated in 1919. A factory was constructed and the production of common salt began in 1924. However, the industrial exploitation of the salt pit began in 1954 after the construction of the first soda plant in the Devnya lowland. The method of boring is applied at present in the salt pits (more than 30 utilized boreholes) and more specifically the method of underground salt leaching through which one major raw material for the Devnya chemical plants is produced – brine.

The deciphering of the tectonic structure in this region shows that it is a real small-blocked mosaic, determined by a significant number of fracture (faults) of different hierarchical and space characteristics. It is considered that this mosaic complexity is due to the impulse raising of salt mass towards the respective central karst structures. The last and the greatest raising, as well as the forming of the present mode of the salt body, has happened long after the Illyrian phase but exact dating has not been done so far.

The reconstruction young tectonic fields are characterized by a sub-horizontal axis of extension in NE-SW direction and an axis of contraction in NW-SE direction. It has been found out that the intensity of these fields is the main factor determining the seismic processes in this region. The fractures active during the neotectonic period are a greater potential seismic danger at present. The fractures of the so-called Provadian beam, by which the salt body last impulse of introduction was realized, are characterized by a maximum amplitude of neotectonic movements. There are data about registered earthquakes of magnitude M4 during the period 1900 – 1901.

In the period 1981 – 1986 the seismic occurrences (became) were rather frequent in this region. This fact gives grounds to suppose that some of them have been the results of the long-term exploitation of the salt deposit which disturbs the tectonic balance. The analysis of the earthquake mechanisms, during the mentioned period, has shown alteration not only in the directions of the local main axis of deformation but in their dips as well.

The activation of the seismic processes brings up the questions how and to what extend every more significant earthquake would be forced to the structures on the salt deposit. Naturally the most dangerous are the earthquakes centered in the vicinity of the salt body, since they would exert influence not only by radiating energy but also by destroying the geological environment in depth. Regardless of the greater plasticity of the salt body, compared to the infitted rocks, each sudden burst of energy in the immediate proximity, would cause disruptive dislocations in the heterogeneous salt-rock mass. That is why periodical geodetic survey is necessary so as to be received authentic information, concerning the horizontal and vertical movements and deformation processes of the earth crust, in order to predict earthquakes.

Geodetic investigation at Mirovo salt deposit has been started since 1980. For measuring the vertical deformations and subsidence a geodesic network of 300 overground leveling bench marks situated on two almost perpendicularly crossing each other in the center of the salt mirror profile lines were build up in 1980. Six cycles of leveling measurements are carried out over this network almost every year. Till 1986 this network was almost destroyed and became unfit for further use. A new special geodynamic network has been created as a combination of horizontal and vertical network, which points are monumentated in a special way and on a suitably chosen places.

It was a geodynamic project which attracted the researchers' interests. A dense network of 26 stable pillars (Fig.1) and 200 leveling bench marks (Fig.2) has been designed and built-up especially to monitor movements and deformations in the deposit area using precise angular and distance measurements, spirit leveling and GPS. Some of the pillars – outside the central zone – control the details of the salt body.

The main purpose of the network is to establish 3D control, and to monitor shifts, deformations and the most significant site displacements. The aim is to examine to what extend the exploitation of rock – salt is connected to the local seismic activity because recently earthquakes have been occurring there more frequently and with increasing strength.

The initial combined geodesic observations were carried out in the period 15 – 30 May 1990. After that the measurements are carried out two times every year.

The angular measurements have completed by the electronic theodolite KERN E2. The slope distances between the points are measured by the precise laser ranger

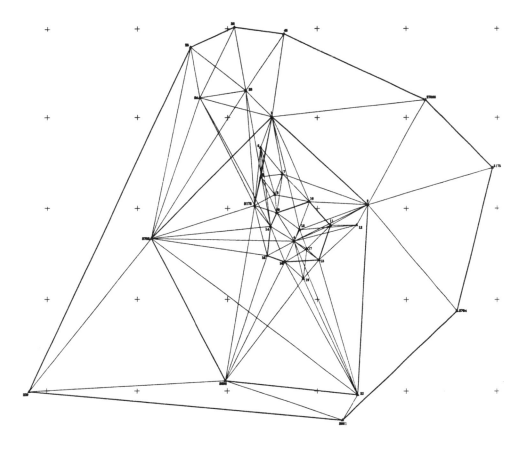

Figure 1. The Geodynamical Network.

/KERN MEKOMETER ME 5000/. Meteorological factors (temperature, pressure, humidity) has been reading out during the measurements. In order to eliminate the indeterminacy in rotation and for integral processing of the network astronomic observations are made. To increase efficacy and reliability of classical geodetic survey GPS takes place every year – measurements on the points of the network. A set of two or more receivers, type Wild System 200 of the Leica Company is used. Precise leveling is accomplished by a set of leveling devices of the KERN Company, levels GK 2 – A, leveling staff, leveling leapfrogs.

A complete gravimetric determination of the network has been accomplished too. The BUGE anomalies and the "free air" anomalies have been computed and a gravimetric map has been worked out for the BUGE anomalies 1 : 10000.

All results from the various geodetic observations are processed separately and after that put under integrated processing.   Horizontal directions, vertical angles, distances, height differences, astronomic azimuths, astronomic coordinates and GPS

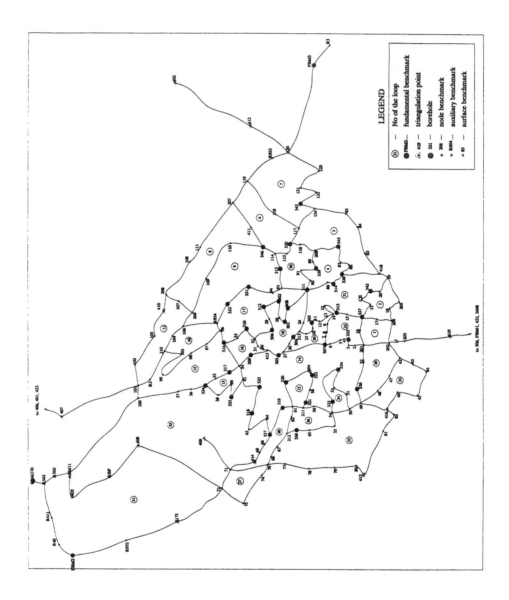

Figure 2. The Geodynamical Network - 200 Leveling Marks.

base lines are included in the common adjustment model. The different kinds of coordinates and heights are calculated with extremely high accuracy.

Eighteen cycles of combine geodetic measurements with very high precision have been accomplished on the network till now.

The accumulation of highly precise geodetic information is a good base for establishing the regularities and the tendencies in the dynamic behavior of the points in

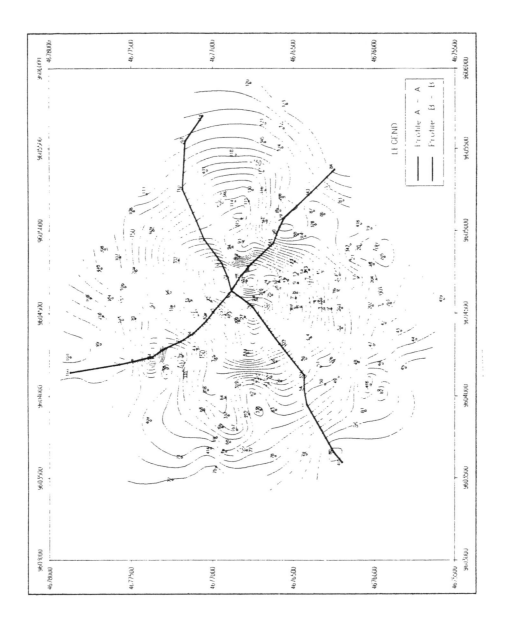

Figure 3.   The Graphic of the Boreholes (Profiles A & B).

the special geodynamic network, as well as   for connecting these regularities and tendencies with the processes in the region having technogenetic, tectonic and seismic character.

Total vectors of displacement of points are calculated by the reliably established displacement at coordinates axis   X,   Y,   and   H, respectively. The established displacements give opportunity for noting some regularity. The shift of net points, located North-East of the Provadija fracture, towards North-West is obvious. On the other hand, points in the region of the salt body, situated to the South of the Provadija fracture, are clearly shifting towards the North-East. The shift decreases with the remoteness of the points from the salt body. One of the causes for this is that salt body has greater plastic properties than the other earth mass in the region. The tendency of shortening the distances between points, situated immediately around the salt body is also evident. This length shortening has been characteristic of all cycles till now and gives an extremely reliable information about the character of the horizontal shifts in the region of the salt body.

Treated together, horizontal and vertical components conform to a certain extend. Subsidence in the central part can stimulate a horizontal shift of the points in the direction of the center of subsidence. The question whether these horizontal motions are caused by a subsidence in the region or there exist some other causes is raised. We are of the opinion that the shift is due to the tectonic component as well, especially at the Provadija fracture.

The components of the isotropic plane deformations of earth blocks in the region of the geodynamic polygon are calculated by the final elements and by the relative linear deformations at suitable sides of the network. The calculation results are the base for a variable analysis of deformation processes in the investigated region.

It is necessary to point out that the deformation intensity for the period investigated is higher than the average critical values of the earth crust. Taking into consideration the shortness of the temporary interval, for which research has been done, the received extreme values of deformations should not ease us, especially in the salt mirror. However, it is necessary to determine what part of the total value of deformation is the one of technogenetic character, and what is the part, which is of tectonic origin. The necessity of such determination in the near future is supported by the fact that a great part of the axes of deformation sections are in NW-SE direction, a direction characteristic of the whole country. Additional specifications are necessary to the activity and the territory detachment of the tectonic processes in the region, and especially to those which could become generators of seismic occurences. A final and definite answer to these questions could not be given for the time. The reason for this opinion is the relatively small amount of information, which prevents us from creating a full concept on the mechanism and the genetic peculiarities of dynamic and deformation processes in the region.

Based on the results of the cyclic leveling survey, annual vertical speed of all kinds of benchmarks has been calculated, and a velocity map has been worked out for the region investigated. It is obvious from the map that the velocities of the drilling bench marks decrease with the remoteness of the bore holes in the central part, which is almost identical with the center of the salt mirror. There is good reason to consider that the behavior of the benchmarks is a reflection of the deformations in the salt body.

It is of great significance that the velocity of subsidence of each of the benchmarks during the years is almost equal. Due to this fact the average year velocity is calculated for each drilling rig, which enables predicting of future subsidence.

The fact that subsidence exists in the regions away from the boreholes, according to the bench marks, could not be explained with the technogenetic activity only. Obviously, causes should be sought elsewhere and probably it is that vague shift that causes the seismic phenomena in the region.

The observed constant and almost regular subsidence is due to the full exploitation of the salt deposit, which runs in a way, not provoking abrupt, unsteady changes. This fact shows that at this environment continues to be permanent. Of course, the present state of balance does not ease us since integral accumulation of stress and deformation could surpass the determined limits and bring about irreversible disaster. Due to these reasons research will continue in future with unceasing interest.

Obviously, the general trend of terrain deformations is a subsidence of the central zone and, as a by-product, shortening of the distance between the center and the outer margin. Increasing residuals with time span are due to the time-dependent components, e.g., site movements, whereas the constant part gives an accurate estimation.

The changes of the distances between the points in the network have been calculated and presented graphically with the purpose to determine the relative movement of points for every side. For instance, the tendency of shortening the distances between the points in the immediate vicinity of the salt body is obvious. The greatest longitudinal changes are observed between the points 11-14 (- 270 mm), 11-16 (-244 mm), 9-11 (-212 mm), 11-15 (-192 mm), 14-18 (-111 mm), 14-18 (-200 mm), 10-15 (- 161 mm), 2-3698 (-131 mm), 9-19 (-149 mm), 11-13 (-201 mm).

Most of these couples of points are situated on both sides of the Provadijan fracture. The shortening the distances in the region of the internal ring, described by points 9, 10, 11, 18, 19, 16, 15, and 14 is characteristic of all cycles to the moment and gives extremely reliable information about the character of the horizontal movements in the region of the salt body. The alteration distances between the points in the horizontal geodynamic network are put to regressive analysis. The alteration of the diagonal sides have been investigated as well as those of the peripheral sides of each ring. There have been obtained reliably the velocities of these alterations, which reach up to 30 mm per year with an average square declination in the framework of 1 mm per year and a coefficient of correlation very close to unity. This presupposes a reliable prognosis of the deformations for a comparatively period of time.

The maximum deformations for the period of investigation are as follows:

Final element	Great semi-axe	Small semi-axe
(11 – 13 – 17)	$A = -21.2 \times 10^{-5}$	$B = - 31.8 \times 10^{-5}$
(4 - 9 - 13)	$A = - 1.0 \times 10^{-5}$	$B = - 40.0 \times 10^{-5}$
(4 – 16 – 17)	$A = 4.9 \times 10^{-5}$	$B = - 32.3 \times 10^{-5}$
(4 - 13 – 17)	$A = -30.6 \times 10^{-5}$	$B = - 41.6 \times 10^{-5}$
(11 – 17 – 18)	$A = 4.0 \times 10^{-5}$	$B = - 22.8 \times 10^{-5}$
(14 - 15 – 3173)	$A = 24.5 \times 10^{-5}$	$B = - 5.3 \times 10^{-5}$

The biggest submergences for the whole period are registered in the region of the boreholes: C12 (-399 mm), C4 (-396 mm), C3 (-392 mm), C6 (-391 mm), C8 (-383 mm), C20 (-376 mm), followed by the drillings C22 (- 337 mm), C7 (- 322 mm), C15 (- 310 mm) and so on. This can be seen in the graphics enclosed (Fig.3).

There have been carried out a regressive analysis of the subsidences in the time. The dependence in the case of all drillings is linear with a very high coefficient of correlation. Reliable values for the year velocities have been obtained. With the help of the dependence obtained it is possible to prognosticate reliably the subsidences for a comparatively long period of time.

The year velocities reach up to 30 mm per year as the vertical year moving shows the following: the great velocities of subsidence are in the central part. The subsidences decrease towards the peripheral part. There is an enduring tendency towards the formation of a kettle/hollow, which is a negative representation of the salt body: its bottom is found under the top of the salt body, its contour coincides with the contour of the deposit. The single graphics of the sinking of the wells very well fit the rheology of the salt-rock mass not only in size, but in distribution as well.

According to the data from the sonar measurements of the drillings the monoaxis vertical deformations of the cameras have been calculated, presupposing that the established sinking of the bench marks on the drilling columns reflected the deformations of the walls of the chambers (caverns).

The following important inferences and conclusions can be drawn:

1. There are deformations present, not only of stretching (dilation) but of compression as well. There are platforms where the deformations are either only of compression, or only of stretching.

2. In the central part of the salt mirror the deformations mainly of compression are included, as the values here are usually great.

3. In the peripheral platforms the deformations vary as a whole and are smaller in value.

4. The greatest deformations, that long-exceeded the critical ones are noticed in the uttermost elements: (4 – 13 – 17), (4 – 9 – 13), (4 – 14- 15), (4 – 15 – 16) and (14 – 15 – 3173), which are situated in the central and the SW part of the deposit.

5. The alteration in the quantitative and qualitative characteristic means that a frequent redistribution of the intensity and the character of the deforming processes in the region occurs. One of the explanations is that this is caused not only by the tectonic activation, but by the technogenic aggression as well. The investigators of the neotectonic movements claim that during the past years these movements have been activated in the earth bowels (i.e., in the salt body) there occurred a disturbance of the equilibrium and the middle part strives to restore the stable state, which is expressed through seismic processes.

6. In connection with the issue of the two components – the tectonic and the technogenic and the technogenic one – the supposition that both are present and act of inter-relation, continues in force. We do not deny that the causes of the deformations are exceptionally technogenic, but taken alone they would hardly display themselves to such an extent, if a lateral tectonic pressure did not exist. The great values of the deformations outside the deposit are in favour of this statement.

7. The values of the deformations in some elements exceed the critical ones. Beyond the critical are also the values of the deformations of the pillars.

8. Probably the chambers deform and in the same time their volumes are reducing, which is proved by the spontaneous increase of the pressure in the ones that are sealed.

9. The results of all cycles show that the deformation process continues developing with almost one and the same rate and with lasting tendency. The deflections that are established from unidirectional process indicate possible seismic (and tectonic) phenomena.

10. As is well known, in the case of the deformations close to the limited ones there is a decrease in the visible module of elasticity and in the periods of anomal increase of the velocity and the deformations it can be made a prognosis of the seismic danger. According to the seismologists tensions are accumulated in the earth crust which cause these deformations and when the latter reach the utmost values, there occur sudden disruptions and earthquakes strike. The values of the deformations in the earth crust vary within the boundaries of 1×10^{-4} to 1×10^{-5}. Here these values are, as we stated above, long-exceeded. Taking into consideration the frequent weak seismic occurrences, although with a small magnitude, it can be supposed that a great deal of the tension is put out by these earthquakes. In these sense the frequent but weak earth tremors are useful.

11. An interesting conclusion that can be come to is that the behavior of the surface layer creates the impression that our measurements are carried out on a living substance, i.e., the diapir continuously creeps.

12. Another inference is that no matter whether some drillings work or not, the values of the vertical and horizontal movements are preserved almost constant. This, however, does not mean that in case of increasing the salt extraction the movements and the deformations will not increase. It is necessary to bear in mind that the reaction of the earth crust occurs with a certain delay.

The established processes in situ have been tested in laboratory conditions. The caused rheological deformations for different varieties of salt diapir (deposit) correspond to the long-term deformation behavior in the massif in analogical conditions. The relationship between the static and the dynamic modules has been estimated. A criterion has been given about the tension-deformation condition.

The static deformation constants, obtained by a dynamic method (by means of supersonic measurements), allow to be classified as indicators giving the "initial" state of the tentative body, i.e., on elastic dynamic module ($E\alpha$) and a deformation dynamic module (Ed). These constants are compared with the static deformation modules, obtained by the classic method (gradual loading and unloading) E and n0. The latter reflect the alteration of the tentative body under the different degrees of the working tensions and that's why we call them constants of "the state of the deforming body". Such a connection between the two states has not been made so far. Jaeger [1975] cursorily points out that the elastic modules (E), defined in a static way are lower than the dynamic ones (Ed).

The methods offered for the comparison of the respective modules, obtained by dynamic and static methods gives the opportunity of assessing the initial state of the rock varieties from different probed points in the rock massif.

The big-small and mixed-crystal salt varieties 100% salt with minor and average defects and with higher modules of elasticity and deformation modules deform according to the scheme Eis<Emin; Mis<Mmin "the modules of the initial state" are always smaller than the minimal modules in the initial stage of loading, since the macrodefects do not mobilize at once. That happens at the end of the deformation process and is accompanied by irreversible disruptions in the structure. The fissuredness never mobilizes completely.

In varieties of salt and marl with mirror contacts, characterized by predetermined surfaces and low maximum modules the scheme Eis>Emax and Mis<Mmax of deformation Emax and Mmax is valid.

In the salt-marl varieties there are observed deformations according to the scheme Emax > Eis > Emin and Mmax > Mis > Mmin. This is conditioned by the differently contents of salt and marl and is valid for salt 100%, but with emphatically acicular and fibrillose crystals and fluid orientation in the mixed-crystal salt. This scheme is valid also for the marls participating in prolonged deformation processes in the massif.

The presented criteria and conditions for deformation as regards "the initial state" have been done for the first time and represent a basis for systematization of a great deal of research and an opportunity of applying laboratory supersonic research to the characterizing of real bodies in broad scope.

The presence of numerous varieties of salts in the diapir massif (100% big-crystal salt, 100% small-crystal salt, 100% acicular salt and so on, different percentage of salt and marl components) determines a difference in the mechanism of creeping and the prolonged ambiguous behavior in the board range of structural distinctions even in the presence of comparatively stable composition. Differentiated rheologic of the most typical varieties of salt and marl has been made under laboratory conditions and under monoaxial pressure in order to carry out an analyses of the observed deformational processes. The scheme of research is consistent with the method of exploitation – a degree of loading until destructing.

It has been found that:
- salt creeps under all tensions from 0.1 of the destructing tension;
- a characterisric feature of the prolonged deformation is that in all cases there appears a mobilization of the contacts among the crystals – a typical macrorheological process. Only in the case of giant crystals in the boundary ot the prolonged strength, the macromechanic defects of the crystal fissuredness mobilize.

Big – crystal salt 100% (Fig.4) deforms in tiers, even under low tensions, at several velocities of creeping (BC - CC' - C'C"), which asymmetrically increase or decrease. The terrace-like activity of the deformation process is determined by the irregular and non-oriented discontinuance of the macrostructure. Besides that, the different arrangement of the macrocrystals in the structure of the lithological diversity creates power local barrages with great inertness for relative shifting and spinning.

Acicular salt 100% (Fig.5) has a peculiar deformation behavior. Acicular salt has been formed in the diapire at its first formation caused by seismical and other processes. In the laboratory the acicular salt also creeps with decreasing or increasing speeds, which are characteristic of the lower tensions (BC – CC' – C'C" – C'C" ' – C" "). Here the process of creeping consecutively mobilizes the microdefects in the preliminary deformed crystals until reaching the prolonged strength.

Creeping in the case of marls is conditioned by two factors:
- mobilization of the fissures at 100% marl (Fig.6)
- simultaneous mobilization of the fissure (at the presence of marl 90%)

and the large salt crystals (at a presence of 10% salt component).

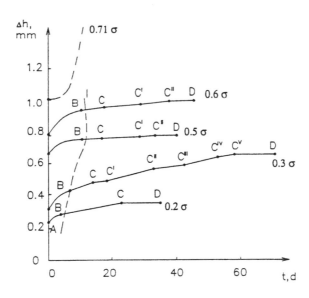

Figure 4.   Creep of Big-Crystal Salt.

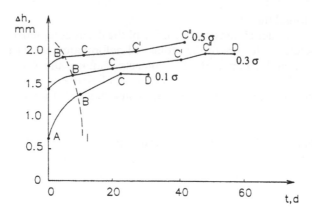

Figure 5.   Creep of Acicular Salt.

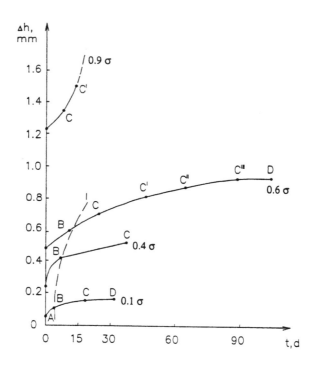

Figure 6.   Creep of Marls.

On Figure 7, regardless of the fact that salt is only 10%, the development of the process of deformation is analogous to that described for 100% salt.

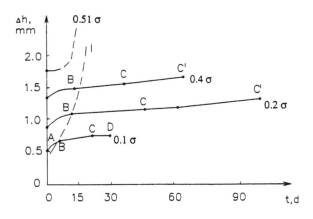

Figure 7.   Creep of Salt.

Part II

Coupled effects and permeability

Basic and Applied Salt Mechanics, Cristescu, Hardy, Jr & Simionescu (eds)
© 2002 Swets & Zeitlinger, Lisse, ISBN 90 5809 383 2

Effects of fluid-rock interaction on mechanical behavior of rock salt

Ph.Cosenza, M.Ghoreychi & S.Chanchole
Groupement pour l'étude des Structures Souterraines de Stockage (G.3S), Ecole Polytechnique, France
UMR 7619 Sisyphe Université Paris VI, France

ABSTRACT

An experimental investigation based on triaxial compression tests, relaxation and creep tests was performed in "wet" condition (without jacket) in order to allow percolation of the confining fluid *(brine or oil)*. Comparison of the results with those obtained in "dry" condition shows that failure and damage criteria in "wet" condition are significantly and systematically lower than those in "dry" condition. Morever, viscoplastic flow is enhanced when damage occurs.

The involved fluid/rock interactions are : a mechanical effect due to pore pressure and a chemico-mechanical effect due to salt solubility in brine. These hydro-chemico-mechanical phenomena may play a significant role in the safety assessment of long-term radioactive disposal.

INTRODUCTION

Salt formations are considered as suitable repository media since they are assumed to be impervious to fluids : rock salt has been selected in many countries for underground storage of hydrocarbons, industrial and radioactive wastes.

Mechanical drilling disturbs rock salt and leads to the permeability increase. This phenomenon associated with damage may induce significant hydro-mechanical coupled effects which may make the safety assessment difficult. Therefore, the following questions arise whether rock salt is permeable :
a- Can the concept of "pore pressure" be applied to rock salt ?
b- If so, in which conditions the concept of "effective stress" may be used ? and what are the effects of pore pressure on the strength of rock salt ?
b- In which way the mechanical behavior (elastic and viscoplastic) is modified when brine-salt interactions are involved ?

In order to answer to these questions and to characterize hydro-mechanical effects induced by damage, an experimental work based on triaxial compression tests, relaxation and creep tests was carried out.

In the first part of the paper, a special triaxial device allowing percolation of confining fluid*(brine or oil)* in the sample is presented. The laboratory results obtained in wet condition (with no jacket) are given and compared with those in dry condition (with jacket). In the second part, the device is used to carry out relaxation and creep tests in "dry" and "wet" conditions. In a third part, the experimental results are discussed in the framework of the safety assessment of storage in rock salt.

TRIAXIAL COMPRESSION TESTS

1. Experimental Set-up

The triaxial tests were performed in "wet" condition, without jacket allowing brine percolation into the sample. Confining fluid is applied using a filling screw between a viton jacket and the sample (Figure 1). Pore pressure is measured by two pressure transducers.

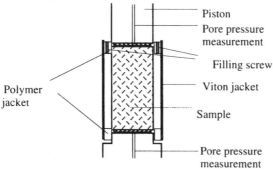

Figure 1- Triaxial Device for "Wet" Condition Tests.

Axisymmetric triaxial tests were made under different confining pressures p (0 to 60 MPa) and at a constant strain rate (about 10^{-5} s^{-1}). Sample dimensions were : diameter $\varnothing_0$= 50 to 60mm, height h_0=120mm. Confining pressure was applied by means of a pressure controller GDS (maximum pressure 60 MPa, accuracy<0,25% of maximum pressure); axial displacement was controlled using a MTS servo-hydraulic machine (maximum load 250 kN or 500 kN), and an extensometer (accuracy<0.5% of the full-scale displacement) located outside the cell. Axial force was measured with a load cell (accuracy<0.5% of the full-scale force).

Confining liquid was oil or saturated brine with dynamic viscosity of 1.95 10^{-3} Pa.s (1.95 cP) and 0.93 10^{-3} Pa.s (0.93 cP) respectively (at room temperature and atmospheric pressure).

Prior to the deviatoric loading, a hydrostatic pressure was applied during at least 24 hours. This first step aimed to "heal" if possible the present microcracks and to estimate bulk modulus and initial porosity. Then, in a second step, axial loading was applied at a constant strain rate. During deviatoric loading, some samples have been subjected to the cycles of axial force in order to measure Young's modulus changes. All the tests were performed at room temperature.

To measure pore pressure during hydrostatic and deviatoric loadings, it was necessary to make sure that contact areas between the sample and the end cap and/or piston were tight. The tightness was assured by little pieces of viton jacket sticked to the tip of sample.

2. Results

Samples have been cored in the purest part of a layer called "S1" belonging to French Potash Salt Mine (MDPA) located in the Mulhouse basin. Many previous laboratory and in situ studies have been performed in this mine especially in S1 layer [Thorel & Ghoreychi, 1996; Cosenza et al., 1999].

Unjacketed triaxial tests were performed on 15 samples : 7 using oil and 8 with saturated brine. Three values of confining pressure have been applied : 20, 40 and 60 MPa. A few tests have been carried out with multi-step creep and relaxation paths (see further).
Whatever the mean stress and the confining fluid are, the deviatoric stress corresponding to failure remains in the range of 27 to 42 MPa (mean value of 33.5 MPa and standard deviation of 4.7 MPa) (Figure 2). The mean value of 33.5 MPa is very close to the uniaxial strength of MDPA rock salt (about 30 MPa).
In "wet" condition (triaxial tests without jacket), confining pressure does not improve the strength as it is observed in dry condition : the failure remains brittle (Figure 3).
On the other hand, the chemical composition of the confining fluid has also a significant effect on the material strength. The average maximum strength is 30.4 MPa (standard deviation of 4.5 MPa) for the tests performed with brine and 37.1 MPa for those using oil (standard deviation of 1.7 MPa).

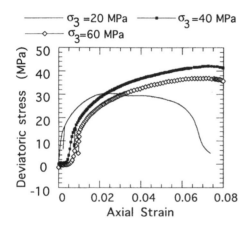

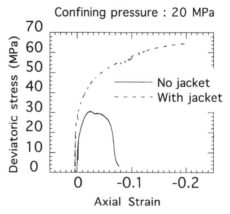

Figure 2 - Deviatoric Stress versus Axial Strain for 3 Different Confining Pressures (Tests Without Jacket).

Figure 3 - Deviatoric Stress versus Axial Stress in "Wet" Condition (with No Jacket) and in "Dry" Condition (With Jacket).

Another investigated aspect is the influence of pore pressure build-up in the sample during deviatoric loading. This aspect is difficult to study for two reasons :
a- the tightness of the experimental device is probably close to the very low permeability of a virgin rock salt sample;
b- the observed failure is associated with a very low number of cracks; the sample volume of the sample is consequently below the Representative Elementary Volume

(REV). Then, the concept of "pore" pressure cannot be used here to interpret the performed tests.

For the most samples, during the hydrostatic loading, the pore pressure sensor reaches quickly the confining pressure value. Two reasons can be invoked to explain this observation :

- in a few tests, a lack of tightness of the contact between the sample and the jacket, provided by a polymer jacket;

- connected cracks in the sample induced by the deconfining process and not closed/healed during the hydrostatic loading.

For one sample (n°41), a pore pressure build-up was observed during the deviatoric loading after the peak stress (Figure 4). This result will be discussed further.

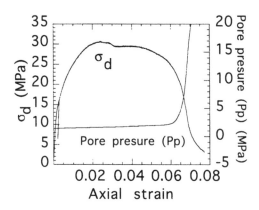

Figure 4 - Deviatoric Stress (Left Side) and Pore Pressure (Right Side) versus Axial Strain for 20 MPa Confining Pressure Test Without Jacket.

In addition to the different values of deviatoric stress corresponding to the sample failure in "dry" and "wet" conditions, there is a significant difference in the failure mode. In "dry" condition, for the values of confining pressure between 20 to 60 MPa, a significant plastic deformation of the individual grains is observed. Macroscopic fractures growth is small but the development of a network of microcracks cannot be excluded (Thorel, 1994; Thorel & Ghoreychi, 1996). In "wet" condition, as it is observed in unconfined compression tests for other rocks (Jaeger and Cook, 1969), irregular longitudinal splitting is observed (Figure 5).

Nevertheless, the observed brittle behavior of rock salt does not seem to modify in a sensitive way elastic behavior. Young's modulus values measured during loading-unloading deviatoric paths at the beginning of the compression triaxial tests do not show significant changes. The calculated average value 13.7 GPa (standard deviation of 1.6 GPa) is comparable to the values given by Thorel (1994) in case of unjacketed compression tests. This result suggests that damage effect is more visible regarding strength and ductility of rock salt than with respect to elastic properties.

Figure 5 - Cylindrical rock salt sample after failure in "wet" condition (without jacket) submitted to a value of 20 MPa confining pressure. Sample dimensions are : diameter $\emptyset_O$= 50 to 60 mm, height h_O=120 mm. Sample displays an irregular longitudinal "splitting", typical in uniaxial compression tests.

3. Discussion

3.1 Mechanical effect of pore pressure

As it was observed in "wet" condition, application of a high confining pressure up to 60 MPa does not improve the strength value when the sample is in direct contact with the confining liquid (see Figure 3). This result is in contradiction with that obtained in "dry" condition.

In "dry" condition and for the confining pressure values up to 12 MPa, the jacketed sample shows a ductile behavior : failure does not occur and large strains up to 20 % are obtained. From this view point, the main feature of the mechanical behavior is the strength increase (hardening). Consequently, the obtained results cannot be explained if it is assumed that the material is strictly impervious (permeability equal to zero) and consequently no pore pressure can be developed.

This difference between brittle and ductile behaviors related to "wet" and "dry" conditions respectively (Figure 3) may remind of the brittle-ductile transition also observed for other rocks. Figure 6 shows stress-strain curves for a limestone (Robinson, 1959). Indeed the material is more permeable and porous than rock salt, at a confining pressure of 10 kpsi (1kpsi = 6.9 MPa) and under various values of pore pressure. The transition between ductile and brittle behaviors is associated with a pore pressure increase. Such a transition is controlled by the effective confining pressure σ_3-p (Jaeger & Cook, 1969) : for the same value of confining pressure σ_3, when pore pressure p increases, effective confining pressure σ_3' decreases leading to an increase in brittleness of limestone.

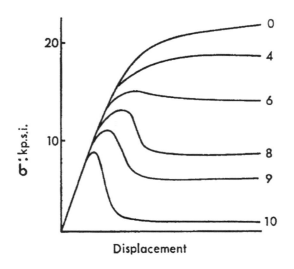

Figure 6 - Influence of Pore Pressure on the Brittle-Ductile Transition in Limestone at a Confining Pressure of 10 kpsi (1kpsi=6.9 MPa).

Numbers on the Curves are Values of Pore Pressure in kpsi (From Robinson, 1959 in Jaeger and Cook, 1969).

One may wonder whether the concept of effective confining stress can also be applied for rock salt to explain the "ductile-brittle" transition observed experimentally. During the deviatoric loading in triaxial unjackcted tests, pore pressure p increases in the existing and growing microcracks to attain the value of the confining pressure σ_3. Consequently, the effective confining pressure σ_3-p (or in a more general case, σ_3-bp with b : Biot's coefficient) drops to a low value :

$$\sigma_3' = \sigma_3 - bp \quad \Rightarrow \quad \sigma_3' = (1-b)p \tag{1a}$$

$$\text{If } b \approx 1, \ \sigma_3' \approx 0 \tag{1b}$$

Since the effective confining pressure value is small, nearly zero, when Biot's coefficient b is close to 1, the effective stress state in the sample is similar to that of uniaxial compression tests (or triaxial tests with very low confining pressure, less than 1 MPa). Therefore, at failure, the value of axial effective stress σ_1' has to be equal to a value close to the Uniaxial Strength R_c (i.e. a bit higher $(1+\epsilon)R_c$):

$$\sigma_1' = (1+\epsilon)R_c \tag{2}$$

For the same reason, when the confining liquid penetrates into the microcracks, the axial effective stress at failure σ_1-p (or σ_1-bp) tends to a value close to the total deviatoric stress σ_1-σ_3, since b approaches 1:

$$\sigma'_1 = \sigma_1 - p \quad \Rightarrow \quad \sigma'_1 \approx \sigma_1 - \sigma_3 \tag{3}$$

Equality of the equations (2) and (3) leads to:

$$\sigma_1 - \sigma_3 \approx (1+\varepsilon) R_c \tag{4}$$

Such an interpretation based on the concept of effective stress allows to understand two important features observed at failure in the experiments conducted on unjacketed samples :

i- a failure mode similar to that observed in uniaxial compression test (i.e. without or with a very low confining pressure);

ii- a nearly constant value of total deviatoric stress at failure close to that of Uniaxial Strength R_C.

In this analysis, the value of Biot's coefficient was supposed to be close to 1. Such a value seems to be very high considering the low porosity, less than 1 %, of rock salt. But it is to be kept in mind that here the considered stress state is the failure state : the sample is greatly dilated by a lot of cracks/microcracks and its stiffness is much smaller than that corresponding to an undisturbed state (undamaged).

3.2 Chemical effects of brine

The strength of rock salt samples subjected to brine percolation is less than that of a sample subjected to oil. This fact has not clearly been explained yet but the difference is suspected to be caused by chemical reactions strongly coupled with propagation and nucleation of microcracks.

This phenomenon could be associated with "stress corrosion" or "stress-aided corrosion process" studied on quartz (Kranz, 1983) and granites (Althaus et al., 1994). It occurs at the crack (or microcrack) tip and induces a very slow crack or microcrack propagation, called "subcritical". This theory indicates that two factors mainly control subcritical crack propagation velocity (Kranz, 1983) :

a- the rate at which "corrosion" (or chemical) reactions proceed in the vicinity of the crack tip ;

b- the rate at which "corrosive" elements, mostly water molecules, can be brought to the crack tip area (i.e. hydraulic permeability and diffusivity of "corrosive" elements).

For a salt/brine system, the slowest factor which governs the kinetics of the whole phenomenon is certainly not the chemical reaction of the salt dissolution which is known to be very fast compared to the current natural chemical reactions. Nevertheless, the "stress corrosion" phenomenon in rock salt/brine system may constitute an important research area, especially when rock salt is the host medium for the storage of industrial toxic wastes.

3.3 Effects of inert or reactive liquid on mechanical criteria

Failure criterion

Failure is supposed to correspond to the peak of the stress-strain plot. Following a classical procedure in Soil and Rock Mechanics, total deviatoric stress (σ_1-σ_3) and mean stress ($\frac{\sigma_1 + 2\sigma_3}{3}$) for the corresponding failure state are reported in the stress invariants plane [Mean stress; Deviatoric stress] (Figure 7). As it was previously mentioned, the yield surface is associated with failure is in the range of 27-42 MPa of deviatoric stress. It is slightly sensitive to mean stress and depends on the chemical composition of the confining liquid (brine or oil). The yield surface is lower, when brine is used.

Compared to the results obtained with jacketed samples in "dry" condition, a much lower internal friction is obtained both for brine and oil. This is mainly due to the percolation of liquid during the damage process leading to the failure. There is no significant chemical effect on the internal friction reduction.

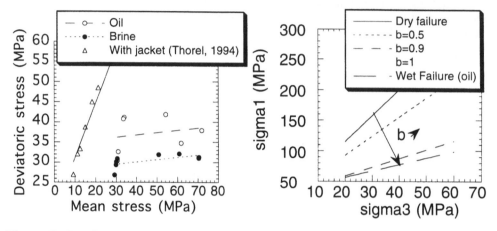

Figure 7- Maximum Deviatoric Stress versus Mean Stress at Failure ("Wet" or "Dry " Condition).

Figure 8- Mohr-Coulomb Criteria Expressed in the $[\sigma_1-\sigma_3]$ Plane.

Another classical way to represent a failure criterion in Rock Mechanics is to use the "axial stress - confining stress" plane or $[\sigma_1-\sigma_3]$ (see for instance, Jaeger & Cook, 1969). Data for "dry" condition have been fitted using a Mohr-Coulomb line :

$$\sigma_1 = R_c + K_p\sigma_3 \tag{5}$$

with R_c : Uniaxial Strength (30.2 MPa); K_p : a parameter related to the angle of internal friction Φ :

$$K_p = \frac{1+\sin\Phi}{1-\sin\Phi} \tag{6}$$

It may be interesting to express this "dry" criterion in terms of effective stress and to compare it with the real wet criterion obtained *directly* by the unjacketed tests. The experimental "wet" criteria may be called "pseudo-wet" criteria since they are obtained using "dry" criterion and are estimated by the following relationship :

$$\sigma_1 - bp = R_c + K_p(\sigma_3 - bp) \tag{7}$$

The "pseudo-wet" criteria are then compared with the real wet criterion in figure 8. It can be observed that the real wet criterion is surrounded by two "pseudo-wet" criteria associated with two Biot's coefficient values, 0.9 and 1.

Consequently, the experimental "wet" criterion and the "dry" criterion expressed in terms of effective stress with a Biot's coefficient between 0.9 and 1 are close This result confirms the validity of the previous analysis given in section 4.1 : near failure, the concept of effective stress can be applied using a Biot's coefficient value close to 1.

Damage criterion

Since, damage criterion is rarely used in Geology and Rock Engineering, it may be useful to explain in which way it has been established. Damage initiation in rock salt is currently attributed to irreversible volumetric strain (dilatancy) since crystal of sodium chloride leads to negligible volumetric strains (of the order of 10^{-7} according to Davidge and Pratt, 1964). Thus, dilatancy is associated with genesis of a microcracks network.

According to some previous experimental works (Thorel, 1994), the onset of irreversible volumetric strain coincides with the loss of linearity of the volumetric strain-mean stress plot : the "linearity limit". The corresponding stress state (characterized by the values of deviatoric stress and mean stress) related to the "linearity limit" can be reported in the plane of stress invariants [Mean stress; Deviatoric stress] (Figure 9) as it was done for the failure criterion.

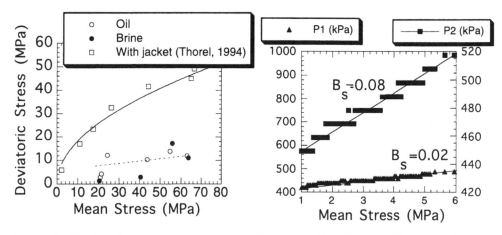

Figure 9 - Deviatoric Stress versus Mean Stress at Damage Initiation ("Wet" and "Dry" Condition).

Figure 10- Pore Pressure Evolution Versus Mean Stress Evolution During Hydrostatic Loading on Sample n°41.

From figure 9, the following results can be pointed out:

a- There is a significant difference between the criterion obtained using jacketed and unjacketed triaxial tests . This means that despite the extremely low permeability of rock samples (typically 10^{-20} m^2) the very little percolation of liquid (oil or brine) into the sample has a real effect on the damage initiation. As for failure criterion, a reduction of angle of internal friction is observed.

b- There is no chemical effect on damage initiation. This point confirms that the drastic drop of the failure criterion obtained using brine is a strong coupled process combining fracturation and chemical reactions (i.e. a mechano-chemical effect): it is not a purely chemical effect.

c- The four lower points in the plane [Mean stress; Deviatoric stress] in figure 9 are those obtained on the sample subjected to loading-unloading paths in order to measure the

Young's modulus value. This result which was not observed for failure criterion suggests that somehow a "fatigue" effect may play a role in reduction of deviatoric stress associated with the damage initiation.

3.4 Evolution of Biot's coefficient during damage

In order to estimate the value of Biot's coefficient related to the intact salt prior to damage, an attempt was made to measure, during the hydrostatic loading, Skempton's coefficient defined as follows :

$$B_s = \frac{\Delta P}{\Delta \sigma_m} \qquad (8)$$

ΔP : pore pressure change; $\Delta \sigma_m$: means stress change.

Poroelastic theory allows to link Biot's coefficient with Skempton's coefficient by means of the following relationship (see for instance, Coussy, 1995) :

$$b = 1 - \frac{1}{1 + \dfrac{B_s}{1 - B_s} \phi \left(\dfrac{K_s}{K_{fl}} - 1 \right)} \qquad (9)$$

with ϕ : porosity; K_s and K_{fl} : bulk modulus of solid and liquid phase respectively.

As it was already mentioned, Skempton's coefficient measurement is difficult since rock salt permeability is very low prior to damage . Besides, the concept of "pore pressure" is not validated in this condition. Thus, the following calculations have to be considered cautiously. But our purpose here is to obtain an order of magnitude for Biot's coefficient and to verify whether the value of 0.1 calculated using Mc Tigue's data (Mc Tigue, 1986) can be used.

On sample n°41, pore pressure evolution has been measured by the external liquid pressure sensors (figure 1). We believe that the two curves of figure 10 cannot be attributed to a "pore pressure" evolution at the sample scale, since the evolutions are different. All the involved microcracks are certainly not connected yet (i.e. below the so called the Percolation Threshold) and the REV is bigger than the characteristic dimension of the sample. Nevertheless, it is assumed that the equivalent "pore pressure" concept can be applied beyond REV using the two values given by the pressure sensors.

Consequently, Skempton's coefficient value which is the ratio of pore pressure change over mean stress change (equation 8) can be estimated by the given upper and lower bands .It is likely in the range of 0.02-0.08 (Figure 10). Using the following values for rock salt ($K_{fl} \approx 2.0$ GPa; $K_s \approx 23.5$ GPa; $\phi \approx 1\%$), the equation (9) leads to Biot's coefficient values in the range of 2.10^{-3}-8.10^{-3}. This result that has to be verified by further experiments. suggests that the extremely low value of 0.1 obtained using Mc Tigue's data seems to be realistic for an undisturbed salt.

Considering a likely low value of Biot's coefficient for undisturbed salt and a value of 1 at failure state (see section 3.1), one may conclude that Biot's coefficient may increases substantially during damage. The real evolution of such a hydro-mechanical parameter requires to perform further experiments. However, it may look like permeability evolution induced by damage as it was observed by Peach (1991) on Asse salt or by Stormont & Daemen (1992) on bedded salt from near Carlsbad at Waste Isolation Pilot Plant (WIPP) site.

CREEP AND RELAXATION TESTS

1. Methodology

To investigate the influence of damage and fluid-rock interaction, 7 multi-step compression triaxial tests with creep and relaxation paths were performed on rock salt samples from the same MDPA site.

During deviatoric loading, relaxation and creep steps were carried out with different values of stress corresponding to pre, or post damage initiation phases. Each step was short (20 minutes). Thus, only the transient viscoplastic behavior was studied : steady-state creep was not investigated.

The aim of this study was not to propose a new viscoplastic law for transient and irreversible strains but to estimate qualitatively the effects of damage and liquid-rock interactions on viscoplastic strain mechanisms.

As a matter of fact, for each type of test (relaxation or creep), viscoplastic strain rate measured at the end of the step, $\dot{\varepsilon}_{vp}$, associated with the deviatoric stress σ is reported in the plane [$\text{Log}\,\dot{\varepsilon}_{vp}$, $\text{Log}\sigma$]. In order to be compared to damage criterion, deviatoric stress value σ is kept at the beginning of each step for relaxation tests. Strain partition is assumed to calculate viscoplastic strain rate as follows:

$$\dot{\varepsilon}_{vp} = \dot{\varepsilon} - \dot{\varepsilon}_e \qquad (10)$$

$\dot{\varepsilon}$: total strain rate; $\dot{\varepsilon}_e$: elastic strain rate.

In case of relaxation test ($\dot{\varepsilon}=0$), viscoplastic strain rate is equal to elastic strain rate given by Hooke's law :

$$\dot{\varepsilon}_{vp} = -\dot{\varepsilon}_e = -\frac{\dot{\sigma}}{E} \qquad (11)$$

E : Young's modulus value, already measured is equal to 14 MPa.

For creep test, the measured strain rate is considered to be equal to the viscoplastic strain rate : there is no elastic strain change since axial stress and confining pressure are constant in a creep test. Associated parameters ($\dot{\varepsilon}_{vp}$, σ) for the different steps of the same test are measured or calculated at the same time in order to be compared easily.

2. Results and discussion

2.1 Multi-step jacketed relaxation tests

The parameters ($\dot{\varepsilon}_{vp}$, σ) corresponding to two jacketed relaxation tests, 20 MPa and 60 MPa of confining pressure, are reported in figure 11.

"Linearity limit" corresponding to the damage initiation obtained in "dry" condition is also given. From figure 11, two comments can be made :
a- Two classical features are found. On one hand, the higher the deviatoric stress, the higher the viscoplastic strain rate. On the other hand, viscoplastic strain rate depends slightly on confining pressure (i.e. mean stress) before damage initiation.
b- A significant increase of viscoplastic strain rate is observed beyond damage initiation, under 20 MPa of confining pressure. This effect is not significant when a high value of confining pressure up to 60 MPa is applied.

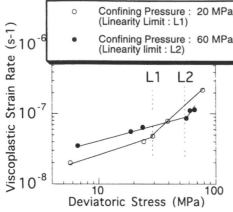

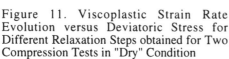

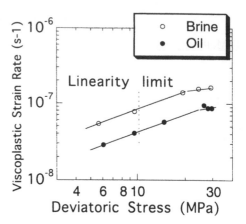

Figure 11. Viscoplastic Strain Rate Evolution versus Deviatoric Stress for Different Relaxation Steps obtained for Two Compression Tests in "Dry" Condition

Figure 12. Viscoplastic Strain Rate Evolution versus Deviatoric Stress for Different Relaxation Steps obtained for Two Compression Tests in "Wet" Condition (Brine and Oil).

2.2 Multi-step unjacketed relaxation tests

The parameters values ($\dot{\varepsilon}_{vp}$, σ) corresponding to two unjacketed relaxation tests with multi-steps, are reported in figure 12. "Lineary limit" corresponding to damage initiation obtained in "wet" condition from figure 9 are also given in figure 12. The two liquids, oil and brine were used as confining liquid at 20 MPa of pressure

The main difference with figure 11 is the continuous aspect of the ($\dot{\varepsilon}_{vp}$, σ) curve beyond the "linearity limit". Micro fracturing associated with damage does not seem to have a significant influence on strain mechanisms involved in relaxation. This looks very surprising and without any clear physical meaning, but it may be related to the pore-pressure build-up followed by a pore-pressure dissipation in the sample during the deviatoric loading.

Meanwhile, an interesting point can be noticed with respect to the influence of the chemistry of the confining liquid : viscoplastic strain rate associated with "oil" is systematically lower than that associated with "brine". This effect can be compared with that observed through the experiments for which a few authors introduced the "pressure solution" process as the main strain mechanism (i.e. Spiers et al., 1986).

2.3 Multi-step unjacketed creep tests

The parameters values ($\dot{\varepsilon}_{vp}$, σ) corresponding to two multi-step unjacketed creep tests are reported in figure 13. In creep test, viscoplastic strain rate is obtained directly from the measured strain curve ; no assumption on elastic parameters is necessary. "Lineary limit" values corresponding to damage initiation obtained in "wet" condition from figure 9 are also given in figure 13. Confining liquid submitted to 20 MPa or 60 MPa of pressure, was brine.

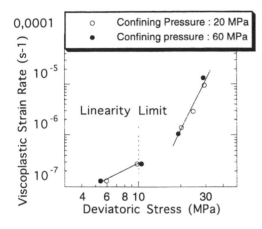

Figure 13. Viscoplastic Strain Rate Evolution versus Deviatoric Stress for Different Creep Steps obtained for Two Compression Tests in "Wet" Condition (Brine).

As for jacketed relaxation tests, the results on figure 13 show :

a- A drastic increase of viscoplastic strain rate $\dot{\varepsilon}_{vp}$ is observed beyond damage initiation.

This increase can be explained by the chemical composition of the confining liquid when it penetrates into microcracks ,or through a "lubrication" effect.

Since viscoplastic strain rate does not depend on the confining pressure value, the second phenomenon seems to be more significant : percolating liquid would have enhanced "sliding" on the lips of microcracks generated during damage.

b- The transient viscoplastic flow depends slightly on confining pressure (and thus on mean stress) *before* and *after* the damage initiation. This seems surprising since dilatancy associated with damage is known to be depend strongly on mean stress. An explanation may be a "threshold effect" : beyond a certain characteristic mean stress value, dilatancy does not depend any more on mean stress. This hypothesis is to be confirmed by complementary tests.

CONSEQUENCES FOR UNDERGROUND STORAGE

Regarding the safety assessment of underground storage in rock salt, this laboratory investigation allows to focus attention on the following points :

a- With respect to a hydrocarbon storage or to a likely flooded storage, it is clear that "dry" (damage and failure) criteria are not the only essential factors to be considered. The mechanical effect of pore pressure and possibly chemical effects in presence of brine have also to be taken into account. According to our experiments, the failure criteria in "wet" condition are significantly and systematically lower than the failure and damage criteria in "dry" condition.

b- In a disturbed salt, when brine percolates into microcracks, a mechano-chemical phenomenon called "stress corrosion" or "stress-aided corrosion process" may play a significant role. The key parameters governing such an effect in rock salt are not clearly identified yet. A better physical understanding based on laboratory and in situ experiments seems necessary.

c- When the damage criterion is over and the confining pressure value is low, dilatancy (i.e. microcracking) takes place and leads to increase the viscoplastic flow rate. Thus, a creep law only based on experiments conducted on undisturbed ("undamaged") samples may not be relevant for the study of the "near field" of storage which includes the so-called DRZ (Disturbed Rock Salt).

This confirms previous observations made by Pérami *et al.* (1993) that emphasized on the essential determination of damage criterion. Such a criterion, the knowledge of the in situ stress state around the storage, and also the appropriate constitutive law of salt (taking account of damage) allow to estimate the extent and the intensity of the Damaged Rock Zone (i.e. the "DRZ") and the resulting long term irreversible strains and displacements.

Let us carry out simple calculations for two underground structures : spherical and cylinder cavities subjected to an initial isotropic stress. The cavities are supposed to be excavated instantaneously. For both conditions ("dry" and "wet"), cavities are assumed to be full of liquid with a density close to that of water : radial stress on the cavities wall is equal to the corresponding hydrostatic pressure.

Failure and damage criteria in "wet" condition are deduced from figures 7 and 9. For both liquids, Tresca's criterion is assumed since the stress state slightly depends on mean stress. Cohesion values for failure criterion (C_e) and damage criterion (C_d) are 15 MPa and 5 MPa respectively (Table I). We are aware that this approach is not very realistic : rock salt in the vicinity of wall cavity is not totally in contact with liquid as it is the case of the unjacketed samples. But, a rough qualitative estimation of the mechanical stability may be done by means of two distinct analyses considering either a "dry" criterion or a "wet" criterion.

Table I : Poroelastic Analysis of Cylindrical and Spherical Analysis.
Constants C_d et C_f are Cohesion Related to Damage and Failure Criterion Respectively (Tresca), Obtained Both in "Wet" Condition.

	Criterion in "dry" condition (no permeability)		Criterion in "wet" condition	
	Cylindrical cavity	Spherical cavity	Cylindrical cavity	Spherical cavity
Evolution of deviatoric stress σ_{eq} with depth (initial stress σ_0)	$\sigma_{eq}=\sqrt{3}(\sigma_0+P_1)$	$\sigma_{eq}=\frac{-3}{2}(\sigma_0+P_1)$	$\sigma_{eq}=\sqrt{3}(\sigma_0+P_1)$	$\sigma_{eq}=\frac{-3}{2}(\sigma_0+P_1)$
Damage criterion	$\sigma_{eq}=(\frac{-\sigma_0}{0.04})^{1/2}$	$\sigma_{eq}=(\frac{-\sigma_0}{0.05})^{1/2}$	$\sigma_{eq}=2C_d$ (C_d=5 MPa)	$\sigma_{eq}=2C_d$ (C_d=5 MPa)
σ_{eq} value at cavity wall	32 MPa	29 MPa	10 MPa	10 MPa
	($\approx$1830 m)	($\approx$1950 m)	($\approx$580 m)	($\approx$660 m)
Failure criterion	$\sigma_{eq}=-1.43\sigma_0+10.3$	$\sigma_{eq}=-1.5\sigma_0+3$	$\sigma_{eq}=2C_f$ (C_f =15 MPa)	$\sigma_{eq}=2C_f$ (C_f =15 MPa)
σ_{eq} value at cavity wall	never	never	30 MPa ($\approx$1730 m)	30 MPa ($\approx$2000 m)

Comparison between the results of calculations on cylindrical and spherical cavities given in Table I shows the following results :

a- Failure and damage risks are strongly depend on the initial stress σ_0 and radial stress P_1 playing somehow a role of "lining" effect in "dry" condition.

b- As it is expected, failure and damage risks are more important in "wet" condition (i.e. when rock salt is supposed to be permeable). The risk is greater at short term when fluid pressure is low and thus deviatoric stress is high. At long term, relaxation of stress due to the viscoplastic behavior of rock salt leads to a decrease in deviatoric stress which improves the mechanical stability.

CONCLUSION

In the framework of safety assessment of underground storage and to study "altered" scenarii for deep storage of radioactive wastes, a laboratory investigation on fluid-effect interactions on mechanical behavior of rock salt was performed on the samples belonging to the Mines de Potasse d'Alsace (MDPA). The results of triaxial compression tests carried out without jacket show a significant effect of pore pressure and chemical composition of percolating liquid on damage and failure criteria.

The results of creep and relaxation tests show an increase of viscoplastic flow when the damage criterion is over under small mean stress. This effect is associated with dilatancy provided by microcracking.

All the experimental results confirm the necessity of further researches, experimental and theoretical as well, on Thermo-Hydro-Chemico-Mechanical (THCM) coupled phenomena in rock salt.

REFERENCES

ALTHAUS E., FRIZ-TÖPFER A., LEMP Ch., NATAU O., [1994], "Effects of water on strength and failure mode of coarse-grained granites at 300 °C", Rock Mech. Rock Engng., vol 27, 1-21.

COSENZA Ph., GHOREYCHI M., BAZARGAN-SABET B., MARSILY (de) G., [1999], "In situ rock salt permeability measurement for long term safety assessment of storage," International Journal of Rock Mechanics and Mining Sciences, vol. 36, 509-526.

COUSSY O., [1995], Mechanics of porous continua, John Wiley & Sons, New York.

DAVIDGE R. W.., PRATT P.L., [1964], "Plastic deformation and work hardenning in NaCl", Phys. Stat. Sol., vol. 6, 759-776.

KRANZ R. L., [1983], "Microcracks in Rocks: A Review", "Microcracks in rocks : a review", Tectonophysics, vol 100, 449-480.

JAEGER C. J., COOK N. G. W., [1969], "Fundamentals of Rock Mechanics", Methuen & Co Ltd, London.

Mc TIGUE D. F., [1986], "Thermoelastic response of fluid-saturated porous rock", J. of Geophys. Res., vol. 91, 9533-9542, .

PEACH C., [1991] "Influence of deformation on the fluid transport properties of salt rocks, " Ph. D Thesis Utrecht University, Geologica Ultraiectina, N°77, Hollande.

PÉRAMI R., CALEFFI, C., ESPAGNE, M. and PRINCE, W., [1993], "Fluage et microfissuration dans les stockages souterrains", Geoconfine '93, Arnould, Barrès et Côme (Eds.), 99-104 Balkema, Rotterdam.

SPIERS C.J., URAI J.L., LISTER G.S., BOLAND J.N. ZWART H.J., [1986], "The influence of fluid-rock interaction on the rheology of salt rock", Final report, Commission of the European Communities, EUR 10399.

STORMONT J.C., DAEMEN J.J.K., [1992], "Laboratory study of gas permeability changes in rock salt during deformation," International Journal of Rock Mechanics and Mining Sciences, vol. 29, 325-342 .

THOREL L., [1994], "Plasticité et endommagement des roches ductiles - Application au sel gemme". Ph. D. Thesis,Ecole des Ponts et Chaussées, Paris.

THOREL L., GHOREYCHI M.[1996], "Rock salt damage - Experiment results and interpretation", 3rd Conference on Mechanical Behavior of Salt , Palaiseau, France, 1993, 175-190, Trans. Tech. Publications, Clausthal-Zellerfeld, Germany.

Basic and Applied Salt Mechanics, Cristescu, Hardy, Jr & Simionescu (eds)
© 2002 Swets & Zeitlinger, Lisse, ISBN 90 5809 383 2

Humidity induced creep and it's relation to the dilatancy boundary

Udo Hunsche & Otto Schulze
Federal Institute for Geosciences and Natural Resources (BGR), Hannover, Germany

ABSTRACT

Numerous uniaxial tests with stepwise variation of the relative air humidity ϕ (%RH) of the surrounding air show that steady state creep rates are increased in average by a factor of 90 between ϕ = 0 and 75 %RH in a nonlinear way. A number of triaxial tests with stepwise variation of the confining pressure p show that steady state creep rates are increased by an average factor of about 30 at p = 0.1 MPa compared to high values of p. It is explained with moisture assisted microscopic recovery at areas with strain hardening at sites of loaded contacts around open grain boundaries and microcracks. This humidity induced creep is observed only in the dilatant stress domain where the salt is permeable for moisture and it is strongly dependent on the distance from the dilatancy boundary. Therefore, it is active only in the damaged zone around underground openings. A reliable empirical function F has been formulated on the basis of 255 creep test phases with constant conditions describing the acceleration of steady state creep rates as a function of ϕ, p and of the distance from the dilatancy boundary.

INTRODUCTION

It is well known that water or water vapour has a large influence on deformation of rocks. Especially rock salt is largely influenced, but this has not been considered in model calculations for mines, repositories or caverns in salt formations. It is the purpose of this publication to supply an equation for the description of the influence of air humidity and confining pressure on creep of rock salt, to give a deeper insight into the controlling mechanism and to serve a guide for application. This publication is based on the paper of Hunsche & Schulze (1996) where the influence of both – relative air humidity ϕ and confining pressure p - on creep rates have already been described by an equation based on a limited number of laboratory experiments. Meanwhile the experimental basis has considerably been extended. In addition our understanding of the effect of dilatancy and the dilatancy boundary on deformation and on humidity induced creep has been increased.

Humidity induced creep has been addressed by a number of publications in the past: Varo & Passaris (1977), Pharr & Ashby (1983), Horseman (1988), Spiers et al. (1989, 1990), Brodsky & Munson (1991), Jockwer et al. (1991), Schulze (1993), Hunsche, Schulze & Langer (1994), Hunsche & Schulze (1996), Le Cleac'h et al. (1996), Cristescu & Hunsche (1998).

The mechanical condition for humidity induced creep is generation, evolution, or healing of microcracks. As described e.g. by Cristescu & Hunsche (1993, 1996, 1998), Hunsche (1993), Hunsche & Hampel (1999), and shown in figure 1 the dilatancy boundary subdivides the stress space into two domains. In the dilatant domain at higher values of octahedral stresses τ microcracks are formed or growing with the result that dilatancy, damage, microacoustic emissions, seismic travel times, and permeability are increasing. Other related effects are becoming possible, e.g. tertiary creep, creep rupture, and also humidity induced creep. In the compressible domain at lower values of the octahedral shear stress τ existing microcracks are closed and practically no new ones are formed. This causes decrease of dilatancy, damage, microacoustic emissions, and seismic travel times. Then the other related effects are no longer possible, e.g. tertiary creep, creep rupture and humidity induced creep. It must be emphasised, that the dilatancy boundary is rather a band then a line where practically no volume change occurs, as indicated in figure 1.

This implies that the dilatancy boundary is a kind of safety boundary: Below it deformation is only due to creep by dislocation movement without damage, above the bouundary creep plus micro-cracking advances with the mentioned unfavourable effects including humidity induced creep.

In the following at first representative examples for uniaxial creep tests are given, then the evaluated and compiled test results. Furthermore, a representative example for the triaxial creep tests, the evaluation, and the compiled test results are shown. On this basis a new equation for the acceleration of the deformation rates due to humidity induced creep is given, which takes into account its dependence on the stress state relative to the dilatancy boundary. Finally the controlling mechanism is discussed.

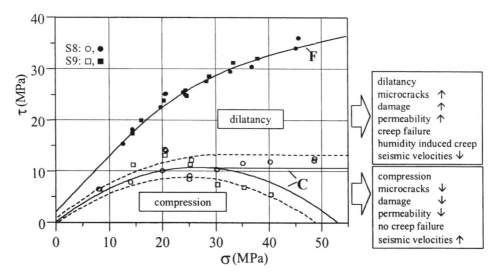

Figure 1: Failure surface F for short term failure and dilatancy boundary C for rock salt. The dilatancy boundary is rather a band, since between the full line for the dilatancy boundary and the parallel dashed lines practically no dilatancy occurs. After Cristescu & Hunsche (1998).

τ: octahedral shear stress, σ: octahedral normal stress.

UNIAXIAL CREEP TESTS

A great number of uniaxial and triaxial creep tests on various types of rock salt have been carried out at BGR for the evaluation of the humidity induced creep. 37 uniaxial tests with 220 test phases (testing steps) under constant conditions have been performed in two rigs. One rig (N7) allows simultaneous testing of five samples (standard size: 10 cm diameter, 25 cm length) at T = 22°C and controlled relative air humidity between ϕ = 0 and 75%RH. The other rig (F1) allows simultaneous testing of four samples (10 cm diameter, 25 cm length) at temperatures of 22 to 80°C and relative air humidity of 0 to 75%RH, see figure 2. Temperature is controlled better than 1K, relative air humidity is controlled better than 3%RH. In our experiments 75%RH was not exceeded, since this is the relative air humidity at the 3-phase-equilibrium (critical humidity) for rock salt over a great temperature range (Greenspan, 1977). Above the critical humidity rock salt behaves hygroscopicly and dissolves.

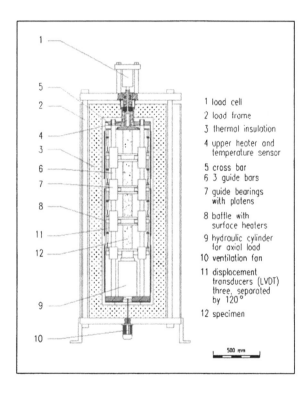

1 load cell
2 load frame
3 thermal insulation
4 upper heater and temperature sensor
5 cross bar
6 3 guide bars
7 guide bearings with platens
8 baffle with surface heaters
9 hydraulic cylinder for axial load
10 ventilation fan
11 displacement transducers (LVDT) three, separated by 120°
12 specimen

500 mm

Figure 2: Uniaxial rig (F1) at BGR for four simultaneous creep tests under controlled climate. Temperature: T = 22 to 80°C, relative air humidity: ϕ = 0 to 75%RH.

Figure 3 presents the creep curves of four uniaxial creep tests with stepwise variation of relative air humidity ϕ on two types of rock salt at room temperature. The duration was about two and a half years. This needs a very stable long term control for all the components of the system. All the great changes of the creep rates in figure 3 are caused by changes in air humidity, the other parameters are kept constant. As already stated in Hunsche & Schulze (1996) the qualitative results of the uniaxial creep tests are the following:

- The relative air humidity RH has a great effect on creep rate
- The specimens react immediately after a humidity change
- Repeated increase and decrease of the relative humidity show that the effect is reversible
- Each change of the relative humidity is followed by a normal transient phase, i.e. with decreasing creep rate.

Nearly steady state creep rates have been determined from the successive tests phases and were normalised by the values at 0%RH. The results of the respective tests in figure 3 are

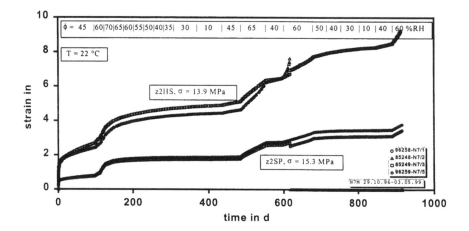

Figure 3: Four long term uniaxial creep tests on rock salt in the same rig with stepwise variation of relative air humidity ϕ. All changes of creep rates are due to changes in ϕ. The two salt types have different creep behaviour and exhibit different sensibility to air humidity. z2HS: Gorleben mine, z2SP: Asse mine. For creep rates see figure 4.

shown in figures 4. It can clearly be seen that the two rock salt types exhibit different sensibility to air humidity. The slower creeping z2SP (Speisesalz from the Asse mine) exhibits a lower sensibility than the faster creeping z2HS (Hauptsalz from the Gorleben mine). The creep rate is increased by a factor of 50 or 300, respectively, between 0 and 70 %RH. This demonstrates that the salt type has an influence on the sensibility to humidity. These acceleration factors are surprisingly large and should be taken into account in precise model calculations. In addition, the dependence on ϕ is very nonlinear, even on the logarithmic scale.

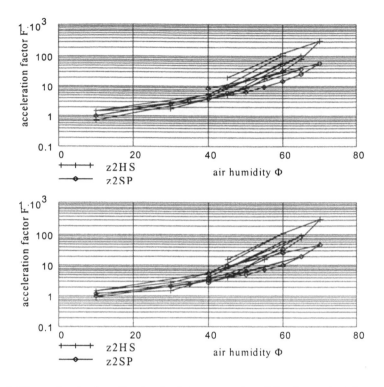

Figure 4: Normalised values for steady state creep rates of the four uniaxial creep tests on rock salt with stepwise variation of relative air humidity φ as shown in figure 3. The two types are differently sensible to the air humidity. F: acceleration function given in (2). σ = 15.3 MPa (z2HS) and 13.9 MPa (z2SP). z2HS: Gorleben mine, z2SP: Asse mine.

Altogether 37 uniaxial multistep long term creep tests on different types of rock salt have been performed and evaluated. This comprises 220 different test phases at constant conditions. All normalised steady state creep rates are compiled in figure 5 on logarithmic and linear scales. The results show a considerable scatter mainly due to different salt types. The great number of tests gives a reliable basis for further conclusions. The line in figure 5 represents the acceleration function F for humidity induced creep given by equation (2).

It could already be shown in Hunsche & Schulze (1996) that no change of the dependence on φ can be observed with respect to temperature between 22 and 50°C.

TRIAXIAL CREEP TESTS

Three multistep triaxial compressive creep tests with varying confining pressures p (0.1 to 15 MPa) and constant stress differences Δσ (14, 16 or 21 MPa) – i.e. 35 test phases – have been carried out at BGR. One of the used rigs is shown in figure 6. Figure 7 presents the representative creep curve of one of the tests with a duration of two years. The large influence of p on creep is very remarkable. The creep rate clearly decreases with increasing values of p. The main results are the following:

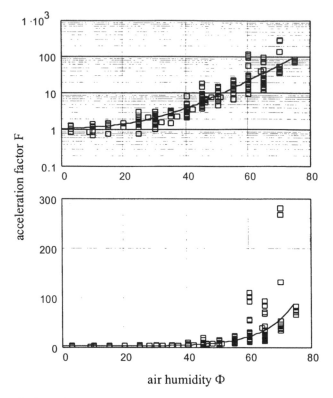

Figure 5: Compilation of 220 normalised steady state creep rates derived from uniaxial creep tests on different types of rock salt with stepwise variation of relative air humidity φ. Logarithmic and linear vertical scales. Lines: acceleration function F given in (2).

- The confining pressure p has a great effect on creep rates
- The sample reacts immediately after a pressure change
- Repeated increase and decrease of p shows that the effect is reversible
- Each change of p is followed by a normal transient creep phase, i.e. with decreasing creep rate
- The results for triaxial tests show the same characteristics as for uniaxial tests.

Again, the nearly steady state creep rates have been determined from the 35 successive tests phases of the 3 tests and then normalised by the rates at high confining pressures as shown in figure 8. It results that the effect of p is very nonlinear even on the logarithmic scale and reaches a factor of about 30 at p = 0.1 MPa for $\Delta\sigma$ = 14 MPa.

Figure 8 shows a very important additional result: The humidity induced creep gradually comes into effect only under dilatant stress conditions and the acceleration factor F is dependent on the distance of p from the dilatancy boundary. Near the dilatancy boundary and at stress states below it, the acceleration factor is F = 1. The values for the stress state on the dilatancy boundary as determined by Cristescu & Hunsche (1998) and given in equations (5), (6), (7) are indicated in figure 8 by triangles for the three used values of $\Delta\sigma$. The three lines in figure 8 represent the acceleration function F for humidity induced

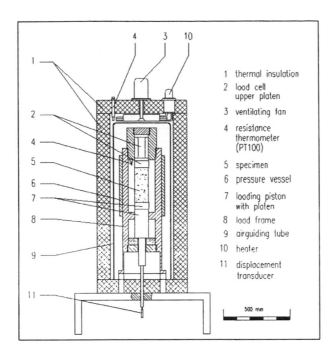

Figure 6: Triaxial rig (D1 to D4) at BGR for creep tests at temperatures up to 70°C.

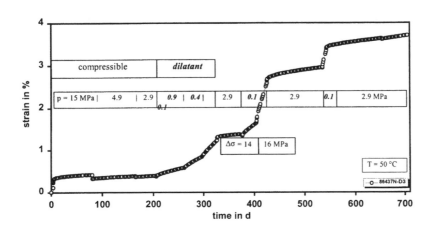

Figure 7: Triaxial long term creep tests with variation of confining pressure p at constant stress difference Δσ. It shows that the creep rate is only increased in the dilatant domain (italic ciphers). This is due to humidity induced creep.

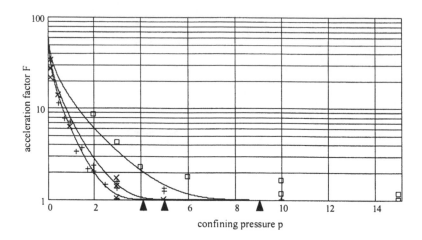

Figure 8: Compilation of 35 normalised steady state creep rates from three triaxial creep tests with variation of confining pressure p. It shows the change of creep rate in the dilatant domain due to humidity induced creep. T = 50°C. $\Delta\sigma$ = 14, 16 and 21 MPa; the dilatancy boundary (▲) is at p = 4.1, 5.1 and 8.6 MPa, respectively. The lines represent the respective acceleration functions F given by (2).

creep for the three values of $\Delta\sigma$ as given by equation (2). The acceleration rapidly increases with increasing distance from the dilatancy boundary. This is also visualised in figure 9 by the stress paths of the three tests.

The triaxial test results denote: increasing confining pressure (or minimum principal stress) suppresses humidity induced creep.

No difference was found between tests were 1 cm^3 of brine was injected behind the rubber jacket and tests performed without added brine. This means that the small amount of natural brine in the German domal salt (ca. 0.1wt%) is sufficient for the induction of humidity induced creep. Moisture works as a catalyst. This effect is totally suppressed in uniaxial as well as triaxial tests on specimens impregnated with a small amount of mineral oil (see Hunsche & Schulze, 1996, and Pharr & Ashby, 1983).
Brodsky & Munson (1991) have carried out triaxial tests on samples which were suddenly flushed with brine during the tests. They found that creep acceleration occurred below p = 3.5 MPa only, which is in good agreement with our tests. However, creep acceleration at higher confining pressures due to pressure solution, as predicted by Spiers et al. (1989, 1990), was not observed.

THE EQUATION FOR HUMIDITY INDUCED CREEP

On the basis of the presented results an average equation was developed for the description of the acceleration of creep due to relative air humidity and confining pressure, named humidity induced creep. The relative air humidity is the appropriate parameter for this equation since the 3-phase-equilibrium at 75%RH is not dependent on temperature. Therefore, humidity effects must be discussed in dependence to the relative air humidity – not to the absolute air humidity or the water content of the rock.

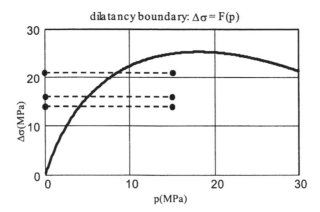

Figure 9: Stress conditions with respect to the dilatancy boundary in the 3 triaxial creep tests on rock salt for the evaluation of humidity induced creep.

Creep of dry salt or salt under high confining pressure is caused by dislocation movement. This has been described by Hunsche & Schulze (1994), Hunsche & Hampel (1999), Hampel & Hunsche (2000). Humidity accelerates this creep mechanism since it reduces strain hardening as explained in the next chapter. Therefore, we introduce the acceleration factor F(f,p).

The basic equation for the effect of relative air humidity and confining pressure on steady state creep is

$$\dot{\varepsilon}_{s,h}(\Delta\sigma, T, \phi, p) = \dot{\varepsilon}_s(\Delta\sigma, T) \cdot F(\phi, p) \tag{1}$$

$$F(\phi, p) = 1 + e(\phi) \cdot f(p) \qquad \text{(acceleration function)} \tag{2}$$

s stands for steady state, h for humidity
$\Delta\sigma$: stress difference, effective stress
T   : temperature in K
ϕ : relative air humidity RH in %
p   : confining pressure
$e(\phi)$: function for the effect of ϕ on creep
$f(p)$: function for the effect of p on creep
$\dot{\varepsilon}_{s,f}(\Delta\sigma, T, \phi, p)$: steady state creep rate with humidity induced creep
$\dot{\varepsilon}_s(\Delta\sigma, T)$: steady state creep rate of dry rock salt

The functions $e(\phi)$ and $f(p)$ have been defined with empirical considerations as follows:

$$e(\phi) = w \cdot \sinh(q \cdot \phi) \tag{3}$$

$$f(p) = \left\langle \left[1 - \left(\frac{p}{p_{dil}} \right)^g \right]^h \right\rangle \text{, with } \langle A \rangle = A \text{ for } A \geq 0 \text{ (or } p \leq p_{dil}) \text{ else } \langle A \rangle = 0 \tag{4}$$

It follows: $0 \leq f(p) \leq 1$.

$p_{dil} = p_{dil}(\Delta\sigma)$ is the function for the dilatancy boundary and p/p_{dil} is a reasonable measure for the distance from that boundary. As explained above it is an important result of our research work that the humidity induced creep is only active in the dilatant region and the acceleration F is strongly dependent on the distance from the dilatancy boundary. Therefore, the function f(p) is rather nonlinear and must incorporate the function for the dilatancy boundary given below.

The four parameters for the average acceleration function were determined as follows:

From triaxial tests: g = 0.31, h = 2.85

From uniaxial tests: w = 0.1, q = 0.1

These parameters must be used in equations (3) and (4); they differ from those given in Hunsche & Schulze (1996), because new test results are included and the dilatancy boundary is involved.

Maximal acceleration in this equation is: F(75%RH, 0 MPa) = 90

For the practical determination of acceleration function one needs the function for the dilatancy boundary p_{dil}. The following function is given by Cristescu & Hunsche (1998):

$$\tau_{dil}(\sigma) = f_1 \cdot \sigma^2 + f_2 \cdot \sigma \tag{5}$$

where τ = octahedral shear stress, σ = octahedral normal stress and
$f_1 = -0.01697 \text{ MPa}^{-1}$, $f_2 = 0.8996$.

For cylindrical symmetry (σ_1, $\sigma_2 = \sigma_3 = p$) with

$$\tau = \frac{\sqrt{2}}{3} \cdot \Delta\sigma, \quad \Delta\sigma = \sigma_1 - p, \quad \sigma = \frac{\sigma_1 + 2 \cdot p}{3} = \frac{\Delta\sigma + 3 \cdot p}{3}$$

it follows:

$$p_{dil}(\Delta\sigma) = -\frac{\Delta\sigma}{3} - \frac{f_2}{2 \cdot f_1} + \frac{1}{2 \cdot \sqrt{3} \cdot f_1} \cdot \sqrt{3 \cdot f_2^2 + 4 \cdot \sqrt{2} \cdot f_1 \cdot \Delta\sigma} \qquad \text{with } \Delta\sigma \leq \Delta\sigma_{dil,max} \tag{6}$$

and

$$\Delta\sigma_{dil}(p) = \frac{3}{\sqrt{2} \cdot f_1} - 3 \cdot p - \frac{3}{2} \cdot \frac{f_2}{f_1} - \frac{3}{2 \cdot f_1} \cdot \sqrt{2 - 4 \cdot \sqrt{2} \cdot f_1 \cdot p - 2 \cdot \sqrt{2} \cdot f_2 + f_2^2} \tag{7}$$

In order to give an impression of the results figure 10 shows the dependence of the acceleration function F on p for four different values of the relative air humidity ϕ. This also represents qualitatively the dependence on the distance from the surface of an underground opening. That means: The large effect of humidity on creep is in effect only in the disturbed zone around underground openings. In figures 5 and 8 the fit of the equation to the measured values is given.

DISCUSSION OF THE MECHANISM OF THE HUMIDITY INDUCED CREEP

The practical result of the evaluation is that air humidity has a large effect on creep deformation but only in the vicinity of underground openings, where the stress condition allows moisture to migrate through the network of microcracks. According to the tests this is the case only for rocks with stress conditions above the dilatancy boundary. This

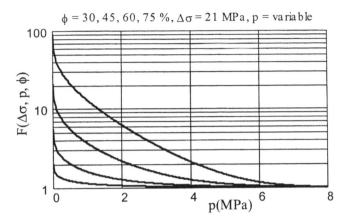

Figure 10: Dependence of the acceleration function F on the relative air humidity and the confining pressure p relative to the dilatancy boundary (p = 8.6 MPa for $\Delta\sigma$ = 21 MPa). Maximum acceleration is F(75%RH, 0 MPa) = 90.

means that there exists a shell of some decimeters thickness near the surface of drifts, shafts, and pillars – as measured in salt – where the rock is considerable more ductile than the rock more distant from the surface. The weakening is highly dependent on the relative air humidity in the adjacent opening and on the distance from the opening. A humidity value of 45 %RH accelerates creep by a factor of about five compared to dry condition or undisturbed salt. This should be considered in precise model calculations. Changes of humidity induced creep is observed in mines caused by changing relative air humidity due to the seasons, see Plischke & Hunsche (1989). The thickness of the excavation disturbed zone (EDZ) around openings in salt mines have been measured by Alheid et al. (1998) with seimic methods.

In accordance with the model of Pharr & Ashby (1983) and the general ideas of Horseman (1988) we explain our observations as follows: If rock salt is permeable, then water vapor can propagate through the network of open grain boundaries and microcracks. The water molecules in the humid air are preferably adsorbed at places of increased energy, i.e. at sharp convex surfaces and in strain hardened regions at contact areas in cracks and pores. At these sites the solubility is locally raised (and the critical relative humidity is decreased) and a thin brine film is formed. The stored mechanical energy, predominantly produced by strain hardening due to pile up dislocations will be dissipated by recovery, thus facilitating creep. This mechanism is strongly influenced by the local relative air humidity. Because moisture acts as a catalyst, only very small amounts are needed. Thus recovery of strain hardened material steadily takes place at varying places causing a great increase in creep rate. It results that a small range around open grain boundaries and cracks becomes ductile and carries the humidity induced creep. It is still unclear how far this effect extends into the grains. The marked transient creep phases after a humidity change show that moisture needs some time to come to a steady state.

Apparently humidity induced creep is possible only if the humid air can migrate through the dilatant rock salt. The results in figure 10 show how the pathways are steadily closed and humidity induced creep is gradually suppressed by the increasing confining pressure unless the acceleration factor F is equal to one below the dilatancy boundary (F = 1). The tests show that this process is reversible.

For a complete description of the humidity induced creep we need a constitutive model which would describe dilatancy and evolution of microcracks during creep, where the knowledge of the dilatancy boundary is of special importance. The constitutive model of Cristescu & Hunsche (1993, 1996, 1998) is an appropriate model. In figure 11 five empirical functions and some measurements for the dilatancy boundary of rock salt are shown together with the graph for the short term failure function. It is again obvious that the stress conditions, where the humidity induced creep starts, must be just above the dilatancy boundary.

Domal salt in Germany as used for the tests generally has a brine content of about 0.1wt%. It is, however, well known that bedded salt usually has a higher brine content of more than 1wt%. Therefore, the question arises: Is there a brine content, above which humidity induced creep is always present, independent of air humidity and confining pressure? There exists a comparative study of Le Cleac'h et al. (1996) on French "milky salt", a bedded rock salt with 1 to 5wt% brine, with German domal salt from the Asse mine with about 0.1wt% brine. Their uniaxial creep tests in a microcell under a microscope show that "milky salt" exhibits increased creep rates under all conditions; change of air humidity has only little effect. Asse salt, however, is considerably affected by air humidity. Our own tests on "milky salt" under triaxial conditions also show increased creep rates. We conclude, that accelerated creep due to high brine content is permanently present in rock salt above about 1wt%, if the brine is finely dispersed.

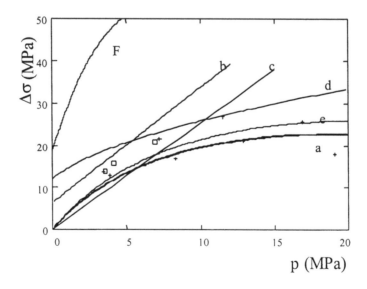

Figure 11: Dilatancy boundaries for rock salt determined by several authors

$\Delta\sigma$: Stress difference σ_1 - p, σ_1 : axial stress, p : confining pressure
a : Cristescu and Hunsche (1993, 1996, 1998),
b : Spiers et al. (1989),
c : Van Sambeck et al. (1993), d: Thorel and Ghoreychi (1996),
e : Nonlinear fit of the data of Van Sambeck (1993) by Hunsche.
F : Short term failure surface for rock salt by Cristescu and Hunsche (1998).
□ : Beginning of humidity induced creep
+ : Measured values for the dilatancy boundary (this publication)

Another arising question is: Does humidity cause crack corrosion and thus enhances the development of microcracks and dilatancy? For the solution of this question we have measured the microacoustic emission rates (AE) during uniaxial creep tests with stepwise variation of relative air humidity. It results that the AE-rates are nearly constant at changing creep rates for all values of humidity. This indicates that humidity does not initiate additional growth of cracks due to crack corrosion, the acceleration is not caused by crack corrosion.

CONCLUSIONS

It can be concluded:

- Humidity induced creep is present under uniaxial stress conditions.

- Humidity induced creep is active under triaxial stress conditions only in the dilatant domain. The acceleration depends on the distance from the dilatancy boundary.

- In average the maximal creep acceleration by humidity is given by a factor of $F = 90$ at 75% relative air humidity.

- Humidity induced creep has been described by an equation including the dependence on air humidity ϕ, confining pressure p (minimum principal stress), and stress difference $\Delta\sigma$. It regards the stress condition relative to the dilatancy boundary.

- Creep acceleration due to humidity is only important in the disturbed zone around underground openings.

- Rock salt with a brine content above about 1wt% exhibits permanent accelerated creep

- A reliable equation for the acceleration due to humidity induced creep is now available.

REFERENCES

Alheid, H.-J., M. Knecht & R. Lüdeling (1998): *Investigation of the long-term development of damaged zones around underground openings in rock salt.*- Proc. NARMS'98, Cancun 1998, Int. J. of Rock Mech. & Min. Sci. 35: 4-5, Paper No. 27.
Brodsky, N. S. & D. E. Munson (1991): *The effect of brine on the creep of WIPP salt in laboratory tests.*- In: Proc. of 32nd US Symp. on Rock Mech. as a Multidisciplinary Science, Norman (USA) 1991; Editor: J.-C. Roegiers; 703 - 712. Balkema, Rotterdam.
Cristescu, N. & U. Hunsche (1993): *A constitutive equation for salt.*- In: Proc. 7. Int. Congr. on Rock Mech., Workshop on Rock Salt Mech. (Aachen) 1991, v. 3, p. 1821 - 1830. Balkema, Rotterdam.
Cristescu, N. & U. Hunsche (1996): *A comprehensive constitutive equation for rock salt: determination and application.*- In: The Mechanical Behavior of Salt III, Proc. of the Third Conf., Palaiseau (1993).
Cristescu & U. Hunsche (1998): *Time Effects in Rock Mechanics.*- Series: Materials, Modelling and Computation. John Wiley & Sons, Chichester (UK). 342 p.

Greenspan, L. (1977): *Humidity fixed points of binary saturated aqueous solutions.*- J. of Research of the Nat. Bureau of Standards.- A. Physics and Chemistry, 81A, No.1.

Hampel, A. & U. Hunsche (2000): *Extrapolation of creep of rock salt with the composite model.*- In: The Mechanical Behavior of Salt V; Proc. of the Fifth Conf., (ME-CASALT V), Bucharest 1999. (This volume)

Horseman, S. T. (1988): *Moisture content - a major uncertainty in storage cavity closure prediction.*- In: The Mechanical Behavior of Salt II, Proc. of the Second Conf., Hannover (Germany) 1984; Editors: Hardy, H. R. Jr. & M. Langer; p. 53 - 68. Trans Tech Publications, Clausthal.

Hunsche, U. (1993): *Failure behaviour of rock salt around underground cavities.*- In: H. Kakihana, H. R. Hardy, Jr., T. Hoshi & K. Toyokura (Editors), Proc. Seventh Symp. on Salt, Kyoto (Japan). April 1992, Vol I, p. 59 - 65, Elsevier, Amsterdam.

Hunsche, U. & A. Hampel (1999): *Rock salt - the mechanical properties of the host rock material for a radioactive waste repository.*- Engineering Geology 52, 271 - 291. Elsevier, Amsterdam.(Special Issue on Nuclear Waste Management in Earth Sciences. Eds.: Ch. Talbot & M. Langer)

Hunsche, U. & O. Schulze (1994): *Das Kriechverhalten von Steinsalz.*- Kali und Steinsalz, v. 11 (8/9), 238 - 255.

Hunsche, U. & O. Schulze (1996): *Effect of humidity and confining pressure on creep of rock salt.*- In: The Mechanical Behavior of Salt III; Proc. Third Conf., Palaiseau (France) 1993; Editors: M. Ghoreychi, P. Berest, H. Hardy, Jr. & M. Langer; p. 237 - 248. Trans Tech Publications, Clausthal.

Hunsche, U., O. Schulze & M. Langer (1994*): Creep and failure behaviour of rock salt around underground cavities.*- In: Der Bergbau an der Schwelle des XXI. Jahrhunderts, Proc. 16th World Mining Congress (WMC). Sofia (Bulgarien) 1994. Vol. 5, p. 217 - 230. Bulgarian National Organizing Committee, Sofia.

Jockwer, N., J. Mönig, U. Hunsche & O. Schulze (1991): *Gas release from rock salt.*- In: Proc. of a Workshop on 'Gas Generation and Release from Radioactive Waste Repositories', organised by NEA in cooperation with ANDRA, Aix-en-Provence (France) 1991; 215 - 224. OECD, Paris.

Le Cleac'h, J.M., A. Ghazali, M. Deveughele & J. Brulhet (1996): *Experimental study of the role of humidity on the thermomechanical behaviour of halitic rocks.*- In: The Mechanical Behavior of Salt III; Proc. Third Conf., Palaiseau (France) 1993; Editors: M. Ghoreychi, P. Berest, H. Hardy, Jr. & M. Langer; p. 231-236. Trans Tech Publications, Clausthal.

Pharr, G.M. & M.F. Ashby (1983): *On creep enhanced by a liquid phase.*- Acta metall., v. 31, 129 - 138.

Plischke, I. & U. Hunsche (1989): *In situ-Kriechversuche unter kontrollierten Spannungsbedingungen an großen Steinsalzpfeilern.*- In: Rock at Great Depth, Proc. ISRM-SPE Int. Symp., Pau (France) 1989; Editors: Maury, V. & D. Fourmaintraux; v. 1, p. 101 - 108. A.A.Balkema, Rotterdam.

Thorel, L. & M. Ghoreychi (1996): *Rock salt damage - experimental results and interpretation.*- In: The Mechanical Behavior of Salt III, Proc. of the Third Conf., Palaiseau (1993).

Schulze, O. (1993): *Effect of humidity on creep of rock salt.*- Geotechnik-Sonderheft 1993, 169 - 172. Essen.

Spiers, C. J., C. J. Peach, R. H. Brzesowsky, P. M. Schutjens, J. L. Liezenberg & H. J. Zwart (1989): *Long-term rheological and transport properties of dry and wet salt rocks.*- Final Report, Nuclear Science and Technology, Commission of the Europian Communities, EUR 11848 EN.

Spiers, C. J, P. M. Schutjens, R. H. Brzesowsky, C. J. Peach, J. L. Liezenberg & H. J. Zwart (1990): *Experimental determination of constitutive parameters governing creep of rocksalt by pressure solution.*- In: Deformation Mechanisms, Rheology and Tectonics; Editors: Knipe, R. J. & E. H. Rutter; Geological Special Publication No. 54, p. 215 - 227.

Van Sambeck, L., A. Fossum, G. Callahan & J. Ratigan (1993): *Salt mechanics: Empirical and theoretical developments.*- In: H. Kakihana, H. R. Hardy, Jr., K. Toyokura and T. Hoshi (Editors), Proc. Seventh Int. Symp on Salt, Kyoto (Japan). April 1992, Vol I, p. 127 - 134. Elsevier, Amsterdam.

Varo, L. & E. K. S. Passaris (1977): *The role of water in the creep properties of halite.*- In: Proc. Conf. Rock Eng., Brit. Geotech. Soc./Univ. Newcastle upon Tyne, p. 85 - 100.

Basic and Applied Salt Mechanics, Cristescu, Hardy, Jr & Simionescu (eds)
© 2002 Swets & Zeitlinger, Lisse, ISBN 90 5809 383 2

Mechanical phenomena and ground water at salnic saline

Ervin Medves
S.C. IPROMIN S.A., Bucharest

Christian Marunteanu, Victor Niculescu & Sorin Mogos
University of Bucharest, Bucharest Romania

Christian Stoicescu
SALRM, Bucharest

ABSTRACT

The salt massif SLANIC-PRAHOVA, including three exploitation fields, respectively Victoria (with small rooms and square pillars on 11 levels), Unirea (with large trapezoidal rooms on 3 levels) and Cantacuzino (with small rooms and square pillars on 6 levels in process of exploitation) is affected by the water infiltration from Slanic stream and from the underground accumulations of meteoric waters. These infiltrations created in time channels, dissolution and partial submersion of some mining works, aleatory actions hard to predict. The effect of these manifestations concreting in geomechanical phenomena that are presented and described in this paper. Our research tried to characterise the phenomena in time and space and allowed the elaboration of some solutions for stopping or limiting the undesirable effects.

SLANIC SALT DEPOSIT

Slanic salt deposit is situated in the southern part of the East Carpathian Bend, being provided by the Miocene formation of the Slanic syncline.

The salt carrying formation is the Salt formation (Middle Miocene), disposed on the tuffs and the Globigerina marls of Slanic tuffs. Shales, siltstones, clayey-marley breccia, salt and gypsum represent salt formation. The salt deposit is developed as a lens with the plane view of triangular shape, 2.7 km long (E-W) and 1.8 km wide (N-S). The profile of the salt lens looks like a pillow with the maximum thickness of 500 m.

MINING WORKS

The salt massif Slanic includes three main exploitation fields, respectively *Unirea* (with large trapezoidal rooms on 3 levels - opened in the 19th century and closed in different stages between 1935 and 1970), *Victoria* (exploited using the method with small rooms and square pillars on 11 levels and closed in 1992) and *Cantacuzino* (with small rooms and square pillars on 6 levels in process of exploitation).

GROUND WATER EFFECTS

The presence and the circulation of ground water in the covering formation of the salt body cause adverse effects in the salt mine. The groundwater derives from percolation and accumulation of meteoric water and from the water infiltration from Slanic stream crossing over the salt deposit. The waters circulating through permeable formations that are in contact with the salt body created in time solution caverns and channels and determined subsidence phenomena and sometimes partial salt mine flooding.

GEOMECHANICAL PHENOMENA IN THE *CANTACUZINO* FIELD

In April and May 1996, following intense precipitation, the salt mine was partially flooded through the shaft *Cantacuzino*. The active solution and erosion of salt by circulating brine were appreciable, resulting in the formation of small and large channels and caverns. A large depression on the back of the salt produced subsidence phenomena accompanied by the fissuring of the ground and of the stream bed. The directions of the drainage of the waters to the shaft are shown on the map of the apparent resistivity few days after the flooding of the mine (Fig. 1).The sections from the Figure 2 represent geological and geoelectrical profiles based on boreholes and VES data. This process determined water inflow and partial flooding of the mine through the shaft in the directions of progressive solution of salt: circulating brine, brine from the old mine *Baia Verde* and water from *Slanic* and *Tulburea* streams.

On the basis of these ascertainments and measurements, measures of remediation have been taken: lining of Slanic stream channel using a clay layer and cemented concrete slabs, cement injections around the shaft location, especially against the water flow main directions, reconstruction of the concrete lining of the shaft from the surface to the first exploitation level (about 100 m depth).

As consequence of these measures, the water inflow in the shaft and the flooding of the mine stopped and the main draining system of the underground waters became again the Slanic valley. The apparent resistivity map with the main underground water flow directions after the impermeabilization measures in December 1996 is presented in the Figure 3.

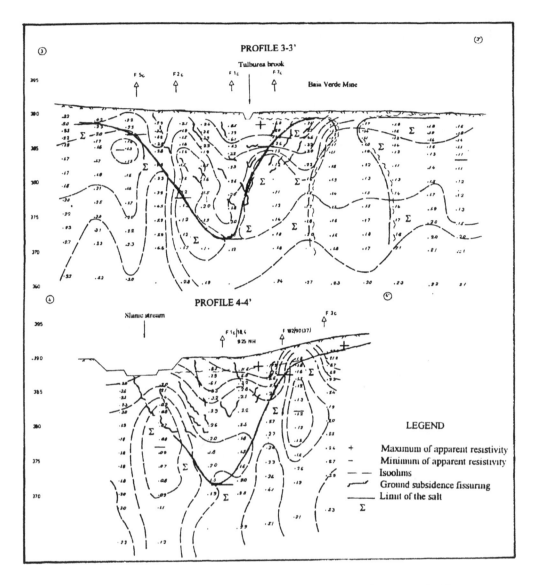

Figure 1. Geological and Geoelectrical Profiles Showing the Subsidence
Phenomena Due to the Dissolution of Salt.

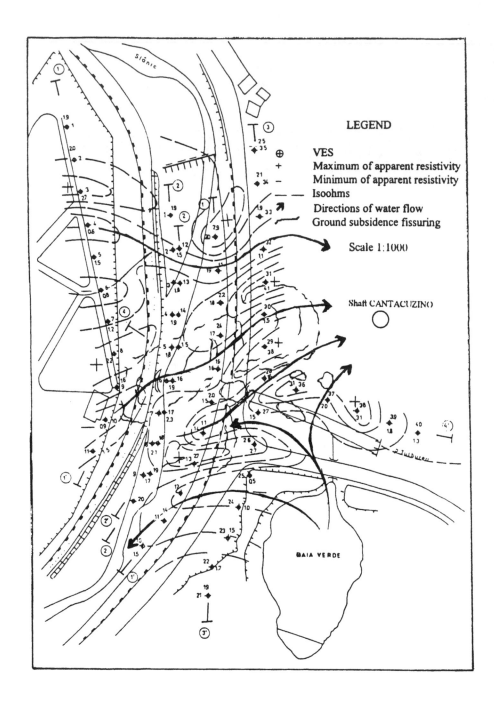

Figure 2.   Apparent Resistivity Map with the Main Underground Water Flow
Directions and Ground Subsidence Phenomena in June 1996.

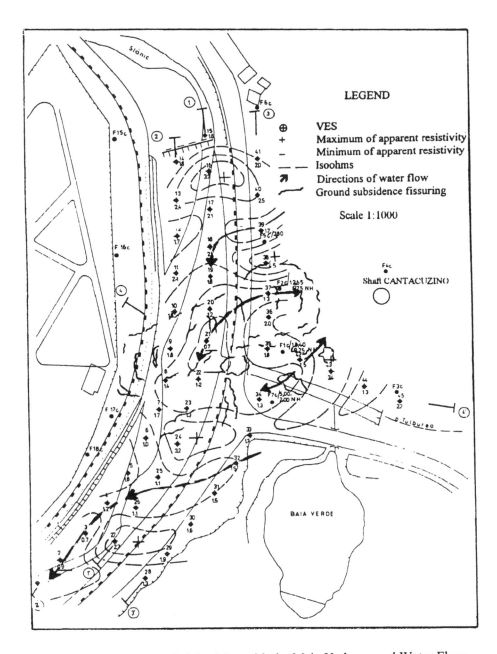

Figure 3. Apparent Resistivity Map with the Main Underground Water Flow
Directions after Impermeabilization Measures in Decemner 1996.

Basic and Applied Salt Mechanics, Cristescu, Hardy, Jr & Simionescu (eds)
© 2002 Swets & Zeitlinger, Lisse, ISBN 90 5809 383 2

Permeation and development of dilatancy in rock salt

T.Popp & H.Kern
Institut für Geowissenschaften, Universität Kiel, Germany

O.Schulze
Bundesanstalt für Geowissenschaften und Rohstoffe, Hannover

Abstract

Crack-sensitivity of ultrasonic wave velocities is used to monitor the development of the micro-structure and to identify the onset of dilatancy. Changes of wave velocities and permeability, which are characteristically related to the generation of micro-cracks, allow the determination of the dilatancy boundary. The equation given by Cristescu & Hunsche (1998) is confirmed. During loading in the dilatant stress domain the patterns of micro-cracks develop anisotropically, depending on the geometry of loading. Cracks are found to propagate predominantly parallel to the direction of the maximum principal stress. Therefore, measurable changes in permeability and the permeation of fluids affecting strength are directly coupled with both crack density and orientation of the micro-cracks. The results are used to assess loading conditions as expected in situ. We conclude that pore pressure effects on strength and integrity due to permeation are of concern only in the dilatant stress domain where dilatancy and permeability develop with strain.

1. Introduction

Rock salt with its low permeability provides an excellent geologic host rock for storage cavities (oil, gas) and waste repositories. Laboratory measurements on intact or re-compacted rock salt samples give very low permeability ($k < 10^{-20}$ m^2). However, experiments on un-jacketed samples (e.g. Fokker, 1995; Cosenza and Ghoreychi, 1998) give evidence that fluids may permeate into the polycrystalline salt microstructure as a result of increasing pore fluid pressure, reducing the effective mean stress. As a consequence marked differences in strength are observed between un-jacketed and jacketed specimens deformed in triaxial testing (see Fig. 1). Also, fluid pore pressure can have a remarkable effect on permeability of rock salt. Kenter et al. (1990) found that rock salt shows a rapid and significant increase in permeability at fluid pore pressures exceeding the confining pressure (see Fig. 2). Both, the increase of permeability and the reduced strength are interpreted as a result of micro-fracturing caused by pore pressure effects (Fokker, 1995).

It is important to note that many samples used for the experiments were disturbed due to drilling and sample preparation (e.g. Peach, 1991; Popp & Kern, 1998). This pre-damaged state of the specimens facilitates permeation of fluids (e.g. Kenter et al., 1990). Fluid permeation is not confirmed when carefully re-compacted rock salt specimens are used as is documented in Fig. 3. Our strength tests which were performed under similar conditions as reported in Fig. 1 demonstrate that the occurrence of pore pressure effects strongly depends on the initial permeability. Differences in permeability were produced by different pre-treatments. In case of "slight pre-

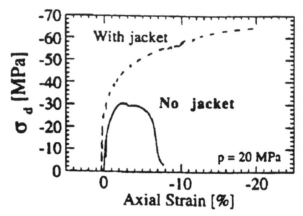

Fig. 1: Deviatoric stress versus axial strain during triaxial testing in compression with and without jacket; confining pressure p = 20 MPa. Without jacket failure strength is reduced to uniaxial strength. Taken from Cosenza & Ghoreychi (1998).

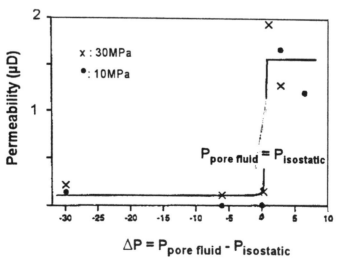

Fig. 2: Permeability of rock salt during isostatic loading p_{iso} (x : 30 MPa, • : 10 MPa) as function of excess pressure $\Delta p = (p_{fluid} - p_{iso})$. As soon as p_{fluid} exceeds p_{iso}, the permeability is increasing step-like. Taken from Kenter et al. (1990).

damage" the specimen was fractured during an isostatical loading at p_{iso} = 10 MPa by increasing the fluid pressure, first, at top of the cylindrical specimen via a small pipe through the axial loading piston, and second, via a corresponding pipe into a small blind hole at bottom ($\phi \approx$ 5 mm, l = 10 mm; for details see Fig. 22). In both cases hydraulic-fracturing is initiated at an excess fluid pressure of $\Delta p \approx$ 2 MPa in accordance with the results of Kiersten (1988), see Fig. 16. Inspection of such a pre-treated specimen revealed a single crack plane, intersecting the specimen in its length with an angle of approx. 15°. The "heavily pre-damaged" specimen was deformed by a reduction of the confining pressure from initial isostatic pressure conditions at p_{iso} = 15 MPa to a final p_c = 0.6 MPa (σ_1 = 15 MPa = const.), creating a marked disturbance of the microstructure. Subsequent triaxial deformation of the "heavily pre-damaged" specimen is associated with a significant reduction of the hardening behavior and failure strength, similar to that observed in the un-jacketed specimen (Fig. 1).

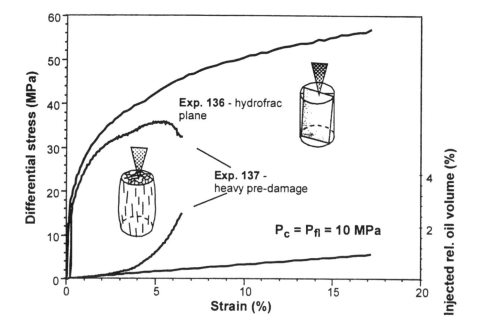

Fig. 3: Strength testing on pre-damaged specimen. Fluid injection via axial loading piston. Fluid pressure, p_{fluid}, is equal to confining pressure, p_c = 10 MPa. The transition from non-dilatant to dilatant deformation is at a flow stress of $\Delta\sigma \approx$ 20 MPa.

Therefore, we conclude that permeability and related pore pressure effects are very sensitive to the state of dilatancy. Dilatancy is defined as the inelastic increase of the volume during deformation (e.g. Cristescu and Hunsche, 1998). Deformation under conditions of differential stress causes dilatation or compaction, which give rise to changed pore space properties. The amount of change in pore space during deformation depends on the loading conditions with respect to the so-called dilatancy boundary. Understanding the effect of pore fluid pressure on the mechanical behavior of rocks requires knowledge of the state of the initial dilatancy and of the evolution of the micro-crack structure. Analysis of the interdependence between stress, strain and hydraulic properties can make use of the fact that petrophysical properties are sensitive to dilatancy, that is crack development or crack healing. This is schematically illustrated in Fig. 4. Exceeding a critical differential stress (i.e. the transition from non-dilatant to dilatant behavior), dilatancy is associated with remarkable changes of volumetric strain, electric conductivity, acoustic emission, permeability, and ultrasonic wave velocities (longitudinal: Vp and transversal: Vs). As shown by Peach (1991) and Stormont & Daemen (1992) the onset of dilatancy is generally associated with an increase of permeability of several orders of magnitude. Nevertheless, the relative low permeability of rock salt makes continuous measurement of transport properties during deformation very difficult. Also the interpretation of variations of electrolytic conductivity during deformation is not simple (Kern, Popp & Takeshita, 1992). Because elastic wave velocities are very sensitive to micro-cracks, monitoring the changes of Vp and Vs provides clues for an indirect estimation of changes in pore space during deformation (Popp and Kern, 1998).

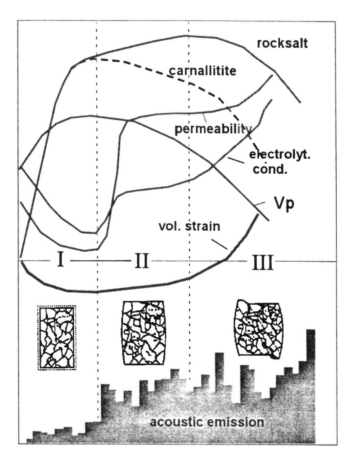

Fig. 4: Syndeformational petrophysical monitoring illustrated as schematic stress-strain-diagram. Properties like volume, electrical conductivity, acoustic emission, permeability, and ultrasonic wave velocity are used to monitor dilatancy.

The aim of this work is twofold. First, we examine the evolution of dilatancy and permeability in natural rock salt deformed during loading in compression (path m = -1) and extension (path m = +1) focusing, in particular, on the determination of the transition from compressive (non-dilatant) to dilatant deformation (i.e. the dilatancy boundary). Measurements of gas-permeability were combined with measurements of the ultrasonic longitudinal (V_p) and transversal wave (V_s) velocities. Second, we present results from the investigation of the effect of pore pressure on the mechanical properties of rock salt which are of importance for the assessment of the stability and long-term integrity of repositories.

2. Monitoring of dilatancy by ultrasonic wave velocities and permeability

The combined measurements of permeability and ultrasonic wave velocities during mechanical testing were performed in a Karman-apparatus. The methods to determine permeability (steady state and pulse peak technique) are described in detail by Popp & Kern (1998). Gas was used as the permeating fluid. The direct determination of volume change (change of porosity) is not possible in this device, but the gas-accessible porosity can easily be measured. The arrangement to measure Vp and Vs is schematically shown in Fig. 5. The device allows simultaneous measurements of permeability and ultrasonic wave velocities parallel to the specimen axis which can be loaded in compression or extension. In accordance to this arrangement, the changes of Vp and Vs during deformation are different depending on loading geometry. In case of loading in compression (load path m = -1) Vs reacts more sensitively to dilatancy and crack propagation than Vp. In case of loading in extension (load path m = +1) the reverse is true, that is, Vp reacts more sensitively than Vs.

A great number of strength tests as well as creep tests were performed to investigate dilatancy. Changes of the microstructure during deformation are documented by characteristic changes of the physical parameters as deformation proceeds. A typical set of experimental data from a strength test is shown in Fig. 6, where the normalized ultrasonic wave velocities Vp, Vs and the gas-permeability are plotted as a function of strain (loading in compression; controlled strain rate $\dot{\varepsilon}$ = 10^{-5} s^{-1}; confining pressure p_c = 5 MPa = const.).

Initial compaction is characterized by a significant increase of both ultrasonic wave velocities and a slight decrease of gas-permeability. In addition to the compaction during the preceding isostatic loading phase, this is clearly due to further closure of micro-cracks and the interruption of pre-existing interconnections at open grain boundaries at the very beginning of triaxial loading. Further deviatoric deformation gives rise to marked changes of the physical properties in the opposite direction. The onset of dilatancy during triaxial deformation in compression is associated with a pronounced velocity decrease and a coeval significant increase of gas-permeability (compare Fig. 5 and Fig. 13), indicating micro-fracturing and opening of grain boundaries. The increase in permeability (during loading in compression, loading path m = -1) is highest at the beginning of the loading phase. Unfortunately, the permeability changes cannot be monitored continuously with the applied pulse technique, which has to be used for measuring low permeability. By contrast, ultrasonic wave velocities which are very sensitive to micro-cracks can be measured continuously.

The experimental results in Fig. 6 convincingly document that the pronounced change in the sign of the slope of a velocity vs. strain curve is a sensitive indicator for the strain-related transition from compaction to dilation. As already mentioned, in case of compressional loading (m = -1) Vs starts to decrease earlier than the corresponding Vp. Thus, the onset of dilatancy is indicated by the maximum Vs. As is shown schematically in Fig. 4 and experimentally confirmed in Fig. 13, further straining of the sample give rise to characteristic changes of permeability and ultrasonic wave velocities which are closely related to changes of the microstructure. After the initial increase, a sharp velocity drop is indicating the onset of dilatancy. This is followed by a phase of nearly constant velocity decrease. A second rapid drop in velocity correlates with the failure of the specimen as indicated by the maximum differential stress (failure strength) at the end of a strength test.

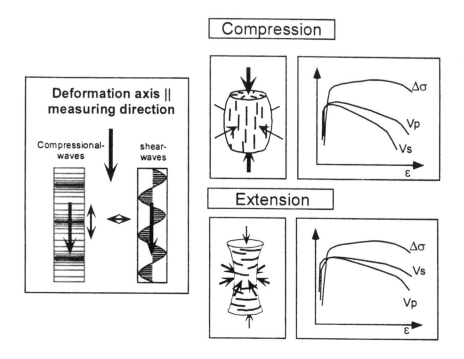

Fig. 5: Ultrasonic wave velocities Vp and Vs (left) and their variation with strain in compression and extension in relation to the stress curves. Note: The crack pattern develops anisotropically in dependence on the load geometry.

The relations between crack closure, crack opening and changes of ultrasonic wave velocities during deformation are caused by the following features. There are two types of micro-cracks developing during loading in compression: (1) cracks oriented perpendicularly to the maximum principal stress (σ_z) which close, and (2) cracks oriented parallel to the maximum principal stress direction which are continuously opening at a differential stress above the dilatancy boundary (see Fig. 6). Following Ayling, Meredith & Murrel (1995), a measure for the density of micro-cracks can be derived from expressions (3) and (4) relating the crack density parameter δ to Vp and Vs (the ultrasonic wave velocities are measured parallel to the z-axis, i.e. the axial direction of the specimen):

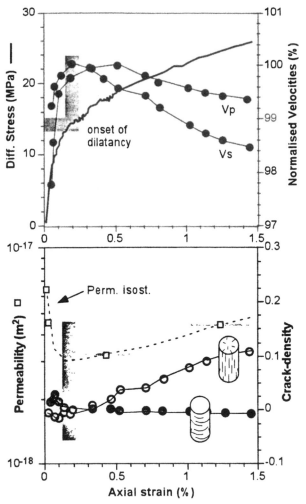

Fig. 6: Changes of physical properties as a function of compressive strain (strength test; controlled strain rate $\dot{\varepsilon} = 10^{-5}$ s^{-1}; confining pressure $p_c = 5$ MPa; temperature T = 30 °C).

Upper part: Diff. Stress and normalised velocities (Vp, Vs)

Lower part: Permeability and crack-density parameter.

Onset of dilatancy is determined by maximum in Vs. Crack-density is derived by the equations given by Ayling et al. (1995).

$$\delta_z = 0.92 \cdot (1 - (V_p/V_{p,max})^2 - 0.57 \cdot (1 - V_s/V_{s,max})^2) \qquad (3)$$
$$\delta_{x,y} = 4.28 \cdot (1 - (V_s/V_{s,max})^2 - 1.77 \cdot (1 - V_p/V_{p,max})^2) \qquad (4)$$

$V_{p.max}$ and $V_{s.max}$ are the velocities in the undisturbed or virgin rock (resp. the maximum velocities at the maximum isostatic stress, $p_{iso} = 30$ MPa). δ_z stands for micro-cracks lying perpendicular to the z-axis, and $\delta_{x,y}$ stands for the micro-cracks lying parallel to the z-axis (that is normal to the x- and y-axes). For details see Ayling, Meredith & Murrel (1995). It has to be stated that these empirical equations have been derived for the investigation of granite which behaves elastically or has no ductile behavior. But, the equations are obviously appropriate also to describe the anisotropic development of crack pattern in rock salt.

The transition from non-dilatant to dilatant behavior is discussed in detail by Cristescu & Hunsche (1998). It has to be mentioned that both types of testing, strength and creep tests, yield to the same result for the onset of dilatancy. Exemplary, Fig. 7 shows the variation of volume change and micro-acoustic emissions (AE) in a true triaxial compressional creep test. By stepwise increase of the octahedral shear stress the passing of the dilatancy boundary (C) at $\tau = 7.3$ MPa is indicated by the transition from a small volume compaction to volume increase and considerable increase of acoustic emission.

In Fig. 8a we have plotted the stress conditions for the dilatancy boundary of rock salt as derived from the transversal wave velocity (Vs) reversals in a differential stress versus confining pressure diagram during loading in compression. This dilatancy boundary as derived from our experimental data corresponds accordingly with the dilatancy boundary determined by Cristescu & Hunsche (1998) – compare Fig. 8b where the dilatancy boundary (C) is plotted with octahedral shear stress τ versus mean stress σ (see Cristescu & Hunsche (1998), page 109 ff.):

$$\tau_{Dil} = 0.8996 \cdot \sigma - 0.01697 \cdot \sigma^2 \tag{5}$$

It should be noted that the equation for the dilatancy boundary (C) is only dependent on the mean stress, and is independent of loading rate, loading geometry, or the type of rock salt. At higher mean stresses the graph for the dilatancy boundary is split. At stresses below the lower part of the graph, a continuous reduction in volume occurs, at stresses above the upper part dilatant behavior occurs. In between these domains the volume remains constant. In addition, the mean

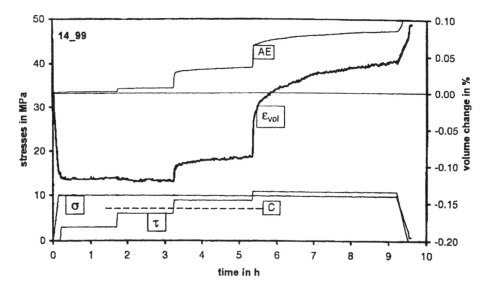

Fig. 7: Volume change and microacoustic emissions (AE) in a true triaxial compressional deformation test (taken from Hunsche & Schulze, this issue). Stepwise increase of the octahedral shear stress τ below and above the dilatancy boundary (C) at $\tau = 7.3$ MPa for constant mean stress $\sigma = 10$ MPa.

failure strength $\sigma_F(\tau)$ is plotted where short term failure is expected to occur (F), compare Cristescu & Hunsche (1998, page 113 ff.):

$$\sigma_F = 0.91 \cdot \tau + 1.025 \cdot 10^{-8} \cdot \tau^6 - 1.82 \qquad (6)$$

At a state of stresses between dilatancy boundary (C) and failure strength (F) tertiary creep can take place finally ending in creep failure.

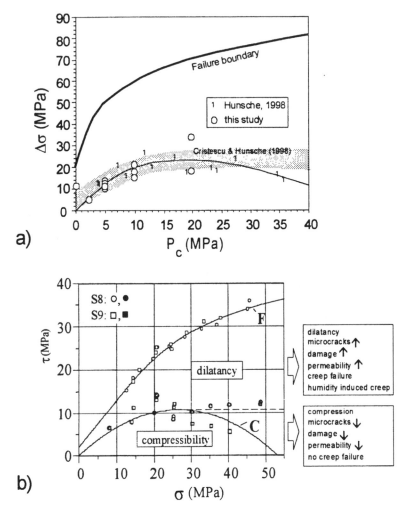

a)

b)

Fig. 8: Dilatancy boundary for rock salt, (i.e. boundary between the dilatant and compressible domains)

a) Differential stress/confining pressure (compression tests, using the Vs maximum as indicator for the dilatancy boundary).

b) Octahedral shear strength τ versus mean stress σ or the dilatancy boundary (C) and the mean failure strength (F). The dilatancy boundary is only depending on mean stress. The behavior of rock salt loaded in the non-dilatant or in the dilatant stress domain.

3. Development of dilatancy and permeability in the dilatant stress domain

In this part, the results of the experimental work are reported which has been performed to investigate how dilatancy and permeability develop while loaded in the dilatant stress domain and to analyze which parameters are important. Of particular interest is the amount of dilatancy and its effect on fluid transport properties in deforming rock salt.

3.1 Effect of confining pressure

The effect of increased or lowered confining pressure during deformation on the development of dilatancy in compression is illustrated in Fig. 9 and Fig. 10. In both figures the development of ultrasonic wave velocities, calculated crack density, and measured permeability is shown. The strain rate-controlled strength test in Fig. 9 was started at constant confining pressure $p_c = 30$ MPa. The ultrasonic wave velocity response shows that the dilatancy boundary is passed at a differential stress $\Delta\sigma_{Dil} \approx 22$ MPa. Although the specimen is deformed to more than 10 % axial strain in the dilatant stress domain reaching a differential stress of $\Delta\sigma \approx 50$ MPa, the expected increase in permeability is not observed. At a strain of approx. $\varepsilon_1 \approx 11\%$ the confining pressure is changed from 30 MPa to 10 MPa. Continuing the deformation in the dilatant stress domain at the lowered confining pressure results in a remarkable velocity decrease due to crack opening, which is accompanied by an increase of permeability.

In contrast, the second experiment was started at the lower confining pressure ($p_c = 10$ MPa), see Fig. 10. During the strain rate-controlled strength testing the dilatancy boundary is passed at $\Delta\sigma_{Dil} \approx 20$ MPa. This time, the ultrasonic wave velocities are more intensively decreasing and permeability is increasing by more than three orders of magnitude. Again, at an axial strain of approx. $\varepsilon_1 \approx 11\%$ the confining pressure, p_c, is changed, from 10 MPa to 30 MPa, which suppresses the development of dilatancy, although the differential stress $\Delta\sigma \approx 50$ MPa is beyond the dilatancy boundary, $\Delta\sigma_{Dil}(p_c = 30 \text{ MPa}) \approx 22$ MPa. Ultrasonic wave velocities are increasing again and permeability is correspondingly decreasing. although the stress remains in the dilatant domain all the time. Therefore, the observed recovery of ultrasonic wave velocities and permeability is a transient effect. The transient recovery is maintained by closure of part of the micro-cracks. Afterwards, the crack density is expected to increase again during the continuous loading in the dilatant stress domain, in accordance with the behavior plotted in the first part of Fig. 9.

The two experiments prove that the amount of dilatancy is reduced at increased confining pressures. At high confining pressures (approx. $p_c \geq 20$ MPa) the gas-permeability is below the resolution of the measuring device ($k < 10^{-20}$ m^2). The dependence of dilatancy on confining pressure is strongly confirmed by the graphs for the calculated crack density, plotted in Fig. 11. Obviously, the deformation-induced increase of the crack density becomes smaller as confining pressure is increased.

The experimental findings convincingly document that both, ultrasonic wave velocities and permeability, are important indicators for dilatancy induced by straining in the dilatant stress domain. At present, we are not able to provide a simple equation relating permeability and ultrasonic wave velocities. However, ultrasonic wave velocities representing crack densities and therefore "porosity", yield a better insight into permeability-porosity relations.

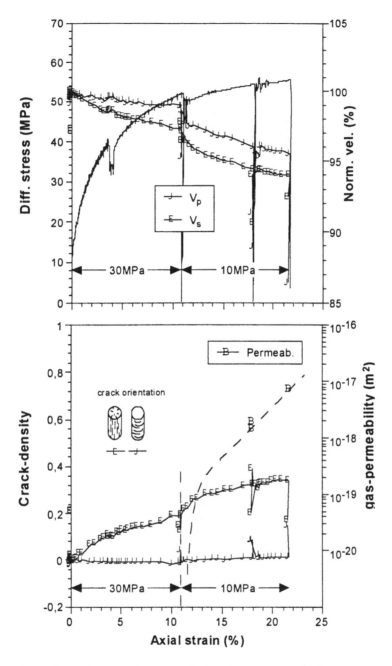

Fig. 9: Strength test (compression; controlled strain rate $\dot{\varepsilon} = 10^{-5}$ s^{-1}; confining pressure p_c = 30 MPa and p_c = 10 MPa ; temperature T = 30 °C). Development of sound velocities, calculated crack density, and permeability in dependence on confining pressure.

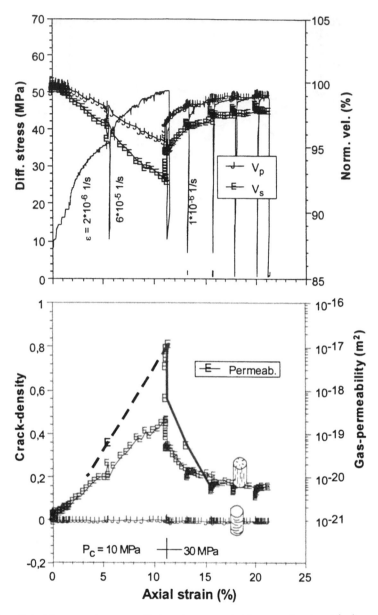

Fig. 10: Strength test (compression; controlled strain rate, varied between $\dot{\varepsilon} = 2 \cdot 10^{-6}$ s^{-1} to $6 \cdot 10^{-5}$ s^{-1}; confining pressure p_c = 10 MPa and p_c = 30 MPa; temperature T = 30 °C). Development of ultrasonic wave velocities velocities, calculated crack density, and permeability in dependence on confining pressure.

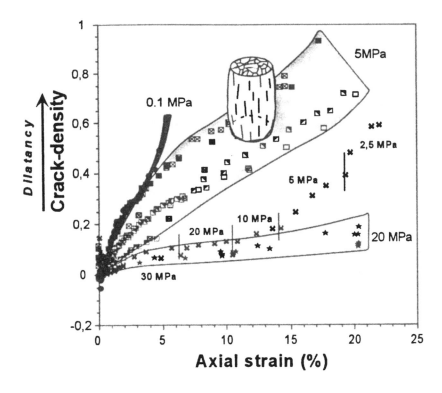

Fig. 11: Dependence of the calculated crack density on confining pressure during deformation in compression (m = -1;). Note: In any case the dilatant stress domain is active since the axial strain is greater than 1 % .

3.2 Permeability-porosity relationship

The prominent influence of confining pressure on the permeability-porosity relation can be seen in Fig. 12 where compiled literature data are plotted. The strength tests on rock salt with low initial porosity were performed in compression (m = -1). Indicated are two areas (results taken from Stormont & Daemen, 1992): the first with the steeper increase in the permeability-porosity relation is typical for low confining pressure (2.4 MPa and 4.4 MPa) and the second for higher confining pressure (7.6 MPa). At low confining pressure (p_c < 5 MPa) dilatant deformation of rock salt causes a noticeable increase of permeability of up to four orders of magnitude while porosity increasing to less than 0.4%. Then, the strain-induced initial slope of the permeability-porosity relation decreases with increasing axial strain approaching a nearly constant value at a permeability in the range of 10^{-16} to 10^{-14} m^2. This suggests that the strain-induced initial rapid increase of permeability is caused by the formation of an interconnecting network of micro-cracks which is most effectively progressing at the beginning of dilatant deformation.

A first approximation for the relationship between permeability, k, and porosity, Φ, can be described by the simple relationship (power law with the exponent χ):

$$k = \Phi^{\chi} \qquad\qquad\qquad (7)$$

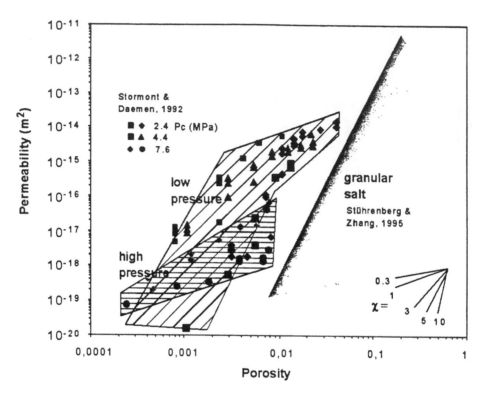

Fig. 12. Permeability vs. porosity of natural rock salt at low and high pressure (compression tests, data recalculated from the figs. 5 - 8 from Stormont and Daemen, 1992) and granular salt (compaction experiments after Stührenberg and Zhang, 1995). A power law is indicated (the slope α is the porosity sensitivity exponent in $k = \phi^{\alpha}$).

From Fig. 12 it is apparent that in the low pressure range the permeability-porosity data of dilating rock salt have to be fitted by a two-domain model for the exponent χ (for details see Stormont and Daemen, 1992). The initial opening of pore space (domain 1) corresponds to an exponent $\chi \approx 6$. In the second domain the increase of permeability is weaker (χ: 1 to 3).

It is important to note that with increasing pressure (i.e. minimum stress) the relationship between permeability and porosity exhibits less scatter and a lower slope ($p_c = 7.6$ MPa; $k_{max} < 10^{-17}$ m² with an exponent of χ: 1 to 3). From the observed dependence of permeability on confining pressure we propose, that the contribution of porosity to the connecting network of micro-cracks reduces as confining pressure increases. We are not aware of any permeability-porosity data from experiments performed at pressures greater than $p_c = 10$ MPa. According to the observed permeability-pressure trend in Fig. 12, our test results demonstrate that at higher pressures ($p_c > 10$ MPa and in general $\sigma_{min} > 10$ MPa, see Fig. 9, 10 &11) the increase of dilatancy-induced permeability is significantly reduced or totally suppressed.

Test data from compaction experiments on highly porous granular salt (i.e. crushed salt) are also plotted in Fig. 12. The data exhibit two main trends: First, compared to rock salt the dependence of permeability on porosity is more sensitive. The data from Stührenberg & Zhang (1995) can be fitted with an exponent $\chi \approx 10$. Second, at the same bulk porosity the permeability of rock salt is much greater than that of crushed salt. For instance, at a porosity $\Phi = 0.03$ with $k \approx 10^{-14}$ m² for deformed rock salt and $k \leq 10^{-16}$ m² for compacted granular salt. Although, the data for rock salt are varying in a broad range, depending on confining pressure or type of salt, they are clearly separated from the results for compacted granular salt.

However, these relations between permeability and porosity are of limited use because they do not reflect the complexity of natural salt. As a result from Fig. 12 it can be concluded that the permeability-porosity relation is not unique. Other parameters have to be considered, e.g. type of salt, humidity, time. From recent investigations on the microstructure of deformed rock salt, we increase of permeability with strain is weaker than in rock salt which is hardened by finely dispersed second phase particles (Popp, Kern and Schulze, subm.) In the latter case, micro-cracks are spreading out from the particle hardened zones disturbing more efficiently the grains of the halite.

3.3 Anisotropic crack patterns

An important parameter which is generally neglected in discussions about the permeability-porosity relation is that of the loading geometry. The micro-cracks and associated permeability develop anisotropically with strain. It has been mentioned before that simultaneous measurements of Vp and Vs provide a valid method to predict spatial micro-crack development in rock specimens during deformation (see Fig. 5 & 6). As a result, the observed velocity signatures can be directly related to characteristic crack patterns. As shown in Fig. 13a, the velocity decrease is more pronounced for Vs than for Vp if loaded in compression, whereas the reverse is true for deformation in the extension mode (Fig. 13b). During loading in compression, a rapid increase of permeability of four orders is observed. But, permeability and porosity often do not pass the limit of resolution or remain nearly unchanged at a low value, compare the corresponding experiment under extensional loading. In this case, the micro-cracks open in the "wrong" direction. The validity of this assumption of anisotropic crack opening corresponding to stress geometry has been confirmed by microstructural observations on the strained specimens (Fig. 14).

Summarizing the results we can state that our experimental investigations confirm that the transition from non-dilatant deformation to dilatant deformation is in agreement with the dilatancy boundary given by Cristescu & Hunsche (1998). The development of porosity (determined by the crack density calculated from the ultrasonic wave velocities or by the gas-accessible porosity) and permeability is more complex than earlier studies claim. Both, permeability and porosity are sensitive to the stress state and the distance from the dilatancy boundary, the minimal principle stress, and the accumulated strain during loading in a certain stress domain. Crack patterns and permeability develop anisotropically and depend on load geometry. The type of salt with regard to the distribution of second phase particles has an influence on the permeability-porosity relation.

With this sophisticated knowledge and continuously improved technical experience we have begun to investigate and to assess loading conditions as expected in situ. The present results are reported in the following section.

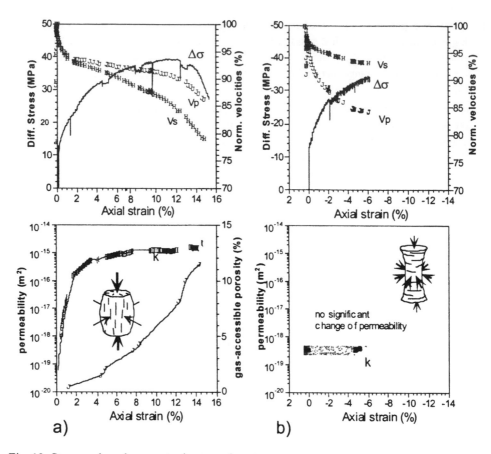

Fig. 13: Stress and crack geometry in strength tests

a) compression: controlled strain rate $\dot{\varepsilon} = 10^{-5}$ s^{-1}; confining pressure p_c = 2 MPa; temperature T = 30 °C) . Behavior of ultrasonic wave velocities indicate developing dilatancy. Dilatancy boundary is at $\Delta\sigma \approx 10$ MPa. Note: Development of permeability reaches saturation although the gas-accessible porosity is increasing.

b) extension: controlled strain rate (1) $\dot{\varepsilon} = 10^{-6}$ s^{-1} and (2) $\dot{\varepsilon} = 2 \cdot 10^{-7}$ s^{-1} ; axial load σ_1 = 2 MPa = const., confining pressure p_c increases; temperature T = 30 °C) . Behavior of sound velocities indicate developing dilatancy. Dilatancy boundary is expected to be passed at $\Delta\sigma \approx 17$ MPa. Note: Permeability remains constant. Gas-accessible porosity is not measurable.

4. Assessment of stability and integrity

In this section, the investigation of the mechanical stability and the hydraulical integrity of a barrier in a salt dome is oriented at the stresses expected to occur at top of the rock salt structure (formation) and around a cavity or repository in the underground. In Fig. 15 the situation is schematically illustrated.

The stress in the vertical direction will be the smallest, because in this direction the stress is reduced by convergence of the rock into the nearby cavity - whereas the pressure gradient is highest in this direction for a fluid at top of salt as well as for a fluid (or gas) in the underground cavity. Therefore, in course of laboratory testing a loading in extension ($\sigma_1 < \sigma_2 = \sigma_3$) and the application of a fluid pressure in the axial direction of the used cylindrical specimens are most appropriate to simulate the in situ situation as illustrated in Fig. 15. The pattern of micro-cracks and, corresponding, permeability develop anisotropically. The results from laboratory testing in another loading geometry will be conservative with respect to the assessment of the functionality of the salt barrier. A loading in compression, for instance, produces micro-cracks which elongate parallel to the axis of a cylindrical specimen. Then, fluid flow is easiest in the axial direction (see Fig. 13).

At top of salt a fluid pressure, p_{fl}, will be at maximum equal to the hydrostatic pressure of ground water under gravity if a rather permeable cap rock is anticipated. In a closed cavity which is filled with brine, crude oil or gas, the fluid pressure, p_{fl}, will rise, due to processes like convergence and gas generation resulting from corrosion or degradation of disposed waste. For all sources of pressure and pressure increase, the question arises which conditions allow "permeation" of the fluid into the salt pore space and what will be the consequences. In this paper the expression "permeation" stands for progressive rock damage by micro-fracturing which is caused by fluids (or gas) under pressure and which leads to an increase of permeability.

4.1 Relationship between effective pressure and pore pressure

A pore pressure which results from permeation will reduce the confining effect of lithostatic pressure in the affected parts of the rock: $p_{eff} = p_{lith} - p_{pore}$. In this paper, an apparent pore pressure relation, $p_{pore} = \alpha \cdot p_{fl}$, is used to discuss its influence on mechanical and hydraulic properties. The Biot-parameter ($\alpha \leq 1$) expresses the fact that only part of the rock mass is affected by the fluid pressure predominantly depending on porosity. In addition, dependence on pore geometry

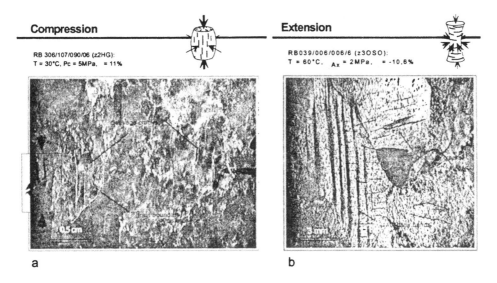

Fig. 14: Macroscopical crack pattern at end of strength tests: a) compression; b) extension (after Popp, Kern and Schulze, subm.).

111

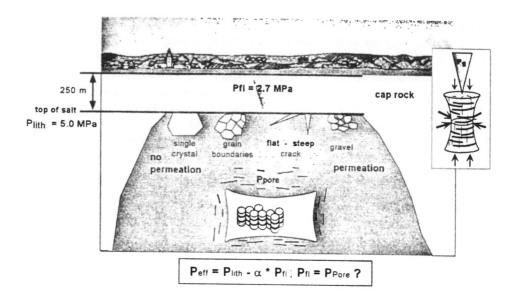

Fig. 15: Schematic of stress situation at top of salt above a repository in a salt dome illustrating the influence of fluid pressure on effective pressure depending on rock permeability.

has to be regarded. To be conservative $\alpha = 1$ is anticipated in most cases. If the apparent pore pressure, p_{pore}, overcomes the minimal principle stress. σ_{min}, then hydraulic fracturing is of a concern. This phenomenon is well known from field and laboratory testing.

In Fig. 16 the results from Kiersten (1988) are re-plotted. Cylindrical specimens were loaded isostatically or in compression, $\sigma_1 \geq p_c$. The fluid pressure was applied via a pipe through the axial piston. The specimens had a small central drill hole. The required excess pressure, $\Delta p = (p_{frac} - \sigma_{min})$, to perform fracturing is found to be $\Delta p \geq 2$ MPa. In Fig. 16 the result from one of our own hydraulic fracturing tests is documented (compare Fig. 22). The specimen which we have used afterwards for strength testing (see Fig. 24) exhibits a comparable excess pressure. Our hydraulic fracturing experiment was performed in two ways. First, the fluid was injected on to the flat end at top side of the specimen. Second, the fluid was injected into a small blind hole at bottom. In all tests a dependence of the excess pressure on the differential stress $\Delta\sigma = (\sigma_1 - p_c)$ or on mean stress, $\sigma = (\sigma_1 + 2 \cdot p_c)/3$, was not found.

In accordance with these results (i.e. fracturing occurs if: $p_{frac} = p_{fl} \geq (\sigma_{min} + \Delta p)$), excess pressure, $\Delta p = 2$ MPa) a "hydraulic criterion" can be formulated. Mechanical stability and hydraulic integrity are reliably confirmed as long as the fluid pressure, p_{fl}, remains smaller than the minimum principle stress, σ_{min}. By contrast, loss of stability and integrity have to be concerned if:

$p_{fl} \geq \sigma_{min}$ hydraulic criterion for loss of stability and integrity.   (8)

In general, the minimum principal stress at top of salt is higher than the fluid pressure at this point and, therefore, hydraulic fracturing will not occur. But pore pressure has to be taken into

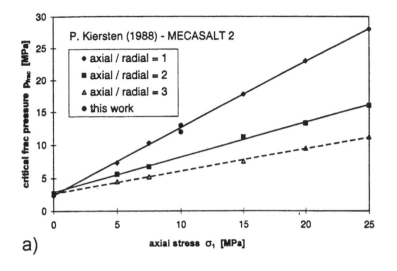

a)

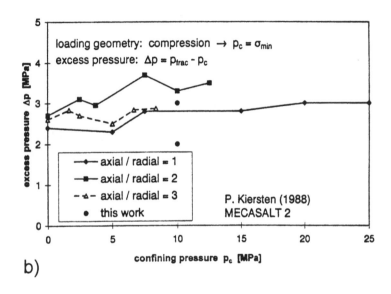

b)

Fig. 16: Results from hydraulic fracturing experiments (Kiersten, 1988). (a) critical frac pressure vs. axial stress and (b) excess pressure vs. confining pressure. In addition, data of our own frac experiments are included (details s. Fig. 22).

account anyway, although the permeability and porosity of un-disturbed rock salt are rather low. The consequences of the pore pressure effect on strength which are possible at maximum (i.e. $\alpha = 1$) are discussed in Fig. 17 (compare Jockwer et al., 1992):

The octahedral shear stress, τ, is not affected by a pore pressure in case of an elastically isotropic rock, because: $\tau(p_{pore}) = \tau$

$$\tau = 1/3 \cdot [(\sigma_1 - \sigma_2)^2 + (\sigma_2 - \sigma_3)^2 + (\sigma_3 - \sigma_1)^2]^{1/2} \qquad [\tau(\sigma_2 = \sigma_3) = (2^{1/2}/3) \cdot \Delta\sigma] \qquad (9)$$

$$= 1/3 \cdot [((\sigma_1 - p_{pore}) - (\sigma_2 - p_{pore}))^2 + ((\sigma_2 - p_{pore}) - (\sigma_3 - p_{pore}))^2 + ((\sigma_3 - p_{pore}) - (\sigma_1 - p_{pore}))^2]^{1/2}$$

But the mean stress, σ, will be reduced:

$$\sigma = 1/3 \cdot (\sigma_1 + \sigma_2 + \sigma_3) \qquad\qquad [\ \sigma(\ \sigma_2 = \sigma_3) = (\Delta\sigma + 3 \cdot p)/3\ ] \quad (10)$$

$$\sigma\ (p_{pore}) = 1/3 \cdot ((\sigma_1 - p_{pore}) + (\sigma_2 - p_{pore}) + (\sigma_3 - p_{pore})) = \sigma - \alpha \cdot p_{fl} = \sigma_{eff}$$

In Fig. 17 the graphs for the mean failure strength (F) and the dilatancy boundary (C) are plotted as given by Cristescu & Hunsche (1998):

(C) $\qquad \tau_{Dil} = 0.8996 \cdot \sigma - 0.01697 \cdot \sigma^2$ $\qquad\qquad\qquad\qquad\qquad\qquad$ (5)

(F) $\qquad \sigma_F = 0.91 \cdot \tau + 1.025 \cdot 10^{-8} \cdot \tau^6 - 1.82$ $\qquad\qquad\qquad\qquad$ (6)

The given example (p_{pore} = 3 MPa) demonstrates that failure and dilatancy are probable to occur at lower stress values. In the diagram this is indicated by the shifted lines (F) to (F') and (C) to (C'). This is the maximal shift in case of $p_{pore} = p_{fl}$ = 3 MPa (i.e. α =1). In the first example (τ = 4 MPa and σ = 10 MPa; triangle) the pore pressure shifts the graphs for the failure strength and the dilatancy boundary in a way which still maintains a non-dilatant state of stress. The second example (τ = 3 MPa, σ = 4 MPa; rhombus) is leading to the result that the shifted graphs indicate a state of stress where instantaneous failure is expected to occur. Generally, for a part of rock which is loaded in the safe area with respect to the safety lines (F) and (C) the distance between these safety lines and the state of stress is reduced by the influence of a pore pressure.

4.2 Re-interpretation of published data

This concept of the influence of a fluid under pressure on the mechanical behavior of rock salt has been investigated by several authors in the field as well as in the laboratory. The first example we want to discuss in this paper results from laboratory tests of Fokker (1995), see Fig. 18. After the application of a specific isostatic pressure the specimen was loaded in extension by reducing the axial stress, σ_1. The specimen was prepared with an open jacket. In Fig. 18 the graphs show where failure (F) and dilatancy (C) are expected to occur if there is no pore pressure. The graphs are shifted to (F') and (C') under the assumption that a pressure, $p_{pore} = p_{fl}$ = 25 MPa, has been applied. In any test the failure was reached very soon at rather low differential stresses. Looking at the loading geometry and the preferred orientation of the micro-cracks the result is not surprising, compare Fig. 13. In addition, we suppose that the specimens were considerably pre-damaged by drilling and preparation. Therefore, nearly no deformation was needed to achieve early failure by hydraulic fracturing.

A respective result for loading in compression (Fokker 1995) is shown in Fig. 19. At the beginning of the strength testing again the specimen was isostatically loaded after preparation with an open jacket, but then the axial stress, σ_1, was increased until failure. The two parallel dotted lines represent the path for loading in compression for uniaxial loading and for loading at a confining pressure, p_c = 10 MPa. The graphs for the mean failure strength (F) and the dilatancy boundary (C) are shifted to (F') and (C') with respect to the anticipated influence of the fluid pressure, see example for $p_{pore} = p_{fl}$ = 10 MPa. The plot for the dilatancy boundary exhibits that right from the beginning of a uniaxial strength test a state of stresses in the dilatant stress domain is valid. At the point where the graph for the loading path is crossing the graph for the mean failure strength the failure of the specimen is expected to occur. The results from Fokker (1995) show that in case of the applicated confining pressure on a specimen with open jacket the failure strength is generally the same as the one which was derived from uniaxial strength tests. Again we propose that these specimens were already sufficiently pre-damaged to offer a network of micro-cracks for the compensation of the confining pressure by the permeated fluid. Therefore,

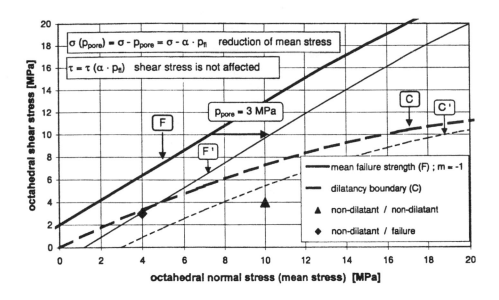

Fig. 17: Shear stress vs. normal stress. Shift of mean failure strength and dilatancy boundary in case of pore pressure effects in porous and permeable rocks ($\alpha = 1$).

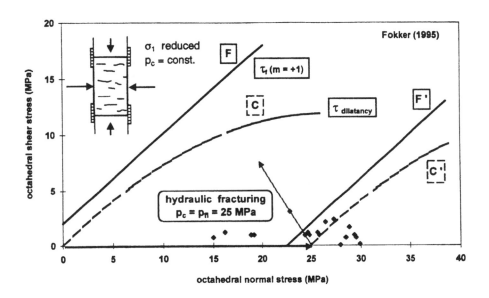

Fig. 18: Strength testing in extension (load path for p = 25 MPa = const.). Pore pressure causes hydraulic fracturing. Reinterpretation of the results from Fokker (1995).

all the strength tests with an open jacket cause a deformation behavior as expected for rock salt which is deformed in an uniaxial strength test.

From these results we conclude that the behavior of rock salt in an initially non-disturbed barrier has to be investigated on very well prepared specimens to avoid artificially facilitated permeation into the specimens. Therefore, an isostatic compaction and healing treatment was applied (in

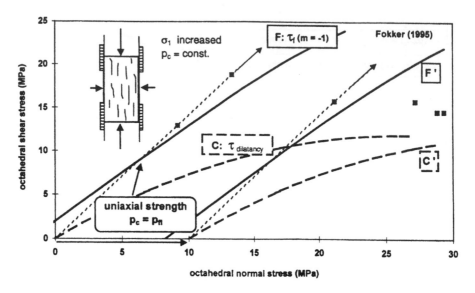

Fig. 19: Strength testing in compression (load path for p = 0 MPa and 10MPa). Pore pressure reduces to uniaxial strength. Reinterpretation of the results from Fokker (1995).

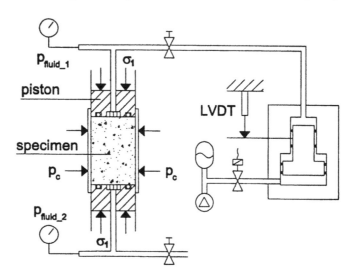

Fig. 20: Testing device for strength testing under the additional influence of fluid pressure.

general: p = 30 MPa; T = 120 °C; t = 3 d) before testing. The aim is to restore the in situ matrix permeability (i.e. $k < 10^{-20}$ m²) or to reduce damage produced during drilling and preparation of specimens as much as possible.

4.3 Measurement of fluid permeation

The experimental work to investigate the mechanical behavior of rock salt under the influence of fluid pressure is performed by applying fluid pressure via pipes through the axial loading pistons.

The testing device is schematically drawn in Fig. 20. During the different types of mechanical testing the amount of the injectable oil volume could be recorded via the LVDT-gauge. In addition, the valves in the supplying pipes could be closed after pressurization to record very sensitively the pressure decay caused by fluid flow into the specimen.

In Fig. 21 one of the test results is documented. After ending the indicated pre-treatment and establishing again the isostatical load condition at p_c = 5 MPa = const. (T = 30 °C) we applied a fluid pressure, p_{fl} = 2.7 MPa. Then the loading in extension was started by decreasing the axial stress. A fluid pressure of p_{fl} = 2.7 MPa (instead of p_{fl} = p_c) was chosen, because at top of the considered rock salt formation this is the hydrostatic pressure which is possible at maximum in the depth of approx. 250 m. The valve at bottom of the specimen was closed to record the increase of pressure in case of noticeable permeation through the specimen. A small blind hole was drilled at the bottom side for fluid collection in the disturbed zone which is expected in the vicinity of this hole.

Before the deviatoric loading was started, the pressure pipe at top of the specimen was also closed to record the decay of the initially applied fluid pressure p_{fl} = 2.7 MPa. The pressure decay gives a measure for the permeability and changes in permeability after the different mechanical treatments (see insert in Fig. 21). The observed slope in the decay, see label (2.1), is similar to the decay in case of an aluminum dummy in place of the specimen. There obviously is a certain unavoidable pressure decay by leakage in the very sensitive system.

At the state of stresses with σ_1 = 2 MPa (p_c = 5 MPa = const.), see label (2.2), the hydraulic criterion is violated: $p_{fl} > \sigma_1$. But no injection of oil was indicated by the LVDT. This observation is in contrast to the result of Fokker (1995), compare Fig. 18. Obviously the deviatoric loading [$\tau = 2^{1/2}/3 \cdot \Delta\sigma = 2^{1/2}$; $\sigma(p_{fl}) = \sigma - p_{fl} = (4 - 2.7) = 1.3$] was too small to cause any changes of the initially very low permeability properties.

Then in this test, the loading in extension was continued by increasing the confining pressure up to p_c = 19 MPa (σ_1 = 2 MPa = const.). Again no injection of oil was monitored by the LVDT. After closing the valve in the pipe for the fluid supply the decay behavior of pressure remained unchanged in comparison with the behavior in the stage under isostatical load, see label (2.3) in the insert of Fig. 21. No effect was observable from the records, although the hydraulic criterion is still violated and, in addition, the dilatancy boundary is nearly reached in this stage of loading:

σ_1 = 2 MPa and p = 19 MPa $\rightarrow$ $\Delta\sigma$ = 17 MPa $\Rightarrow$ $\Delta\sigma_{Dil}(\sigma_1$ = 2 MPa) = 21.5 MPa

$\tau = \sqrt{2/3} \cdot \Delta\sigma$ = 8 MPa and σ = 1/3 $\cdot (\sigma_1 + 2 \cdot p)$ = 13.3 MPa $\Rightarrow$ $\tau_{Dil}(\sigma$ = 13.3 MPa) = 9 MPa

It should be stressed again that in case of dilatancy at extensional loading the micro-cracks propagate perpendicular to the axis of the cylindrical specimen. Therefore, measurable changes in permeability are not occurring during a measurement in the axial direction, compare Fig. 13. As mentioned at the beginning of this section, this geometry is representative for the conditions in the rock salt formation under consideration.

After re-establishing the initial isostatical loading (p_c = 5 MPa) the testing of this specimen was continued by loading in compression. At the differential stress $\Delta\sigma$ = 24 MPa (p_c = 5 MPa = const.) the strength test was stopped (τ = 11.3 MPa; σ = 13 MPa). In Fig. 21, label (3.1) corresponds to this state of stress. Since the differential stress has passed the level of the dilatancy boundary at $\Delta\sigma_{Dil}(p_c$ = 5 MPa) = 16 MPa as derived from equation (5), permeability is expected to increase. (The decay in the ultrasonic wave velocities Vp, Vs indicates that dilatancy starts at $\Delta\sigma \approx$ 20 MPa.) After closing the valve in the pipe for the fluid supply at top again, the decay of pressure is obviously changed this time, compare label (3.1) in insert of Fig. 21.

The strength test was continued in compression by increasing the axial stress. At a differential stress of $\Delta\sigma \approx 32$ MPa the fluid pressure in the closed volume at bottom of the specimen starts to increase. Finally, this strength test was ended at a differential stress of $\Delta\sigma = 47$ MPa. Then the loading had nearly reached the mean failure strength $\Delta\sigma_B(p_c = 5$ MPa$) = 51$ MPa, but the stress versus strain behavior under the influence of the fluid pressure $p_{fl} = 2.7$ MPa is generally identical to the behavior of an unaffected specimen from same origin under same testing conditions.

4.4 Criteria for mechanical stability and hydraulic integrity

From the results reported in the previous section we conclude that the mechanical stability and the hydraulic integrity are not a concern if only the hydraulic criterion is violated. Although in certain parts of the salt barrier the condition $p_{fl} \geq \sigma_{min}$ may be indicated by model calculations, porosity and permeability will remain too low for establishing a predicted fluid pressure in the pore system of the rock salt, as long as the rock salt is loaded in the non-dilatant stress domain. Besides the hydraulic criterion, $p_{fl} \geq \sigma_{min}$, which allows loss of mechanical stability and hydraulic integrity by fracturing to occur, the state of stress must remain in the dilatant domain. Permeation (i.e. rock damage by hydraulic fracturing and increase of permeability) can only occur under the following two conditions:

1. $p_{fl} \geq \sigma_{min}$ hydraulic criterion for loss of stability and integrity   (11)

and

2. $\tau \geq \tau_{Dil}$ dilatancy criterion for loss of stability and integrity   (12)

With help of two additional experiments the anticipated validity of this combined safety-integrity criterion was investigated further. First, the two specimens taken for these experiments were

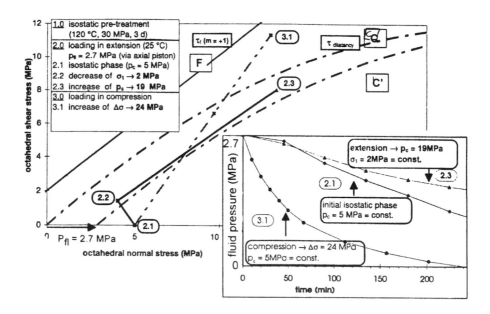

Fig. 21: Permeability testing (fluid pressure decay) in dependence on loading: 1. isostatic load 2. In extension 3. In compression.

prepared in different ways. The procedure for one of the specimens using hydraulic fracturing (see Fig. 22) and the resulting microstructure are already described in the first section of this paper. One has to keep in mind that after re-injection of the fluid at the bottom the pressure at the top side of the specimen is in equilibrium with the pressure at the bottom after approx. 10 hours.

The details of the pre-treatment resulting in "heavy pre-damage" are as follows. At the beginning, the specimen was isostatically loaded up to $p_c = 15$ MPa (T = 30 °C = const.). Then the axial stress was kept constant, $\sigma_1 = 15$ MPa, and the confining pressure was reduced causing the most effective increase in noticeable dilatancy and permeability, because the micro-cracks are propagating in axial direction of the specimen. This is the direction for the measurements of ultrasonic wave velocities and permeability. The reduction of the confining pressure, p_c, was stopped three times, at $p_c = 10$ MPa, at $p_c = 5$ MPa, and at last at $p_c = 0.6$ MPa to perform measurements of the pressure decay after closing the valve in the supply pipe for the fluid pressure at top of the specimen. Fluid pressure was again $p_n = 2.7$ MPa = const. as long as the supply was active. The derived results are plotted in Fig. 23. In the upper part of Fig. 23 the normalized ultrasonic wave velocities and the amount of the injected oil (normalized by the volume of the specimen) are plotted. The decrease in the ultrasonic wave velocities and the corresponding start of noticeable fluid flow coincide with the range where dilatancy is expected to occur. The dilatancy boundary as given by equation (5) is marked: $\Delta\sigma = 11.8$ MPa $= \Delta\sigma_{Dil}$ (p_c=3.2 MPa). In the lower part of this Fig. 23 the fluid pressure decay after closure of the supply pipe is plotted. The decay is caused by flow into the specimen. Although again a certain leakage into the system is recorded (as can be seen from the testing on the aluminum dummy) the general change in the decay behavior is very pronounced. The transition in the decay behavior occurs in a rather narrow band around the stress $\Delta\sigma_{Dil}$ (p_c).

These experimental results confirm that the initial small permeability is only increasing if the loading condition is approaching the dilatant stress domain. Depending on the stress conditions,

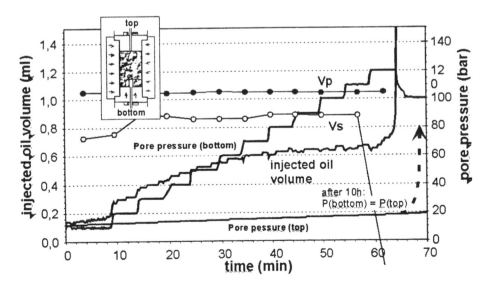

Fig. 22: Hydraulic fracturing experiment. The graph result from the second injection run into the small blind hole at bottom side of the specimen (compare Fig. 16) - pre-treatment for strength test, affected by fluid pressure [Fig. 2; Fig. 24: test 136].

porosity, Φ, and permeability, k, develop with strain [k = f (σ_{min}, loading geometry, strain ...)]. A fluid pressure (if present) is enhancing this process, because at the end of this pre-treatment failure and hydraulic fracturing will become more probable at a very low minimum principle stress ($p_{fl} = 2.7$ MPa > σ_{min} = p_c = 0.6 MPa).

After the pre-treatment, strength testing was performed on the two specimens in compression at a constant confining pressure, p_c = 10 MPa, and under the influence of a fluid pressure of same magnitude, p_{fl} = 10 MPa, which again was applied via the supply pipe to the top of the specimen. The axial stress was increased by strain rate control ($\dot{\varepsilon}$ ~ 10^{-5} s^{-1}). The resulting behavior is plotted in Fig. 24 (compare Fig. 3). Only in case of the "heavy pre-damage" does the specimen behave similar to those subjected to uniaxial compressive loading conditions. Only in this very

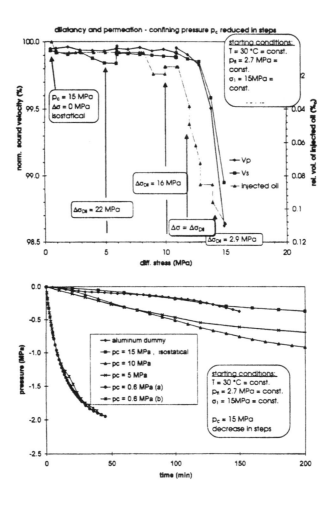

Fig. 23: Ultrasonic wave velocities and injected oil volume as recorded during strength testing (compression, stepwise reduction of p_c). At labeled points decay of fluid pressure was monitored (lower part of figure) after closure of supplying pipe.

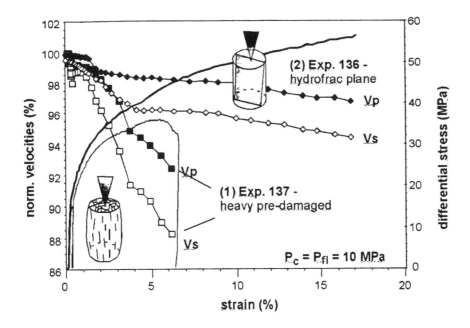

Fig. 24: Strength test - comparison of hardening behavior and failure strength for two specimen after different pre-treatment corresponding to changes of Vp and Vs: (1) slight pre-damage by hydrofrac (see Fig. 22) and (2) hard pre-damage.

permeable specimen the applied fluid pressure can produce a pore pressure which compensates the confining pressure. The specimen which has undergone the "slight pre-damage" is not affected by the applied fluid pressure, because permeability has to develop with strain during loading in the dilatant stress domain. But, the effect of hydraulic fracturing has been too low in case of p_c = 10 MPa.

5. Summary and conclusions

First, the results from our investigation on dilatancy are reported. Dilatancy and dilatancy boundary were determined using the sensitive dependence of ultrasonic wave velocities and permeability on dilatancy. The dilatancy boundary as derived from our data corresponds with the dilatancy boundary given by Cristescu & Hunsche (1998).

In the next section our results with regard to the development of dilatancy and permeability during loading in the dilatant stress domain are documented and discussed. The complex permeability-porosity relation has been analyzed with respect to minimal principal stress, loading geometry, strain, and type of salt. No satisfactory formula can be given yet.

In the last section our results regarding pore pressure effects on mechanical strength and hydraulic integrity are reported. They are of concern only in case of a load in the dilatant stress domain and, in addition, only in the case of low minimum principle stresses where porosity and permeability can develop with strain. Our results are qualitatively summarized in the following schematic diagram:

cap rock

fluid pressure p_{fl}

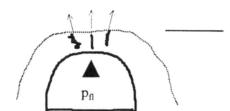

top of salt

$\tau < \tau_{Dil}(\sigma)$ no dilatancy

$p_{fl} < \sigma_{min}$ no frac risk

but pore pressure effect: $\sigma(p_{fl}) = \sigma - \alpha \cdot p_{fl}$

$\rightarrow$ reduction of mean stress in dependence of permeability and hydraulic gradient

un-affected barrier

$\tau < \tau_{Dil}(\sigma)$ no dilatancy

$p_{fl} < \sigma_{min}$ no frac risk

no pore pressure effects

excavation disturbed zone

① $\tau > \tau_{Dil}(\sigma)$

② $p_{pore} = \alpha \cdot p_{fl} \geq \sigma_{min}$ depending on the apparent pore pressure

① The conditions of stress in the vicinity of a cavity are time dependent; they are mainly controlled by depth, geometry and creep properties of the type of rock salt.

② The value of the apparent pore pressure in the rock, p_{pore}, is mainly controlled by:

- pressure increase by volume reduction - which is caused by convergence (creep)
- pressure increase by gas production in case of corrosion and degradation of waste
- pressure decrease by compensating flow: pressure decay depends on permeability and hydraulic gradient (Berest, Brouard & Durup. 1996).
- pressure decrease by permeation - if: $p_{pore} = \alpha \cdot p_{fl} \geq \sigma_{min}$

   ($\alpha \leq 1$ depending on loading geometry, pore geometry, porosity, ...)

To assess the mechanical stability and the hydraulic integrity of the salt barrier between top of salt and a cavity in underground, the conditions of stress have to be analyzed. On basis of our results we conclude that only in the case where both criteria, the hydraulic criterion and the dilatancy criterion, are of a concern at the same time and place, mechanical stability and hydraulic integrity are considerably affected.

Parameters

p_{lith} lithostatic pressure (in situ)

p_{iso} isostatic pressure (laboratory)

p_{fl} fluid pressure

$p_{eff} = p_{lith} - \alpha \cdot p_{fl}$	effective pressure (in situ)
$p_{eff} = p_{iso} - \alpha \cdot p_{fl}$	effective pressure (laboratory)
α	Biot parameter
p_{pore}	pore pressure: $p_{pore} = \alpha \cdot p_{fl}$
p , p_c	confining pressure
σ_1	axial stress
$\Delta\sigma = \lvert \sigma_1 - p \rvert$	differential stress
σ_{min}	minimal principle stress
$\sigma = 1/3 \ (\sigma_1 + 2 \cdot p)$	octahedral normal stress, mean stress
$\tau = 2^{1/2}/3 \cdot \Delta\sigma$	octahedral shear stress
m	loading geometry: $\quad m = -1 \quad$ compression ($\sigma_1 > p$)
	$m = +1 \quad$ extension $\quad (\sigma_1 < p)$
ε_1	axial strain $\quad (\varepsilon_2, \varepsilon_3$ lateral strain)
$\dot{\varepsilon}_1$	axial strain rate
$\varepsilon_{vol} = \varepsilon_1 + \varepsilon_2 + \varepsilon_3$	volumetric strain
Φ	porosity
V_o	volume of specimen (undeformed)
V_{inj}	volume of injected fluid
V_p	ultrasonic compressional wave velocity
V_s	ultrasonic shear wave velocity
δ	(calculated) crack density
δ_{XY}	- normal of cracks vertical to axis of cylindrical specimen
δ_Z	- normal of cracks parallel to axis of cylindrical specimen
k	permeability $\quad$ with χ in $\quad k \sim \Phi^{\chi}$
T	temperature

Literature

Ayling, M.R., Meredith, P.G. & Murrel, S.A.F. (1995): Microcracking during triaxial deformation of porous rocks monitored by changes in rock physical properties. I. Elastic wave propagation measurements in dry rocks.- Tectonophysics 245, 205 - 221.

Berest, P., Brouard, B. & Durup, G. (1996): Behavior of sealed solution-mined caverns.- Proc. EUROCK '96 on Prediction and performance in rock mechanics and rock engineering. Torino, Italia, 1127 - 1131.

Consenza, Ph. & Ghoreychi, M. (1998): Effect of added fluids on mechanical behavior of rock salt.- Proc. 1998 Euroconference on Pore pressure, scale effects and the deformation of rocks. Aussois, France.

Cristescu, N. & Hunsche, U. (1998): Time effects in rock mechanics.- Series: Materials, modeling and computation. John Wiley and Sons, Chichester, 342 pp.

Fokker, P.A. (1995): The behavior of salt and salt caverns.- Proefschrift, Technische Universiteit Delft, 143 pp.

Hunsche, U. and Schulze, O. (this issue): Humidity induced in rock salt and its relation to the dilatancy boundary.

Jockwer, N., Mönig, J., Hunsche, U. & Schulze, O. (1992): Gas release from rock salt.- In: Proc. of NEA-Workshop on Gas Generation and Release from Radioactive Waste Repositories. Aix-en-Provence, France, 23.- 26. Sept. 1991.

Kenter, C.J., Doig, S.J., Rogaar, H.P., Fokker, P.A. & Davies. D.R. (1990): Diffusion of brine through rock salt roof of caverns.- SMRI Fall meeting. Paris.

Kern, H., Popp, T. & Takeshita, T. (1992): Characterization of the thermal and thermo-mechanical behaviour of polyphase salt rocks by means of electrical conductivity and gas permeability measurements.- Tectonophysics, 213. 285 - 302.

Kiersten, P. (1988): Laboratory hydraulic fracturing experiments in rock salt.- In: Hardy, H. R. Jr., Langer, M. (eds.): The Mechanical Behavior of Salt II, Proc. of the Second Conf., Hannover (FRG) 1984; 223 - 233. Trans Tech Publications. Clausthal.

O'Connell, R.J. & Budiansky, B. (1974): Seismic velocities in dry and saturated cracked solids.- J. Geophys. Res. 86, 5412 - 5425.

Peach. C.J. (1991): Influence of deformation on the fluid transport properties of salt rocks.- Geologica Ultraiectina. No.77. Utrecht, pp. 238.

Peach, C.J. & Spiers, C.J. (1996): Influence of crystal plastic deformation on dilatancy and permeability development in synthetic salt rock. Tectonophysics. 256, 101 - 128.

Popp, T. & Kern, H. (1998): Ultrasonic wave velocities, gas permeability and porosity in natural and granular rock salt.- Phys. Chem. Earth 23 (3), 373 - 378.

Popp, T., Kern, H. & Schulze, O. (submitted): The evolution of dilatancy and permeability in rock salt during hydrostatic compaction and triaxial deformation.- J. Geophys. Res.

Stormont, J.C. & Daemen, J.J.K. (1992): Laboratory study of gas permeability changes in rock salt during deformation.- Int. J. Rock Mech. Min. Sci. & Geomech. Abstr. 29, 325 - 342.

Stührenberg, D. & Zhang. C.L. (1995): Results of experiments on the compaction and permeability behavior of crushed salt.- Proc. 5th Int. Conf. on Radioactive Waste Management and Environmental Remediation, Berlin. 1995, (ICEM '95), Vol.1, pp.797-801, ASME, New York, 1995.

Part III

Creep/damage and dilatancy

Basic and Applied Salt Mechanics, Cristescu, Hardy, Jr & Simionescu (eds)
© 2002 Swets & Zeitlinger, Lisse, ISBN 90 5809 383 2

Application of numerical optimization search techniques in field tests to the determination of creep parameters of salt rocks

A.J.Campos de Orellana
Consulting Mining Engineer, Avda. De los Reyes Catòlicos, 23, El Escorial, Madrid 28280, Spain

INTRODUCTION

In the early 80's, Munson and Dawson (1981) presented for the first time constitutuve equations to model the flow of salt rocks within the framework of deformation mechanism maps, based on theoretical thermodynamic studies, microstructural work and laboratory creep tests.

During the following years, creep deformation mechanisms have been a matter of tremendous interest and considerable controversy and debate. Mechanisms such as cross-slip (Wawersik and Zeuch, 1986; Skrotzki and Haasen, 1988), dislocation climb (Carter and Hansen, 1983) and diffusion-controlled pressure solution (Spiers et al., 1988), have been recently widely accepted (Spiers and Carter, 1998), as those deformation models controlling, depending on steady-state strain rate, state of stress, temperature, grain size and water content, the creep behavior of salt rocks in mining and industrial applications (underground storage of oil and natural gas, nuclear waste repositories, oil well, etc.); that is, under equivalent stress (σ_e), temperatures and strain rates, below $25 - 30$ MPa, 200^0 C and 10^{-7} s^{-1}, respectively.

In these three creep mechanisms, the constitutive equations or creep laws, relating creep strain (ε) with stress (σ_e) and time (t), are of the power type:

$$\varepsilon = K \sigma_e^m t^n$$

where, K (creep constant), m (stress-exponent) and n (time-exponent) are specific for each particular saline material and functions of the stress, temperature and strain rate prevailing in the material.

In the determination of creep parameters by laboratory creep tests, equivalent stresses are easily controlled but, on the other hand, tremendous difficulties are encountered in controlling strain rates below $10^{-9} - 10^{-10}$ s^{-1}, which are the rates observed in underground openings and cavities. This make in-situ creep tests preferable for the determination of creep laws, in spite of the disadvantage of not knowing, beforehand, the acting equivalent stresses.

This drawback of the field creep curves can be overcome by means of the application of numerical optimization search techniques (Beveridge and Schechter, 1974) and regression analysis to field creep data.

This work deals with the developments of the analytical procedures of this technique and its validation by numerical simulation of underground opening in a potash operation.

ANALYTICAL PROCEDURES

If i field creep curves obtained from multiple point borehole extensometers (MPBX) are available for a particular salt rock. an average value of the time-exponent (n) and its corresponding standard deviation. can be easily defined by means of regression analysis applied to log-log plots of strain-time data.

Creep curves with a correlation coefficient lower than. let's say 0.9, should be disregarded as nonrepresentative curves. Reading mistakes. extensometer mal-functions, defective installations and damaged heads. may lead to excessively low correlation coefficients. below acceptable values.

In the regression analysis. a set of i values of $(K \; \sigma_e^m)$ are also obtained, being the equivalent stress value (σ_e) unknown and different from one creep curve to another.

The quotient of any two values. out of the i values of $(K \; \sigma_e^m)$ available and derived in the regression analysis of the i creep curves, allows the definition of stress-exponent (m) values. providing that the corresponding equivalent stresses (σ_e) for each creep curve, are known or assumed. Unfortunately. these stresses are never known beforehand.

However, any equivalent stress (σ_e) existing in a salt rock-mass ranges from a minimum value of zero up to a maximum one limited by the Octahedral Shear Strength (K_o) of the material.

Octahedral Shear Strengths (K_o) intervene in the definition of the double Drucker-Prager/Von Mises Plasticity Criteria (Yield Functions) of salt rocks that have beeb reported by many investigators (Serata. 1968. Serata and Fuenkajorn, 1992; Horseman et al. 1992; Dreyer. 1974; Russell. 1979; Russell et al., 1988: Campos de Orellana. 1996. 1998. and Jin et al., 1998) and with values rarely exceeding 15 MPa. Equivalent (σ_e) and Octahedral Shear Stresses (τ_0) are related by the expression:

$$\sigma_e = \frac{3}{2^{1/2}}\tau_0 \quad .$$

For a saline material with an Octahedral Shear Strength (K_0) of 10 MPa, the equivalent stresses (σ_e) acting in the salt rock-mass, under my circumstances, will range from a minimum of zero to a maximum of 21 MPa.

By means of numerical optimization search technique, j random values of equivalent stresses (σ_e), within the established range of $0 - 21$ MPa, are chosen any desired stress-interval (σ_e) (0.5 , 1 , etc. MPa).

For each set of equivalent stress (σ_e) assigned to each of the i $(K\,\sigma_e^m)$ values of a particular saline material, C_2^1 stress-exponents (m) are calculated and therefore an average mean value $(\overline{m})$ and its standard deviation can be derived (s).

Equivalent stresses (σ_e) are modified, within the mentioned range, in a repetitive computing process, until all possible equivalent stresses (σ_e) are assigned to each set of $(K\sigma_e^m)$ values, that is, to each of the i creep curves.

The computing optimization process will continue until j^i mean values of the stress-exponents $(\overline{m})$ with their corresponding standard deviations, are obtained. The mean stress-exponent $(\overline{m})$, out of the j^i computed values, that shows the lowest standard deviation, is considered the stress-exponent (m) that better represents the real value of that parameter for the particular salt rock under study, in other words, the value that best fits the field creep data.

This so-defined stress-exponent (m) is bound to a set of fixed and settled values of equivalent stresses (σ_e) for each of the i creep curves or $(K\sigma_e^m)$ values. This makes possible to determine the creep constant (K) and its standard deviation.

We can express all these concept in mathematical terms, by taking the standard deviation (s) of the stress-exponents (m) as the objective function to be optimized (minimized):

$$s = \left(\frac{\sum (m_k - \overline{m}_1)^2}{k} \right)^{1/2}$$

being:

$$m_k = \frac{\log A_k}{\log\left(\sigma_{ex}/\sigma_{ey}\right)}$$

$$\overline{m}_l = \frac{\sum m_k}{k}$$

$$A_k = \left(K\,\sigma_{ex}^{m_k}\right)/\left(K\,\sigma_{ey}^{m_k}\right) = \left(\sigma_{ex}/\sigma_{ey}\right)^{m_k}$$

$$\sigma_{ex},\sigma_{ey} = 0,\ \Delta\sigma_e,\ 2\Delta\sigma_e,\ 3\Delta\sigma_e,\ldots\left(3\,K_o/2^{1/2}\right)$$

$$1\leq k \leq C_2^i\ ,\quad 1\leq l\leq j^i$$

$$1\leq x \leq i\ ,\quad 1\leq y\leq i\ ,\quad x\neq y\ .$$

C_2^i denotes the number of ways in which 2 values of $(K\,\sigma_e^m)$ can be chosen from the available i values of $(K\,\sigma_e^m)$ of the order in which they are arranged.

The standard deviation function (s) is therefore, function of the equivalent stresses (σ_e) acting in the i creep curves, which has to be optimized (minimized) in the i-dimensional equivalent stress space.

RESULTS AND VALIDATION

The above described analytical procedure was applied to the seven saline horizontal layers (halite, carnallite, sylvinite and rock-salt) of the Saline Formation (Catalonia Potash Basin) existing in an underground mine operated by IBERPOTASH (Dead Sea Works Ltd.) and sited in the township of Sallent, Barcelona, Spain.

Details of the geology and elastic-plastic parameters of these seven fine-grained layers (1-2 mm.) are reported by Campos de Orellana (1996, 1998).

Octahedral Shear Strengths (K_o) for the seven beds range from 14 MPa to 21 MPA and the available number of creep curves (i) with correlation coefficients greater than 0.9, between 1 and 28.

Application of regression analysis and numerical optimization search techniques led to creep parameters for the seven salt rocks, in the range, 0.36 − 0.90 for the time-exponents (n). 1.03 − 1.71 for the stress-exponents (m) and $5.473426\times10^{-4} - 1.024864\times10^{-2}$ for the creep constants (K), measuring strains (ε) in mm/m, equivalent stresses (σ_e) in MPa and time (t) in days.

These creep parameters utilized in a numerical finite element simulation of four control stations, where roof sags, floor heaves and horizontal converge were monitored for periods between 102 and 250 days.

Control stations were set up in four galleries with different cross sections, cut in distinct geological configurations and located, one to each other, at horizontal distances of 1.000 – 1.500 meters apart.

Creep parameters were modified until differences between calculated and measured roof, floor heaves and horizontal convergence in the steady-state creep, were of the order of 10^{-3} mm. During the first 5 days after the gallery was opened, this difference reached an average value of 15% and from the fifth day to the transition time (change from temporary to steady-state), 7%.

Modification introduced during the numerical simulations, in the creep parameters of the seven analyzed salt rocks, with respect to the values determined by optimization techniques, were in the following ranges:

$$\text{Creep Parameter} \quad (K): 0.59\,\% \ / \ 0.90\,\%$$
$$\text{Stress-exponent} \quad (m): 4.63\,\% \ / \ 7.45\,\%$$
$$\text{Time-exponent} \quad (n): 0.00\,\%$$

except for the Lower Salt layer, where these modifications reached up to 62.84 %, due to the fact that only one creep curve was available for this material and consequently optimization techniques could not be applied. A minimum of three creep curves are required for this technique. Therefore, creep parameters for the Lower Salt layer were estimated, not derived by optimization techniques, which explains the high modifications introduced in the creep parameters during the numerical simulation.

The excellent approximation and good fit between initial and final creep parameter values, that is before and after the numerical match simulation, support the validity and accuracy of numerical optimization search techniques in the determination of creep parameters of salt rocks and the outstanding efficiency obtained by this method in terms of reliability.

The utilization of laboratory creep tests, as an alternative method to the determination of creep parameters of salt rocks, introduces serious short-comings due to the already mentioned technical limitations of conventional testing devices at strain rates below 10^{-10} s^{-1} and to the non-representative values of creep parameters obtained in laboratory tests compared to those of the rock-mass in-situ (effects of sample sizes, sample disturbances, stress-strain history, etc.).

CONCLUSIONS

Application of regression analysis and numerical optimization search techniques to in-situ field creep tests, for the determination of creep parameters of salt rocks, has been proved to be of a tremendous assistance, precision and accuracy, when field creep curves are obtained from multiple point borehole extensometers (MPBX).

A minimum of three field creep curves with correlation coefficients greater than 0.9 are needed for the application of this technique. This implies reliable extensometer read-out equipments, careful readings and good installations.

This technique allows to analyze in deep and to obtain reliable creep parameters of salt rocks, without the need of troublesome, doubtful and misleading laboratory tests, making possible to forecast and to predict long-term behavior of underground salt rock openings with confidence.

REFERENCES

Beveridge, G.S.G. and Schechter, R.S. (1974) "Optimization: Theory and Practice"; Mc Graw-Hill, New York; pp.355–508.

Campos de Orellana, A.J. (1996) "Pressure Solution Creep and Non-Associated Plasticity in the Mechanical Behavior of Potash Mine Openings"; Int. J. Rock Mech. & Mining Sci.; vol.33. pp.347-370.

Campos de Orellana, A.J. (1998) "Non-Associated Pressure Solution Creep in Salt Rock Mines"; 4th Conf. on The Mechanical Behavior of Salt, Université de Montreal, June 1996. Trans Tech Publ., 1998; pp.429-444.

Carter, N.L. and Hansen, F.D. (1983) "Creep of Rocksalt"; Tectonophysics, 92, pp.275-333.

Dreyer, W. (1974) "Results of Recent Studies on the Stability of Crude Oil Storage in Salt Caverns", 4th Salt Symp.; The Northern Ohio Geological Society. Cleveland; vol.2; pp.65-92.

Horseman, S.T., Russell, J.E., Handin, J., Carter, N.L. (1992) "Slow Experimental Deformation of Avery Island Salt", 7th Salt Symp., Tokyo, vol.1; pp.67-74.

Jin, J., Cristescu, N.D., Hunsche, U. (1998) "A New Elastic/Viscoplastic Model for Rock Salt", 4th Conf. on the Mechanical Behavior of Salt. Université de Montreal. June 1996; Trans Tech Publ., pp.249-262.

Langer, M. (1984) "The Rheological Behavior of Rocksalt", 1st Conf. on The Mechanical Behavior of Salt. Pennsylvania State University, Nov. 1981; Trans Tech Publ., pp.201-240.

Munson, D.E. and Dawson, P.R. (1981) "Salt Constitutive Modeling Using Mechanism Maps", 1st Conf. on The Mechanical Behavior of Salt, Pennsylvania State University. Trans Tech Publ. 1984. pp.717-737.

Russell, J.E. (1979) "A Creep Model for Salt", 5th Salt Symp., Hamburg, vol.1, pp.349-353.

Russell, J.E., Handin, J., Carter, N.L. (1988) "Modified Mechanical Equations of State of State of Salt", 2nd Conf. on The Mechanical Behavior of Salt, BGR, Hannover, Trans Tech Publ., pp.409-430.

Serata, S. (1968) "Application of Continuum Mechanics to Design of Deep Potash Mines in Canada" , Int. J. Rock Mech. & Mining Sci., vol.5, pp.293-314.

Serata, S., Fuenkajorn. K. (1992) "Formulation of Constitutive Equations for Salt" , 7th Salt Symp., Tokyo, vol.1, pp.483-488

Skrotzki, W., Haasen, P. (1988) "The Role of Cross-Slip in the Steady-State of Creep Salt" 2nd Conf. on The Mechanical Behavior Salt, BGR, Hannover, Trans Tech Publ., pp.69-82.

Spiers, C.J., Urai, J.L., Lister, G.S. (1988) "The Effect of Brine (inherent or added) on Rheology and Deformation Mechanisms in Salt Rocks" 2nd Conf. on The Mechanical Behavior of Salt. BGR. Hannover, Trans Tech Publ.. pp.89-102.

Spiers, C.J., Carter. N.L. (1998) "Microphysics of Rocksalt Flow in Nature", 4th Conf. on The Mechanical Behavior of Salt, Université de Montreal, Trans Tech Publ.. pp.115-128.

Wawersik, W.R., Zeuch. D.H. (1986) "Modelling and Mechanistics Interpretation of Creep of Rock Salt Below 200° C", Tectonophysics. 121, pp.125-152.

Basic and Applied Salt Mechanics, Cristescu, Hardy, Jr & Simionescu (eds)
© 2002 Swets & Zeitlinger, Lisse, ISBN 90 5809 383 2

Creep of rock salt under small loading

J.P.Charpentier & P.Bérest
Laboratoire de Mécanique des Solides, École polytechnique, 91128 Palaiseau, France

P.A.Blum
Institut de Physique du Globe de Paris, 4, place Jussieu, Paris, France

Abstract

Rock salt cylindrical samples have been tested through small (0.1 MPa) axial loading. Creep strain is measured through high resolution (approximately a nanometer) extensometers. The test was performed in a mine gallery, which favors stable hygrometry and temperature conditions. Relative rotations of the plates are probable, and the average measured creep rate is of the order of 10^{-12} s^{-1} (sample contraction). When the applied stress is reduced by 30%, inverse creep (sample expansion) takes place at a strain rate of - 5 10^{-13} s^{-1}.

Introduction

Study of rock salt creep under small (typically, $\sigma = 0.1$ MPa) mechanical loading is of great importance both for a better understanding of geological deformation features and for the analysis rock mass behavior in the vicinity of underground openings. A small amount of experimental data is available; Hunsche [1988] has described lab tests involving very small strain rates which were obtained through a high-resolution laboratory facility. In fact, constitutive equations for very small mechanical loading are most often extrapolated from constitutive equations validated under higher mechanical loading (say, $\sigma > 5$ MPa).

Sensors

We have conducted long-term uniaxial creep tests on standard cylindrical rock samples (diameter = 7 cm, height =16 cm). High resolution displacement sensors, designed by P.A. Blum, were used. The sample was set between two circular silica plates (width 0.5 cm, diameter 16 cm). A vertical silica string was fixed to the upper plate; at its lower tip, a silica drop encapsulates a ferritic element that moves up and down between the walls of a silica tube fixed to the lower plate and bearing a magnetic coil. In the central part of the coil, a 1-micron displacement of the silica drop (which can easily be measured) generates an approximate 0.7 millivolt tension. Because tensions of the order of a microvolt can be accurately measured, one can expect to be able to measure a 10^{-8} strain. In fact, during each test, four such sensors, set in two vertical planes at a 90° angle, were used, allowing measurements of both the relative rotation of the two plates and their average relative displacement. Tests have been performed on salt samples from the Etrez salt formation that had previously been studied, under standard stress levels, by J.P. Charpentier [1988] and A. Pouya [1991]. The current tests were performed in a salt mine gallery to take advantage of its very stable thermal and hygrometric conditions.

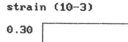

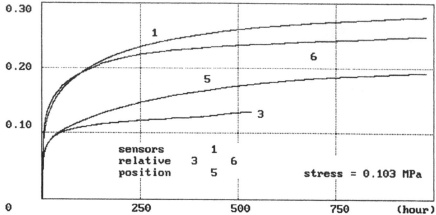

Figure 1 – Test N°1 strain-versus-time curves of the four sensors

Testing Program

On November 17, 1997, test n° 1 started on a phenoblastic salt sample submitted to a 0.103 MPa initial loading. This applied stress was calculated by dividing the weight of steel cylinders set upon the upper plate (three 12.5-kg cylinders, diameter = 20 cm, overall height = 20 cm) by the sample initial cross section area. The applied stress was modified twice during the test.

At the same time, a second test (test n°0) began on a cylindrical duralumin sample whose geometrical dimensions were similar to those in test n° 1, and plays the role of a reference sample (no loading was applied on this sample). On June 8, 1998, 4900 hours after the earlier tests began, a third identical creep apparatus was set on the same table and a second salt sample, significantly purer than the former, was loaded to a stress of 0.102 MPa (test n° 2). On October 26, 1998, a third salt sample, whose quality was intermediate between pure and phenoblastic, was substituted for the reference sample. This third sample was loaded to a stress of 0.02 MPa (test n°3). Tests n° 2 and n° 3 stopped respectively after 9000 hours and 5700 hours, test n° 1 was in progress at the time when the present paper was written (August 1999).

General Results

1. During each of the four tests, one (or two) of the four sensors failed (either temporarily or permanently). The results described below concern the unbroken sensors.

2. Before the tests, displacements induced by temperature variations were thoroughly measured; for each test the measurements provided by the four sensors were consistent and the measured thermal expansion coefficient α belongs to the expected range of 4.4 / 4.6 10^{-5} °C^{-1}. As soon as the load was applied, and later during the creep phase, the four sensor records begin to differ significantly as proved by the four displacement-versus-time curves shown in Figure 1 (test n° 1). This discrepancy may prove either that some relative rotation of the two silica plates, due to non-uniform loading of the upper and lower faces of the sample, takes place; or that the table itself, on which the creep devices are set, rotates. This rotation can also originate in geometrical defects of the sample faces; indeed if the faces are correctly planed (i.e., the defects height is smaller than 10 microns), they are not perfectly parallel (samples heights are not constant by several hundredths of millimeters). Reversible rotations are clearly visible when lights are switched on, they are due to uneven heating of the lateral surface of the samples by the lamps, which induce non-homogeneous thermal expansion. The test table lays on the gallery salt wall and wedges distribute the load, but the three creep devices are set on the same table : for example, the set up of test n° 2, at hour 4900, triggers a visible discontinuity of the test n° 1 displacement-versus-time curve, due to additional weight on the table.

3. The duralumin reference test (test n° 0) sample, which bears no load except for the upper plate weight, exhibits an apparent mean deformation that is four time smaller than the deformation of the loaded salt samples. With the exception of possible sensors drift, the explanation for this non-zero deformation may lay in the crushing of small glue dots left between plate and sample.

Temperature and hygrometry

The rock salt thermal expansion coefficient is approximately $\alpha = 4.4 \ 10^{-5} \,°C^{-1}$. Since the goal of the test is to measure sample deformations as small as 10^{-8}, a stable ambient temperature must be reached. This is why the tests have been performed (with support provided by Compagnie des Salins du Midi et Salines de l'Est) in a 700-meter long, 160-meter deep salt mine gallery remote from present salt extraction and ventilation system. Figure 2 displays temperature and hygrometry, continuously monitored from November 1997 to February 1999. Temperature – as for all other measured quantities – is measured every two minutes during the test. Daily temperature variations, which are recorded with a one thousandth Celsius degree resolution, appeared to be of the order of +/- 0.01°C ; the average temperature value is 13.5°C. These daily variations are clearly correlated to ground temperature variations, and are probably linked to heat transport through the gallery atmosphere. Seasonal variations of a few hundredths of a Celsius degree can also be observed. Human presence in the gallery, or power cuts that switch off electronic and software devices, also create thermal perturbations. For instance, the influence of the (as-measured) temperature can be observed in Figure 2, during

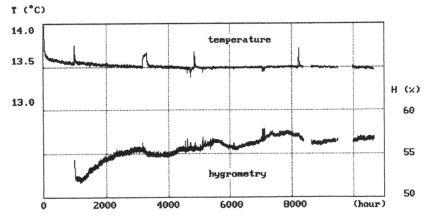

Figure 2 – ambient temperature and hygrometry from November 1997 to February 1999

137

scheduled operations at hours 1000, 4200 and 8200 ; and, during unexpected power cuts, at hours 4700 and 8200. At approximately hour 3150, a lamp was mistakenly switched on and remained on for 6 days when the lamp burned out. Depending upon thermal perturbation intensity and duration, return to thermal equilibrium could last from a few hours to several days.

The striking correlation between temperature variations and displacement variations is a clear evidence of salt thermoelastic behavior (Figure 3). The empirical correlation coefficient is quite close from the thermal expansion coefficient given above. When corrected for temperature variations, the strain-versus-time curves are much smoother (Figure 3); a few discontinuities remain, for instance at the time when a lamp inadvertently remained on. During this period, temperature was kept significantly higher (by 0.15°C) than usual, probably leading to creep acceleration. Surprisingly, however, during this incident (3150-3350 hours), the overall strain, when corrected from thermoelastic variations, is smaller at the end of the heating period than it should have been, had a continuous evolution taken place. A possible explanation of this paradoxical results can probably be found in a concomitant hygrometry decrease. Ambient hygrometry is known to deeply influence salt creep strain rate [Horseman, 1988]. The mine hygrometry experiences small fluctuations (the relative hygrometry lays in the range 52%-57% ; see Figure 2), but a significant reduction of its earlier increase is clearly visible during the time the lamp was left on.

Study of creep under small load

In test n°1, the initial average uniaxial stress (0.103 MPa) was increased to 0.108 MPa (+ 5%) after 1000 hours, then decreased to 0.076 MPa (- 30%) after 8200 hours. Figure 4 displays the strain-versus-time curve, as corrected for thermoelastic variations. Discontinuities of the curve are clearly visible : they are related to the two loading changes and to the set-up of test n°2 (at hour 4900). Strain experiences a long initial transient period up to hour 3000. At that time, the average strain rate stabilizes to $1.4 \ 10^{-12} \ s^{-1}$ (sample contraction), with long period fluctuations of amplitude +/- 20%, possibly related to hygrometry fluctuation.

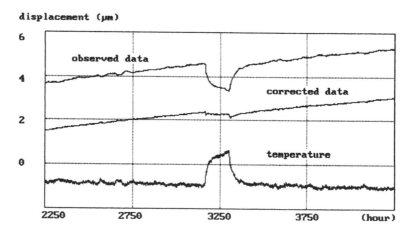

Figure 3 — an example of thermoelastic correction — Test N°1

strain (10-3)   strain rate (10-9 h-1)

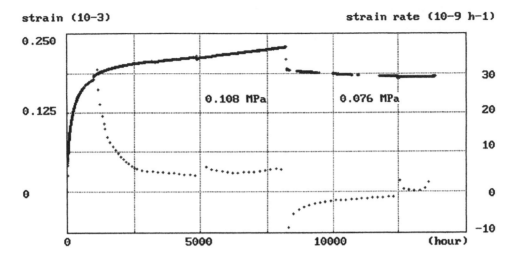

Figure 4 – strain and strain rate during Test N°1

The 30% unloading performed at hour 8200 led to an instantaneous axial expansion of approximately -4 10^{-5} followed by a long period of time during which a delayed axial expansion takes place at a -5 10^{-13} s^{-1}. The delayed expansion can be observed up to hour 12500. During this period, recording was frequently interrupted due to failures of the measurement system. Then creep rate vanishes to zero (hour 13000) and later on contraction of the sample at very low strain rate (0.8 10^{-12} s^{-1}) takes place. This transient inverse creep lasts much longer than that currently observed during the test involving larger stresses.

Test n°2 which began later, also had an initial applied stress of 0.103 MPa. Creep was much faster than during test n°1 : for example during the 2400 to 3200 hours after the beginning of each test, the cumulated strain was 2.5 microns for test n°2 ("pure" salt) and1.8 microns for test n°1 ("phenoblastic" salt). This difference is quite consistent with what is known from previous tests performed at larger stresses on these two salt types.

Test n°3 was performed on an "intermediate" salt under a 0.021 MPa load ; no creep was perceived even 5000 hours after the test began.

Comparison with tests performed under larger stress

Qualitatively, salt behavior under small stress exhibits the same global features as observed under larger stresses (say, 5 to 15 MPa). Rapid stress build up leads to transient creep characterized by slow rate decrease, but this transient phase is much larger in the case of small applied stress. Then creep rate becomes roughly constant (steady-state) or, more precisely, experiences long period fluctuations possibly influenced by slow changes in hygrometry. A 30%-loading decrease triggers a transient inversed creep, at a constant rate, which lasts much longer than that triggered during tests performed under larger stresses.

A. Pouya [1991] has proposed a so-called "Norton-Hoff" rheological law to describe Etrez salt creep behaviour. This law has been fitted to a series of creep tests performed in the 5 - 20 MPa stress range and 293 - 423 K temperature range. Its uniaxial expression is

$$\dot{\varepsilon} = A \exp(- Q / RT) \sigma^n$$

with the following parameters value : $A = 0.64$ an^{-1} MPa^{-n}, $Q/R = 4100$ K, $n = 3.1$. If this expression is extrapoled to the conditions in test n°1 ($\sigma = 0.108$ MPa, $T = 286.5$ K) the calculated strain rate, or $\dot{\varepsilon} = 10^{-17}$ s^{-1}, is smaller by 5 orders of magnitude than the observed strain rate, which is $\dot{\varepsilon} = 1.4 \ 10^{-12} s^{-1}$.

Conclusion

Uniaxial compressive creep tests have been performed under very small (0.1 MPa) mechanical loading. Average creep rates of approximately $10^{-12}s^{-1}$ (sample contraction) have been observed during a one year long period. These results can be considered as preliminary under some respects ; relative rotation of the two plates between which the sample is set cannot be explained fully. The observed strain rates are much larger, by several orders of magnitude, than rates predicted through extrapolation of standard creep laws based on the results of tests performed under much larger mechanical loading.

REFERENCES

CHARPENTIER J.P., [1988], "Creep of Rock Salt at Elevated Temperature", 2nd Conf. Mech. Beh. Salt, Trans Tech. Pub., 131-136.

HORSEMAN S.T., [1988], "Moisture Content – A Major Uncertainty in Storage Cavity Closure Prediction", 2nd Conf. Mech. Beh. Salt, Trans Tech. Pub., 53-68.

HUNSCHE U., [1988], "Measurement of Creep in Rock Salt at Small Strain Rates", 2nd Conf. Mech. Beh. Salt, Trans Tech. Pub., 187-196.

POUYA A., [1991], "Correlation between Mechanical Behavior and Petrological Properties of Rock Salt", 32th US Symp. on Rock Mech., 385-392.

Basic and Applied Salt Mechanics, Cristescu, Hardy, Jr & Simionescu (eds)
© *2002 Swets & Zeitlinger, Lisse, ISBN 90 5809 383 2*

Dilatancy as related to evolutive damage

N.D.Cristescu
Department of Aerospace Engineering, Mechanics & Engineering Science, University of Florida, Gainesville, USA

INTRODUCTION

Dilatancy of several materials as sandstone and several metals, has been studied experimentally in uniaxial tests, by Bauschinger [1879], (see Bell [1973] for the early studies on irreversible volume change during plastic deformation of various materials). For granular materials dilatancy was mentioned by Reynolds [1985]. The word "dilatancy" was used by both these authors. Much later on. it was Bridgman [1949] who has found that in uniaxial compression tests the volume of soapstone. marble and diabase is dilatant at high applied stresses. He has found also that at high stresses dilatancy is produced by "rapid creep". He also suggested that irreversible compressibility is due to closing of pores. while dilatancy (called "retrograde volume change") to the opening of pores.

We know today that dilatancy, characterized as a volume expansion when a shear loading is applied (superposed or not on a hydrostatic pressure) is exhibited by most geomaterials. Careful experimental studies of dilatancy and general constitutive equations describing dilatancy as well as time dependent properties of various rocks, are of recent venue. For those walking on a saturated beach sand it is a "curiosity" but this concept is related to the increase of porosity, of permeability, of the creep characteristics, of the evolutive damage and of prediction of ultimate failure. Thus. this concept it is of fundamental importance for a correct description of the mechanical behavior of rocks.

COMPRESSIBILITY/DILATANCY BOUNDARY

For some stress state the rock salt (and generally most rocks) is compressible, while for other stress states it is dilatant. The stress-states marking the transition from compressibility to dilatancy belong to the compressibility/dilatancy boundary. This concept was used for the formulation of a constitutive equation for granite (Cristescu [1985a,b]) and afterwards for many other rocks as andesite, several kinds of coal, cement concrete, etc. (see Cristescu [1989]). Since the passage from compressibility to dilatancy is quite smooth, the shape of the compressibility/dilatancy boundary is not easy to be found. For rock salt, very careful experiments performed by Hunsche have revealed in detail the main characteristics of this boundary (see Cristescu and

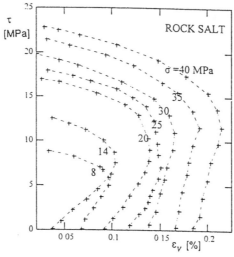

Figure 1. Variation of volumetric strain with octahedral shear stress in true triaxial tests.

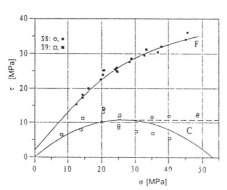

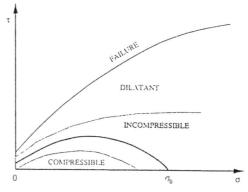

Figure 2.a Measured values of the failure surface and the compressibility/dila- tancy boundary for rock salt (Cristescu and Hunsche [1998]).

Fig.2.b Domains of compressibility and dila tancy, separated by a strip of incompressi- bility; the thick line is a convenient approxi- mation for the modeling of the C/D bounda- ry, to be used in a constitutive equation.

Hunsche [1998]). The variation of the volumetric strain in true triaxial tests on rock salt is shown in Figure 1. The maxima of irreversible volumetric strain ε_v^I are defining the passage from compressibility to dilatancy for each confining pressure. It has been found in true triaxial tests that in reality the compressibility/dilatancy boundary is a strip of incompressibility (Figure 2.a). This strip is shown also in Figure 2.b. It is quite

142

difficult to pinpoint the exact upper and lower boundaries of this strip, mainly at higher confining pressures. For small to medium levels of the mean stress σ, the rock salt is irreversible compressible, while for higher pressures it is incompressible. For higher values of the octahedral shear stress τ, the rock is irreversible dilatant. For higher mean stresses the rock salt is incompressible for small values of τ and dilatant for higher values. Microcracks and pores are closing and opening in all domains shown in Fig.2.b. However, in the compressibility domain most microcracks are closing so that the volume is compressible. It is vice-versa in the dilatant domain. In the strip of transition from compressibility to dilatancy, the closing and opening of microcracks do not produce a significant irreversible volumetric change. Thus the upper boundary of this strip is the boundary from where, for increasing τ, dilatancy starts, while the lower boundary shows up to where, again under increasing τ, the compressibility stops.

For convenience of the formulation of a constitutive equation, the C/D boundary is approximated in the $\sigma\tau$-plane by a curve of equation

$$\frac{\tau}{\sigma_*} = f_1 \left(\frac{\sigma}{\sigma_*} \right)^2 + f_2 \frac{\sigma}{\sigma_*} \tag{1}$$

where σ_* is the unit stress and f_1 and f_2 are two material constants. In Figure 2.b equation (1) is represented as a thick line for $f_1 = -0.01697$ and $f_2 = 0.8996$. The mean stress σ_0 shown in Figure 2.b has the meaning of the smallest pressure closing all pores and microcracks. This limit pressure can not be determined very precise, but by extrapolation, and its exact meaning is the "range of pressure" which is closing all microcracks and pores.

CONSTITUTIVE EQUATION

In order to describe both compressibility and/or dilatancy, as well as time effects, the constitutive law is written in the form

$$\dot{\varepsilon}_{ij} = \frac{\dot{\sigma}_{ij}}{2G} + \left(\frac{1}{3K} - \frac{1}{2G} \right) \dot{\sigma} \delta_{ij} + k_T \left\langle 1 - \frac{W_T(t)}{H(\sigma)} \right\rangle \frac{\partial F}{\partial \sigma_{ij}} + k_S \frac{\partial S}{\partial \sigma_{ij}} \tag{2}$$

where K and G are the non-constant elastic parameters, $H(\sigma)$ is the yield function, $F(\sigma)$ and $S(\sigma)$ are viscoplastic potentials for transient and steady-state creep respectively, all depending on stress invariants, k_T and k_S are viscosity coefficients, δ_{ij} is the unit tensor, $\dot{\varepsilon}_{ij}$ - the strain rate tensor, σ_{ij} the Cauchy stress tensor, and $\langle A \rangle = \frac{1}{2} (A + |A|)$. All the above constitutive parameters and functions are determined by laboratory tests (see Cristescu [1989], Cristescu and Hunsche [1998]). Dilatancy and compressibility are described by the procedure to determine the functions $F(\sigma)$ and $S(\sigma)$, i.e., the equation of the C/D boundary as determined from tests, is incorporated in these

143

two functions. The irreversible volumetric strain rate $\dot{\varepsilon}_v^I$ is obtained from (1) as

$$\dot{\varepsilon}_v^I = k_T \left\langle 1 - \frac{W_T(t)}{H(\sigma)} \right\rangle \frac{\partial F}{\partial \sigma} + k_S \frac{\partial S}{\partial \sigma} \qquad (3)$$

where the right hand side derivatives are with respect to the mean stress. Both these functions F and S are determined starting from the requirements:

$$\frac{\partial F}{\partial \sigma} > 0 \ , \quad \frac{\partial S}{\partial \sigma} = 0 \qquad \text{in the compressibility domain;}$$

$$\frac{\partial F}{\partial \sigma} = 0 \ , \quad \frac{\partial S}{\partial \sigma} = 0 \qquad \text{in the C/D boundary strip;} \qquad (4)$$

$$\frac{\partial F}{\partial \sigma} < 0 \ , \quad \frac{\partial S}{\partial \sigma} < 0 \qquad \text{in the dilatancy region.}$$

Since in the compressibility domain the irreversible volumetric changes cannot be but transient, the condition (4_1) must be satisfied, while in the dilatancy domain (4_3) is showing that microcracking (and thus evolutive damage) must be described by both transient and steady-state terms. Due to this peculiar behavior of the irreversible volumetric strain it is difficult to describe both transient and steady-state creep by a single term in the constitutive equation, if we would like to have this constitutive equation simple enough, in order to be able to use it in mining applications.

For simplicity, if a viscoplastic potential cannot be determined, one can replace it with a strain-rate orientation tensor (Cleja-Tigoiu [1991], Cazacu *et al.* [1997]). For instance, for the transient creep term, the derivative $(\partial F)/(\partial \sigma_{ij})$ is replaced by a tensor valued function $\mathbf{N}(\sigma)$, so that $\text{tr}\,\mathbf{N}$ is satisfying the inequalities (4). Obviously, the determination of $\mathbf{N}$ as function of the stress invariants is certainly simpler than that of $F(\sigma)$.

Evolutive damage is described by the volumetric part of the irreversible stress per unit volume W_V (Cristescu [1986], [1989]):

$$W(T) = \int_0^T \sigma(t) \dot{\varepsilon}_v^I \, dt + \int_0^T \sigma'_{ij}(t) \dot{\varepsilon}_{ij}'^I(t) \, dt = W_V + W_D \qquad (5)$$

i.e., by the first term in the right hand side of (5). In the compressibility domain this term is increasing (closing of microcracks and pores), while in the dilatancy domain it is decreasing (opening of microcracks and pores). Thus as reference configuration for damage must be chosen a state on the C/D boundary. An absolute reference configuration for damage would be a state on the hydrostatic axes for $\sigma \geq \sigma_0$. Therefore, for a loading path leading ultimately (in time) to failure, damage is the energy released by microcracking in the dilatancy domain, having the reference configuration on the C/D boundary (not on the stress-free--strain-free state). In the compressibility domain W_V is an energy stored by the rock.

144

Creep, evolutive damage and ultimate failure taking place in time, depend on the overstress above the C/D boundary. The dependency of ultimate failure (and thus on the evolutive damage) on the overstress above the C/D boundary has been shown for triaxial tests (Cristescu [1993]; see also Rokahr and Staudtmeister [1983]). For stress states approaching the C/D boundary the time to failure increases to infinity. Thus, it is to be expected that steady-state creep is also depending on the overstress above the C/D boundary.

Let us denote by $X(\sigma, \tau) = 0$ the equation of the upper boundary of the C/D strip shown in Figure 2.b. This equation is a relationship between octahedral shear stress τ and mean stress σ: $\tau_D = D(\sigma)$. For rock salt this relationship could be just (1). The subscript D stands for state on the boundary where "dilatancy" starts. Let us write the irreversible strain rate for steady-state creep (last term in (2)) as (Cristescu and Hunsche [to be published]):

$$\dot{\varepsilon}_{ij} = k_S \left[S_V(\sigma, \tau) \delta_{ij} + S_D(\tau) \frac{\sigma'_{ij}}{\tau} \right] \tag{6}$$

if the steady-state creep depends on the invariants σ and τ only. For the volumetric part in (6) we can choose

$$S_V(\sigma, \tau) := \left\langle 1 - \frac{\tau_D(\sigma)}{\tau} \right\rangle \left(\frac{\sigma}{\sigma_*} \right)^n \left(\frac{\tau}{\sigma_*} \right)^m \tag{7}$$

where in the bracket $< \ >$ is just the overstress above the boundary $X = 0$, and n and m are material constants. If we choose S_V in this form, we do not have any volumetric changes under the line $X = 0$ while creep is accelerated for higher overstresses. Starting from (7) one can write the general form of the constitutive equation for the steady-state creep as described in Cristescu and Hunsche [1998], § 5.3. Various other variants of (7), where the bracket $< \ >$ is involved in a nonlinear form could also be considered, if necessary.

DYNAMIC MEASUREMENTS OF DAMAGE

The evolution of damage by dynamic procedures is revealed by measuring the travel time of elastic waves crossing either the specimens in the laboratory or the rock in situ. Results of such measurements have been reported by very many authors (see literature mentioned by Cristescu and Hunsche [1998]). In most cases it was shown that the travel times and thus the elastic parameters are increasing with increasing pressures and that they tend asymptotically towards limit constant values, when pressure becomes high enough to close all pores and microcracks. The octahedral shear stress τ has also an influence on the values of the elastic parameters, but this dependency is not simple and it was less studied. The variation of elastic parameters due to the stress variation only was considered by many authors; see Volarovich et al. [1974], Popp and Kern [1994],

145

Niandou *et al.* [1997], Nawrocki *et al.* [1999], Cazacu [1999], besides many others.

The problem of the elastic parameters depend directly on the stress state or on some other parameters describing damage of rock salt was considered by Matei and Cristescu [1999 a,b]. In a first series of tests a rock salt specimen was subjected to a uniaxial loading and simultaneously two types of elastic waves were propagated through the specimen. From the travel times of the longitudinal and transverse waves the "dynamic" values of K and G were determined. Figure 3 shows the results. As the axial stress is monotonically increased with the loading rate 0.06 MPa/s, the elastic parameters are increasing in the compressibility domain, stay practically constant in the transition zone from compressibility to dilatancy and finally decrease in the dilatancy domain. On the same figure is shown the corresponding variation of the irreversible volumetric strain. This variation is absolutely similar to the variation of K and G. Thus, it seems that direct relationship can be established between the value of G and K and the state of damage existing in the specimen.

Moreover, uniaxial creep tests have also been performed. The specimen was loaded in steps of 0.87 MPa each and after each step increase the stress was kept constant for two days. During this time interval the variation of the strains and the travel times of the elastic waves were continuously recorded. The results are shown in Figure 4. The loading was expressed as a relative loading ratio $\Delta = \sigma_1 / \sigma_c$ where $\sigma_c = 17.88\,\text{MPa}$ is the conventional uniaxial compression strength of the rock salt (in these tests one has used rock salt specimens whose mechanical properties were influenced by humidity). From this figure follows that in the compressibility domain the elastic parameters are increasing every time the stress is increased, but they continue to increase in time even if the stress is constant. Again in the transition zone from compressibility to

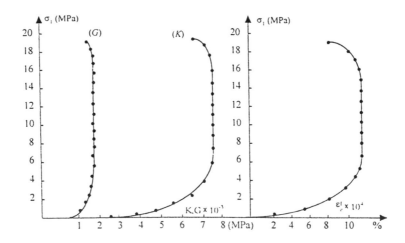

Figure 3 Variation of the elastic parameters K and G and of the irreversible volumetric strain in uniaxial tests.

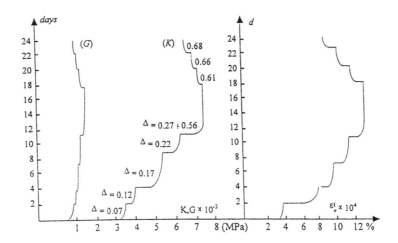

Figure 4. Variation in time of the elastic parameters and of the irreversible
volumetric strain in uniaxial creep test.

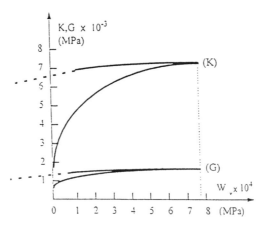

Figure 5. Dependency of elastic moduli on irreversible volumetric stress work
per unit volume w_v.

dilatancy, K and G are essentially constant. In the dilatant domain at each stress
increase, K and G are decreasing, and this decrease continues in time even if the stress
is constant.

The variation of K and G are quite similar to that of ε_v^i, showing that
evolutive damage is the one which governs the variation of the elastic parameters. Since
w_v is a good isotropic parameter to characterize damage, Figure 5 shows the

dependency of elastic parameters on the irreversible stress work per unit volume W_V. With increase of W_V in the compressibility domain, damage decreases and K and G increase. W_V is essentially constant in the transition zone and so are K and G. In the dilatant domain W_V decreases, i.e. damage increases, while K and G decreases.

CONCLUSIONS

We can conclude that the elastic parameters are depending on the stress state and evolution in time of the damage. Also the correct description of dilatancy and/or compressibility, and the correct formulation of the C/D boundary are instrumental in understanding the damage of rock salt. The equation of the C/D boundary, as determined from tests is to be incorporated in the constitutive equation, if we would like to describe accurately evolutive damage. As reference configurations for damage should be chosen states on the C/D boundary and not at the stress-free -- strain-free state.

REFERENCES

Bauschinger, J. [1879], "Ueber die Quercontraction und –Dilatantion bei Laengen-ausdehnung und – Zusammendrueckung prismatisher Koerper," Civilinginieur, Leipzig, 25, 81-124.

Bell, J.F. [1973], "The Experimental Foundations of Solid Mechanics," Handbuch des Physik, Vol. VI a/1, Springer Verlag, Berlin.

Bridgman, P.W. [1949], "Volume Changes in the Plastic Stages of Simple Compression," J. of Applied Physics, 20, 1241-1251.

Cazacu, O., Jin, J., and Cristescu, N.D. [1997], "A New Constitutive Model for Alumina Powders," KONA, Powder and Particle, No.15, 103-112.

Cazacu, O. [1999], "On the Choice of Stress-Dependent Elastic Moduli for Transversely Isotropic Solids," Mechanics Research Communications, 26, no.1, 45-54.

Cleja-Tigoiu, S. [1991], "Elasto-Viscoplastic Constitutive Equations for Rock –type Materials (Finite Deformation)," Int. J. Engng. Sci., 29, 1531-1544.

Cristescu, N. [1985a], "Plasticity of compressible/dilatant rock - like materials," Int. J. Engng.Sci., 23, 10, 1091-1100.

Cristescu, N. [1985b}, "Rock Plasticity," Plasticity Today: Modeling, Methods And Applications. Eds. A. Sawczuk and G. Bianchi. Elsevier Applied Sci.Publ. Ltd. 643-655.

Cristescu, N. [1986], "Damage and Failure of Viscoplastic Rock-Like Materials," Int. J. Plasticity, 2, 2, 189-204.

Cristescu, N. [1989], "Rock Rheology," Kluwer Academic, Dordrecht.

Cristescu, N. [1993], "A General Constitutive Equation for Transient and Stationary Creep of Rock Salt," Int. J. Rock Mech. Min. Sci. & Geomech. Abstr., 30, 2. 125-140.

Cristescu, N.D., and Hunsche, U. [1998], "Time Effects in Rock Mechanics," John Wiley & Chichester, New York, Weinheim, Brisbane, Singapore, Toronto.

Matei, A., Cristescu, N.D. {1999a], "The effect of volumetric strain on elastic Parameters for rock salt, " Mechanics of Cohesive-Frictional Materials, 6.

Matei, A., Cristescu, N.D. [1999b] , "Variation in time of the elastic parameters of rock salt," 1999 International Congress of the International Society for Rock Mechanics, Paris, August 25-28, 635-639.

Nawrocki, P.A., Cristescu, N.D., Dusseault, M. B., Bratli, R. K. [1999], "Experimental Methods for Determining Constitutive Parameters for Nonlinear Rock Modeling,", Int. J. Rock Mech. Min. Sci. & Geomech. Abstr., 36. No.5. 659-672.

Niandou, H., Shao, J. F., Henry, J. P., and Fourmaintraux, D. [1997], "Laboratory Investigation of the Mechanical Behaviour of Tournemire Shale", Int. J. Rock Mech. Min. Sci. & Geomech Abstr.. 34, 3-16.

Popp, T., and Kern, H. [1994], "The influence of dry and water saturated cracks on seismic Velocities of crustal rocks – A comparison of experimental data with theoretical model. Surveys in Geophysics. Kluwer Academic Publ., 15, 443-465.

Reynolds, O. [1885], "On the Dilatancy of Media composed of Rigid Particles in Contact. Phil. Mag.. 20, (5). 469-481.

Rokahr, R.B., and Staudtmeister, K. [1983], "Creep Rupture Criteria for Rock Salt," Sixth International Symposium on Salt. Salt Institute. vol.1 455-462.

Volarovich, M. P., Bayuk, E. I., Levykin, A. I. and Tomashevskaya, I.S. [1974], "Physico-Mechanical Properties of Rocks and Mineral at High Pressures and Temperatures. Nauka, Moscow.

Basic and Applied Salt Mechanics, Cristescu, Hardy, Jr & Simionescu (eds)
© *2002 Swets & Zeitlinger, Lisse, ISBN 90 5809 383 2*

A material model for rock salt including structural damages as well as practice-oriented applications

Z.Hou & K.-H.Lux
Clausthal University of Technology, Clausthal-Zellerfeld, Germany

ABSTRACT

This article compiles the factors to be generally considered from today's point of view as well as the still existing deficits in the field of material model developments for plastic-viscous salt rocks. Consequently, a new material model *Hou/Lux* is introduced, based on the *Lubby2* material model and on the fundamentals of continuum damage mechanics. The article then shows first possible applications.

INTRODUCTION

Geotechnical proofs of safety for underground systems in saliniferous formations, such as mines, storage cavities, subsurface repositories for hazardous waste or final repositories for radioactive waste, require material models, which adequately describe the material characteristics corresponding to the respective requirements. The bearing characteristics have been mathematically calculated for the respective mechanical framework, by means of these material models, together with prognosis models, which have been determined in respect to the location and which conservatively reproduce reality. These bearing characteristics, characterized by states of stresses and deformation, in the course of (future) time, constitutes, as part of the geotechnical framework planning together with corresponding evaluation criteria, the basis for identifying acceptable and unacceptable mechanical framework conditions as well as for the derivation of the design parameters (e.g. framework geometry, operating conditions or safety means).

In view of deformation-related losses of load-bearing capacity, the time-dependent strength of salt rocks and thus also the strength under permanent stress, which are closely related to the softening and creep rupture behavior of the material, are decisive for proofing the load-bearing capacity of saliniferous mechanical frameworks and thus the long-term operational safety of e.g. storage caverns, subsurface repositories for hazardous waste or final repositories for radioactive waste.

Almost all previous material models assume that salt rocks such as rock salt, hard salt or sylvinite have ductile, viscoplastic material characteristics and that fractures can be ruled out by means of an adequate, individually determined limitation of stress e.g. by means of a permissible coefficient of utilization and a limitation of (creep) deformations, *Lux (1984)*. The result is the introduction of stress and deformation related safety coefficients into the proofing process. Rock zones characterized by fracture processes are thus identified by state variable evaluations following the continuum mechanical calculation. On the other hand, the development of rock formations characterized by ruptures is not directly integrated into the actual calculation process. However, such mathematical analyses including the softening process into the calculation are known from other areas of underground constructions, such as e.g. tunneling. Here, the failure of parts of the rock formation, which are then called pseudoplastic zones, are included in the mathematical simulation as part of elasto-viscoplastic calculations. *Yin (1996)* and *Lux & Yin (2000)* provide examples. *Minkley (1997)* shows a similar procedure; he analyses the load-bearing behavior of pillars and mining structures in carnallite formations by means of elastoplastic material models, assigning the carnallitite a softening behavior with post-fracture dilatancy.

The material behavior is viewed differently, assuming that a material a priori has more or less intensively developed structural defects or damages. Depending on the type and intensity of stress, the deformation rate and state of deformation or environmental conditions, these structural damages develop further (->loosening, softening) or they disappear (-> healing).

As part of this point of view, and for now limited to loosening and softening processes, it shall in the following be assumed that the bearing capacity of the stressed salt rock successively decreases over time under the acting stress at different rates, depending on the type of rock and stress as well as stress intensities. At first microfissures develop followed later by single or crosslinked fractures until finally collapsed roofs, wall spalling or floor breaking occur.

In order to quantify these softening processes as well as the long-term strength and the load-bearing behavior of mechanical frameworks as they are reached locally or regionally, sufficient knowledge and suitable statements are necessary, e.g. to describe the tertiary creep phase, the effective fracture mechanisms and the creep rupture.

Especially regarding the above mentioned problems concerning the time-dependent load-bearing capacity and the rupture behavior of rock salts, many questions still remain unanswered today. The rough description of the failure of load bearing elements has shown that the time-dependent load-bearing capacity of the salt rocks is closely related to the development of fractures. The obvious place to look for a statement describing fractured mechanical framework deformations is therefore the area of material science. One tool to solve these problems in a more or less phenomenological point of view could be 'continuum damage mechanics', which allows the description of the material damage as a result of external stresses and thus the 'damage process' until the actual rupture. This theory has been developed over the last 30 years, especially in the fields of metallic materials to describe phenomena such as fracture development, rupture, fatigue, ductility or creep. 'Damage' here means the damaging of the material structure, which could already exist inside a material as an original structural damage prior to any stress. However, under certain conditions this damage could also be generated under acting stress and develop further to finally cause the functional failure of a load bearing element due to e.g. excessive deformation including fissures, fractures and finally rupture processes with the loss of stability or due to increasing permeability and the loss of tightness.

With this background, this article describes the basic factors to be considered from today's point of view as well as the still existing deficits regarding the development of material models for plastic viscous salt rocks. Based on this description, a new material model is introduced, *Hou (1997), Hou & Lux (1998)* – based on the *Lubby2* material model according to *Lux (1984)* and based on continuum damage mechanics, e.g. *Kachanov (1986)* or *Lemaitre (1992)*. Finally, some initial application possibilities will be shown.

MINING AND LABORATORY EXPERIENCE WITH SOFTENING PROCESSES CHARACTERIZED BY RUPTURE

Mining experience shows that mine-working contours are characterized by loosening and softening processes of different intensities. Depending on the type of rock, the stress intensity and the time-to-rupture, ruptured loosening of rock develops, starting at the contour and progressing into the rock mass, with different intensities but, at a first phenomenological glance, increasing over time. This loosening can then further develop to areas of softening and finally ruptured spalling or cracks.

Figure 1 examplarily shows wall spalling and floor breaking in a mine. The intensity of the crack development can be clearly seen, decreasing away from the contour and indicating a stress and time dependent process originating from the contour.

Such softening and fracture processes can also be observed in laboratory investigations, e.g. in common strength tests or in pillar model tests, but also in special investigations with axially perforated large rock salt samples.

Figure 1 - Wall spalling and floor softening in the drift of a salt mine

Figure 2a shows the spalling contour of the axial bore hole in the large rock salt sample. In this example, the almost completely destrengthened area is limited to the contour itself, in spite of the contrary visual impression.

In addition, figure 2b shows the destrengthened areas of such a large rock salt sample, subjected to triaxial stress over a longer period of time, originating from the bore hole contour and visualized as lighter areas by means of tracing.

More information about such experiments will follow.

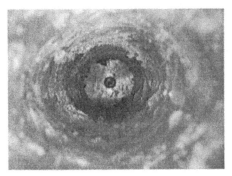

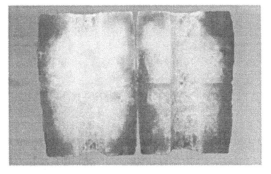

Figure 2a - Axially perforated rock salt sample with fracture zone near the contour (view inside the bore hole)

Figure 2b - Axially perforated rock salt test piece with destrengthened area (vertically cut after removal of the salt rock of the fracture zone near the contour)

FACTORS AND DEFICITS TO BE CONSIDERED CONCERNING MATERIAL MODELS FOR ROCK SALT

From today's point of view, the following applies on an international level:

- Extensive new knowledge about the mechanical behavior of salt rocks and about material models has been gathered, particularly over the last 10 years.

- The behavior of salt rocks during the stationary creep phase has been extensively investigated. Based on these results, several formulations for material models have been developed for their quantification, which in parallel are applied practically, e.g. *Hunsche & Schulze (1994), Langer (1991), Lux (1984),*

- The behavior of salt rocks in the transient creep phase is widely known, but the proposed material models still need to be developed further, especially for the unloading/loading path.

- There are good prospects of physically explaining and mathematically describing the behavior of salt rocks in the tertiary creep phase with the introduction of continuum damage mechanics.

No material model currently meets all demands, and probably none will be developed in the near future, since the corresponding laboratory basics are not yet available in their necessary width and depth. The following list briefly identifies the deficits, which could also be the content of future research activities, *Hou (1997)*. In this regard, the term 'stress' does not only include loads but also temperature changes:

(1) Description of the fracture strength in dependence of the stress geometry, because the strength of salt rocks not only depends on the facies and on the load rate or the load duration, but also on the stress geometry. Therefore, the stress geometry and thus the stress state must be taken into consideration for the determination of the strength allocated to the material. If and how the stress geometry influences the creep rate remains an unanswered question. It is determined, however, that the stress geometry does have an influence, if the stress state is within the semi-ductile and brittle stress range, *Hou (1997), Hou et al. (1998)*;

(2) Analysis of the deformation mechanisms acting in the micro range and derivation of physically based internal state variables to describe the deformation mechanisms;

(3) Derivation of a relation between creep rate and internal state variables;

(4) Analysis of the effects of diffusion and dislocation as the main mechanisms in the stationary creep phase and the secondary mechanisms in the transient phase;

(5) Analysis of the effects of hardening and recovery with a definition of internal state variables to describe hardening and recovery as well as the development or more precise description of a kinetic equation for the internal state variables (for hardening and recovery respectively);

(6) Derivation of a relation between hardening and recovery: equal or unequal internal state invariable to describe hardening and recovery; recovery during hardening as well as hardening during recovery;

(7) Analysis of the effects of damage processes: definition and quantification of damage; conditions for the development of damages; original damage inside the material as well its effect on the newly developing damages and on the further development of the damages; kinetic equation to describe the damage; effects on the creep rate (-> damage induced creep rate); relation between damages for short-term and creep tests regarding the possibility of transfer or extrapolation of the damage behavior of short-term stress and deformation processes onto a large observation span (-> long-term extrapolation);

(8) Effects of damage healing: definition and quantification of damage healing; conditions for the healing of damages; kinetic equation to describe damage healing; effects on the creep rate (-> healing of damage-induced creep rate); healing behavior in case of decreasing stress and in case of relaxation;

(9) Consequences of diffusion, dislocation, hardening and healing: validation of statements regarding the material behavior without damage and without new fracture formation such as isovolume, no changes in porosity and permeability, no moisture effect and effects only on the transient and stationary creep behavior or on the relaxation behavior;

(10) Consequences of damaging: activation and development of dilatancy as well as the effect of moisture; development of pores and cracks; increase in porosity, permeability, creep rate and 'real stress" active inside the solid structure; identification of a damage-induced creep rate; tertiary creep and creep rupture;

(11) Consequences of damage healing: closing and healing of pores and cracks; reduction of damage intensity; reduction of damage-induced consequences such as dilatancy, moisture effects, porosity, permeability, creep rate and effective 'real stress';

(12) Relation between damage development and damage healing: equal or unequal internal state invariable to describe damaging and healing (e.g. damage parameters);

(13) Short-term behavior and fracture strength: portrayal of fracture strength taking into account the stress geometry, the temperature and the deformation or stress rate; flow law; existence of internal state variables, allowing the derivation of a relation between the short-term behavior and the long-term behavior; derivation of a uniform material model for the short-term and long-term behavior;

(14) Creep behavior and long-term strength ->

- *Transient creep*: internal state variable for the hardening after a stress increase and for the recovery after a stress decrease; transient share of recovery in overall hardening; duration of transient creep (mainly estimated to be relatively short); effects of the known deformation mechanisms and implementation of mechanisms into a material model;

- *Stationary creep*: physical explanations of stationary creep e.g. as the result of a compensation between deformation mechanisms; effects of diffusion, dislocation, damaging and healing as well as implementation of these mechanisms into a material model; stationary creep behavior especially in case of little stress; extrapolation of stationary creep rates, gained over a period of weeks or months from creep tests, onto large periods of observation; conservativeness of current statements regarding deformation prognoses, because the actual stationary creep phase is not yet measurable during the common and a relatively short periods of observation, at least in case of not significantly increased temperatures, with simultaneous non-conservativeness regarding the compaction effect;

- *Tertiary creep*: causes and conditions for tertiary creep; effects of damages and healing of damages; damage-induced tertiary creep rate; softening; dilatancy strength; permeability changes;

- *Creep rupture and long-term strength*: Criteria to evaluate the development of creep damage and creep rupture; relation between continuous stress and time until creep rupture; physical reason and mathematical description of the long-term strength and the strength in dependence of the time;

(15) Relaxation behavior: effects of the mechanisms of recovery and damage healing on the relaxation behavior; derivation of the time-dependent strength from the relaxation behavior; existence of a creep limit; mathematical description of the relaxation behavior;

(16) Relation between the creep behavior and the relaxation behavior: uniform model for creep and relaxation; creep during relaxation as well as relaxation during creep; testing of the following statement: $\sigma_{relax}/E_{sys} = \varepsilon_{creep}$.

HOU/LUX MATERIAL MODEL INCLUDING STRUCTURAL DAMAGE

The above mentioned descriptions show that the time-dependent strength and thus the long-term strength as well as the creep rupture behavior can only be estimated using material models, which include a suitable statement for tertiary creep, and that the tertiary creep is closely related to the development of cracks and pores, i.e. the structural damage, and that it can therefore be described by means of continuum damage mechanics. According to this knowledge and in differentiation to existing material models, a new material model named *Hou/Lux* has been developed, based on the *Lubby2* material model according to *Lux (1984)* and based on the fundamentals of continuum damage mechanics. The advantages particularly of the material models by *Cristescu & Hunsche (1993)* as well as *Glabisch (1997)*, by *Chan et al. (1994)* and by *Aubertin et al. (1996)* have been included, *Hou (1997)*.

The *Hou/Lux* material model is a material model, which includes in a phenomenological way the effects of various deformation mechanisms: diffusion and dislocation, hardening and recovery as well as damage and damage healing and which is thus capable of fulfilling the most important demands made on material models, at least in principal. The above mentioned mechanisms contribute directly to the development of inelastic strains/strain rates and thus to the correspondingly current state of deformation. They also contribute indirectly to the state of stress via the rigidity of the load bearing element. The fundamental structure and the most important constituents of this material model are introduced below:

(a) Hou/Lux material model for the total strain rate
- total strain rate: $\qquad \dot{\varepsilon}_{ij} = \dot{\varepsilon}_{ij}^{e} + \dot{\varepsilon}_{ij}^{i}$ $\hfill$ (1)

- elastic strain rate: $\quad \dot{\varepsilon}_{ij}^{e} = \dfrac{1}{2G} \cdot \dfrac{\dot{s}_{ij}}{1-D} + \left(\dfrac{1}{9K} - \dfrac{1}{6G} \right) \cdot \dfrac{\dot{I}_1}{1-D} \cdot \delta_{ij}$ $\hfill$ (2)

- inelastic strain rate: $\dot{\varepsilon}_{ij}^{i} = \dot{\varepsilon}_{ij}^{vp} + \dot{\varepsilon}_{ij}^{d} + \dot{\varepsilon}_{ij}^{h}$ $\hfill$ (3)
with

σ_{ij} $\qquad$ stress tensor
s_{ij} $\qquad$ $= \sigma_{ij} - I_1/3 * \delta_{ij}$, deviatory stress tensor
δ_{ij} $\qquad$ *Kronecker* standard tensor
ε_{ij}^{vp} $\qquad$ viscoplastic strain tensor induced by the deformation mechanisms of diffusion, dislocation, hardening and recovery
ε_{ij}^{d} $\qquad$ viscoplastic strain tensor induced by damages
ε_{ij}^{h} $\qquad$ viscoplastic strain tensor induced by the healing of damages
D $\qquad$ damage variable, $0 \leq D \leq 1$

The statement according to equations (1) - (3) shows that apart from the common fundamental differentiation between an elastic and an inelastic part of the strain rate, there is a further differentiation of inelastic strain rates into a viscoplastic part without direct structural damage and furthermore into a viscoplastic part induced by structural damage and a viscoplastic part resulting from structural healing. The individual parts of the strain rate are additively linked, so that a corresponding laboratory identification of the individual parts becomes necessary on the one hand and must also be possible on the other hand.

Below, the statements for the inelastic parts of the strain rates are introduced. These statements are treated as independent material models (-> partial models).

156

(b) *Hou/Lux-ODS material model (without direct damage) for the viscoplastic strain rate induced by the deformation mechanisms of diffusion, dislocation, strain hardening and recovery*

The viscoplastic strain rate resulting from the deformation mechanisms of diffusion, dislocation, strain hardening and recovery can in principal be calculated using the material models applied in salt mechanics, which account for these mechanisms more or less explicitly, but mostly implicitly. One of these material models is the *Lubby2* material model with deformation hardening. However, the *Lubby2* material model does not account for the differences between hardening after a stress increase or recovery after a stress decrease. Neither does it include an explicit statement for the determination of the maximum transient creep deformation $\max\varepsilon_{tr}$. One modification of this *Lubby2* material model, which eliminates these two deficits by including the reference area for the stress calculation (dA -> (1-D)·dA) reduced as a result of material damage, following *Kachanov (1986)*, is introduced in equations (4) - (7), applying the flow model according to *Mises*. This material model is called the *Hou/Lux-ODS* material model and results in:

$$\dot{\varepsilon}_{ij}^{vp} = \frac{3}{2}\left[\frac{1}{\overline{\eta}_K^* \cdot \exp\left(k_2 \cdot \dfrac{\sigma_v}{1-D}\right)} \cdot \left(1 - \frac{\varepsilon_{tr}}{\max\varepsilon_{tr}}\right) + \frac{1}{\overline{\eta}_m^* \cdot \exp\left(m \cdot \dfrac{\sigma_v}{1-D}\right) \cdot \exp(l \cdot T)} \right] \cdot \frac{s_{ij}}{1-D} \qquad (4)$$

$$\max\varepsilon_{tr} = \frac{1}{G_K} \cdot \frac{\sigma_v}{1-D} \qquad (5)$$

$$G_K = \begin{cases} \overline{G}_K^* \cdot \exp\left(k_1 \cdot \dfrac{\sigma_v}{1-D}\right) \cdot \exp(l_1 \cdot T) & \varepsilon_{tr} < \max\varepsilon_{tr}, \text{hardening} \\[3mm] \overline{G}_{KE}^* \cdot \exp\left(k_{IE} \cdot \dfrac{\sigma_v}{1-D}\right) \cdot \exp(l_{IE} \cdot T) & \varepsilon_{tr} < \max\varepsilon_{tr}, \text{recovery} \end{cases} \qquad (6)$$

$$\overline{G}_{KE}^* \cdot \exp\left(k_{IE} \cdot \frac{\sigma_v}{1-D}\right) \cdot \exp(l_{IE} \cdot T) \le \overline{G}_K^* \cdot \exp\left(k_1 \cdot \frac{\sigma_v}{1-D}\right) \cdot \exp(l_1 \cdot T) \qquad (7)$$

with

$\max\varepsilon_{tr}$ maximum strain of transient creep, in –

G_K^*, k_1, l_1 parameter only for hardenning, correspondingly in MPa, 1/MPa, 1/K

G_K, k_{1E}, l_{1E} parameter only for recovery, correspondingly in MPa, 1/MPa, 1/K

The other parameters correspond to the material parameters in the *Lubby2* material model, *Lux (1984)*.

For the determination of the recovery part of the transient strains it is assumed that only a part of the total transient creep deformation can recover after a stress decrease. This is implemented in equation (7).

(c) *Damage material model Hou/Lux-MDS (with direct damage) for the strain rate induced by damage and damage healing*

The strain rate damage-induced is calculated by means of the *Hou/Lux-MDS* damage material model according to equation (8), which contains a non-associated flow model in order to be able to adapt the calculated volume changes to the laboratory results. The healing variable h and the strain rate induced by damage healing according to equation (9) are not investigated as part of this study, due to a lack of own knowledge and a lack of knowledge to be found in international literature. The following results for the material model:

$$\dot{\varepsilon}_{ij}^d = \dot{\varepsilon}_{ij}^{ds} + \dot{\varepsilon}_{ij}^{dz} = a3 \cdot \frac{\left\langle \frac{F^{ds}}{F^*} \right\rangle^{a1}}{(1-D)^{a2}} \cdot \frac{\partial Q^{ds}}{\partial \sigma_{ij}} + a3 \cdot \frac{\left\langle \frac{F^{dz}}{F^*} \right\rangle^{a1}}{(1-D)^{a2}} \cdot \frac{\partial Q^{dz}}{\partial \sigma_{ij}} , \langle x \rangle = \begin{cases} 0 & x \leq 0 \\ x & x > 0 \end{cases} \tag{8}$$

$$\dot{\varepsilon}_{ij}^h = \dot{\varepsilon}_{ij}^h (D, h, \sigma_v, \sigma_3, \theta, T) \tag{9}$$

with

F^{ds}, F^{dz} flow function for the structural damage (pore and crack formation) due to shearing stresses and tensile stresses, in MPa

Q^{ds}, Q^{dz} potential function for the structural damage (pore and crack formation) due to shearing stresses and tensile stresses, in MPa

$\varepsilon_{ij}^{ds}, \varepsilon_{ij}^{dz}$ strain tensor induced by damage resulting from shearing stresses and tensile stresses

$\varepsilon_{ij}^h, \varepsilon_{ij}^{dz}$ strain tensor induced by healing of damages

F^* = 1 MPa (standardization stress)

a1, a2, a3   material parameters

h   healing variable

(d) Flow and potential functions for the detection of structural damages (pore and crack formation) resulting from shear stresses and tensile stresses

The flow and potential functions contained in equation (8) are made more precise in equations (10) - (16). The flow functions F^{ds} following equation (10) and F^{dz} following equation (11) to describe the effects of structural damages integral for a rock element, represent the mechanical conditions for the formation of pores and cracks, correspondingly due to shear stresses and tensile stresses. The potential functions Q^{ds} according to equation (12) and Q^{dz} according to equation (13) on the other hand describe the corresponding flow directions in the material model:

$$F^{dx} = \sqrt{3J_2} - \beta_{Dil}(\sigma_3, \theta) = \sqrt{3J_2} - \eta_{Dil}(\sigma_3) \cdot \beta(\sigma_3, \theta) \tag{10}$$

$$F^{dz} = a11 \cdot \left[\langle -\sigma_3 \rangle + \langle -\sigma_2 \rangle + \langle -\sigma_1 \rangle - a12 \right]^{a13} \tag{11}$$

$$Q_{ds} = \sqrt{3J_2} - a0 \cdot F^*(\sigma_3, \theta) \tag{12}$$

$$Q^{dz} = a14 \cdot \left[\langle -\sigma_3 \rangle + \langle -\sigma_2 \rangle + \langle -\sigma_1 \rangle \right] \tag{13}$$

$$\eta_{Dil}(\sigma_3) = 1 - a4 \exp(-a5 \cdot \sigma_3) \tag{14}$$

$$\beta(\sigma_3, \theta) = \beta^{TC}(\sigma_3, \theta) \cdot k(\sigma_3, \theta), \beta^{TC}(\sigma_3, \theta) = a6 - a7 \exp(-a8 \cdot \sigma_3) \tag{15}$$

$$k(\sigma_3, \theta) = \left[\frac{1}{\cos\left(\theta + \frac{\pi}{6}\right) + a9 \cdot \sin\left(\theta + \frac{\pi}{6}\right)} \right]^{\exp(-a10\,\sigma_3)} \tag{16}$$

with

$\beta(\sigma_3, \theta)$ strength function from short-term test, in MPa

$\beta_{Dil}(\sigma_3, \theta)$ dilatancy function, in MPa

$\beta^{TC}(\sigma_3, \theta)$ strength function under TC conditions, in MPa

$k(\sigma_3, \theta)$ correction function to account for stress geometry, in -

$\eta_{Dil}(\sigma_3)$ function, describing the conditions for the beginning of dilatancy, i.e. the border conditions for the beginning of crack formations, in the form of the coefficient of utilization

a0, a4 - a13   material parameter

The first term on the right hand side of equation (10) represents, in the form of the effective stress, those kinds of stresses, which cause the damage induced by shear stresses. The second term of equation (10) describes the strength of the material structure and thus the resisting forces, counteracting structural damage. This part of the strength can be calculated from the material strength and the border condition for dilatancy in the form of the coefficient of utilization η_{Dil}. The material strength is, among other things, a function of the facies, the stress rate, the minimum stress and the stress geometry, *Hou (1997)*.

Equation (11) also says that cracks develop, if the tensile stresses of all directions together exceed the tensile stress $\beta_z = a12$. The differentiated treatment of shear and tensile stresses makes it possible to consider tensile stresses and their effects separately.

The derivations of the potential functions can be found in earlier publications, *Hou (1997)* as well as *Hou & Lux (1998)*.

(e) Kinetic equation for the damage development (pore and crack formation)

The kinetic equation for the damage development is described by means of a flow function F^{ds} and F^{dz} according to equation (17):

$$\dot{D} = a15 \cdot \frac{\left[\left\langle \frac{F^{ds}}{F^*} \right\rangle + \left\langle \frac{F^{dz}}{F^*} \right\rangle\right]^{a16}}{(1-D)^{a17}} \qquad (17)$$

with
a15, a16, a17 material parameters

The major difference as compared to the kinetic equation known from *Kachanov (1986)* or *Lemaitre (1992)* is that the stresses in the kinetic equation are not the reference stresses to be found in the corresponding literature, but the damage stresses F^{ds} and F^{dz} according to equation (10) and equation (11).

The reason for this statement is the fact that the structural damages are only activated and further developed, if the strength of the structure, known as dilatancy strength $\beta_{Dil}(\sigma_3, \theta)$, is exceeded. The larger the acting damage stresses, the larger the development of structural damages and the larger the strain rate induced by damage.

If the stress remains constant – a situation, which is for example the case in classic creep tests – equation (17) can be integrated according to the damage variable D. Thus the damage development in dependence of the time is explicit according to equation (18):

$$D(t) = 1 - \left\{ 1 - a15 \cdot (1+a17) \cdot \left[\left\langle \frac{F^{ds}}{F^*} \right\rangle + \left\langle \frac{F^{dz}}{F^*} \right\rangle\right]^{a16} \cdot t \right\} \qquad (18)$$

Assuming that the creep rupture only occurs as the damage variable assumes the numerical value D = 1, the time-to-rupture t_{BK} at a constant stress and until creep rupture can be determined by means of equation (19):

$$t_{BK} = t_{BK}(\sigma_3, \theta) = \frac{1}{a15 \cdot (1+a17) \cdot \left[\left\langle \frac{F^{ds}}{F^*} \right\rangle + \left\langle \frac{F^{dz}}{F^*} \right\rangle\right]^{a16}} \qquad (19)$$

Equation (19) has the following characteristics:

- If the damage stresses F^{ds} and F^{dz} remain negative, there is and will be no damage over time, because the damaging stress is smaller than the long-term strength of the structure (= dilatancy strength). In this case, the possibly already existing pores and cracks cannot be activated. The formation and further development of new pores and cracks is impossible. The creep behavior of salt rocks can thus be satisfactorily be described with known material models such as *Lubby2* on the condition that the stationary creep rate has been correctly determined.

- If the damage stresses F^{ds} and F^{dz} are positive, but still relatively small, damage will develop very slowly. The time until creep rupture could in this case assume very large values.

- If the damage stresses F^{ds} and F^{dz} are positive and also relatively large regarding the level of short-term strength, damage develops very quickly. As a result, tertiary creep and creep rupture will also occur relatively fast.

- The special cases of damage stresses F^{ds} or F^{dz} equaling zero represent the border conditions or the criteria for the fundamental occurrence of creep ruptures and for the long-term strength without damage. That means the strength condition characterizing an intact structure, here called dilatancy strength, is identical to the long-term strength (fatigue strength) of a material. This statement is the fundamental mechanical starting point for the *Hou/Lux*-MDS damage material model. It still has to be quantified with respect to the location by means of laboratory tests and/or by means of in situ observations. The long-term strength itself cannot be measured directly. It is, however, possible to derive the strength conditions of an intact structure from laboratory tests. Thus is would be possible in a first, and because of the relatively high rates also conservative statement, to gain the conditions for the long-term strength, characterized by a structure without damage, from short-term tests.

- First there is a physical reason for the above-mentioned conclusion: In case of creep formation without structural damage, there is no stress state, which could convert stationary creep into tertiary creep. Thus the primary integrity of a material remains the same. The primary crystal defects and micro fissures are in this case suppressed and not activated. The other active deformation mechanisms (without the collaboration of the mechanical structural damage) do produce creep deformations, which, however, only lead to viscoplastic deformations of the same volume but not to dilatancy and which therefore do not cause failure due to rupture. Proofs of this statement are the observations in laboratory tests under triaxial compression and extension stress. According to this, the test samples do not suffer significant losses of strength (and no lasting permeability changes) at very large minimal stresses (e.g. $\sigma_3 > 25$ MPa) in the range of the so-called ideal plastic state of deformation, even in case of very large deformations.

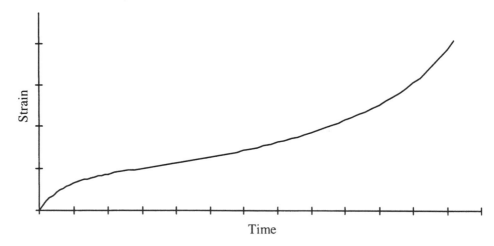

Figure 3 -   Typical creep rupture curve for rock salt

- Together with equation (17) the conclusion can be drawn that the damage develops very slowly as the dilatancy strength is first exceeded, but then increases more rapidly and extremely fast shortly before the creep rupture. This tendency also corresponds to the course of the phenomenologically observed tertiary creep from the beginning of accelerated creep until creep rupture. Figure 3 shows a characteristic creep curve with tertiary creep phase and creep rupture.

- The stress geometry is taken into account by the *Lode*-angle θ.

(f) Dilatancy as a result of damage

The dilatancy ε_{Vol} is here considered to be a consequence of the structural damage. The dilatancy rate can be calculated with equation (20):

$$\dot{\varepsilon}_{Vol} = -a3 \cdot \frac{\left\langle \frac{F^{dh}}{F^*} \right\rangle^{a1}}{(1-D)^{a2}} \cdot \left\{ a0 \cdot a7 \cdot a8 \cdot \exp(-a8 \cdot \sigma_3) \cdot k(\sigma_3, \theta) - a10 \cdot \exp(-a10 \cdot \sigma_3) \cdot \ln \left[\frac{1}{\cos\left(\theta + \frac{\pi}{6}\right) + a9 \cdot \sin\left(\theta + \frac{\pi}{6}\right)} \right] \cdot \left(\sqrt{3J_2} - Q^{dh}\right) \right\}$$

$$-a3 \cdot a14 \cdot \frac{\left\langle \frac{F^{dz}}{F^*} \right\rangle^{a1}}{(1-D)^{a2}} \cdot \left[H(-\sigma_3) + H(-\sigma_2) + H(-\sigma_1) \right] \quad (20)$$

Equation (20) shows that there is dilatancy only, if the dilatancy strength of the structure is exceeded or in case of tensile stresses. At a constant stress, the dilatancy develops together with the damage. As long as the damage variable D is still very small, the dilatancy can be neglected. However, as the damage variable D assumes relatively large values, the dilatancy will increase drastically. Shortly before creep rupture, the dilatancy rate is at its high, because at this point in time, the damage reaches its critical value. Furthermore, the stress geometry is taken into account when calculating the dilatancy.

SOME SELECTED EXAMPLE CALCULATIONS WITH THE *HOU/LUX* MATERIAL MODELS

(1) Material characteristics

In order to present the characteristics and the efficiency of the *Hou/Lux* material model, and also to compare it to experimental results, some exemplary calculations will be introduced in the following, showing the different functions of the material model. The parameters contained in the material model must be numerically determined. Table 1 shows a compilation of the parameters for the *Hou/Lux-ODS* material model for rock salt of a location A, determined during our own laboratory tests in the geomechanical laboratory of the professorship for disposal technology and geomechanics. It also shows estimated parameters for the *Hou/Lux-MDS* damage material model for rock salt, based on literature, e.g. *Aubertin (1993)*. The parameters for the recovery process during transient creep are estimated based on the data published by *Munson et al. (1993)*.

Due to the lack of suitable test data, especially of creep tests with tertiary creep until creep rupture, some of the test data for rock salt must be data found in literature, making it impossible to use a uniform set of parameters. Currently, the parameters e.g. for the *Lubby2* material model, are not determined according to the actually necessary strict condition that there be no damage in the material structure, because the applied creep tests were mainly performed as unaxial compression tests and thus without confining pressure at medium to higher stresses $(10\,\text{MPa} < \sigma_v < 20\,\text{MPa})$, *Lux et al. (1997)*. Structural damage under such stress conditions are obviously unavoidable, so that the creep deformation induced by the damage is implicitly included in the measured data and thus also in the material parameters. That means, the parameters provided in Table 1 generally overestimate the creep rate. Therefore, the following calculations only show an exemplary solution to the tasks closely related to tertiary creep. Parameters especially fitted to the *Hou/Lux* material model can only be determined by means of corresponding laboratory tests. These tests are currently being performed in the laboratory. It must be mentioned that the extend of the investigation as compared to previous laboratory programs will not necessarily be larger. However, additional parameters such as stress and deformation-dependent dilatancy must be determined. Furthermore, the structure of the test program must be modified.

It must also be mentioned that new and better knowledge of the load bearing behavior of saliniferous structures as a basis for solutions to or processing of new questions without extended theoretical instruments cannot be expected. It must also be mentioned though that the corresponding necessary instruments must be determined and used according to the object. The planning of brine extraction caverns and the proof of integrity for a technological weakness in the saliniferous barrier of a final repository are definitely extreme practical examples regarding the effort for the geotechnical proof as part of the mechanical framework planning.

Table 1 - *Exemplary numeric values for the parameter of the Hou/Lux material model for rock salt*

Model	Hou/Lux-ODS			Model	Hou/Lux-MDS	
				a0	1/35	-
η_{k*}	8,94E+04	MPa.d		a1	6	-
k_2	-0,168	1/MPa		a2	4	-
				a3	2,00E-10	1/d
G_{k*}	5,08E+04	MPa		a4	0,8	-
k_1	-0,191	1/MPa		a5	0,055	1/MPa
				a6	67	MPa
η_{m*}	2,03E+07	MPa.d		a7	41	MPa
m	-0,247	1/MPa		a8	0,25	1/MPa
l	0	1/K		a9	1	-
T	295	K		a10	0,25	1/MPa
l_1	0	1/K		a11	-	-
				a12	-	MPa
parameters for recovery				a13	-	-
G_{kE}	3,05E+04	MPa		a14	-	-
k_{lE}	-0,191	1/MPa		a15	1,67E-08	1/d
				a16	5	-
				a17	5,50	-

(2)   Short-term and long-term strength including the stress geometry

Figure 4 shows the short-term and long-term strengths under the special stress conditions of triaxial compression (TC) and extension (TE) in the invariant plane. First of all, the strength under TE conditions is always less than the strength under TC conditions. If a material has the same strength under TC and TE conditions at an equal minimum stress, this is a special case, which can also be determined by means of the developed general strength function with a9 = $\sqrt{3}/3$, *Hou (1997)*. That means by varying the parameter a9 in equation (16) it is possible to set the difference between the strength under TC and TE conditions at an equal minimum stress according to the laboratory results.

It is also evident that the long-term strength at higher minimum stresses as compared to the short-term strength under otherwise equal conditions is far less reduced than at lower minimum stresses. This is the case because the ideal-plastic state is only reached at very high minimum stresses and because the strength reduction caused by damage is pronounced less the more the state of stress approximates the ideal-plastic state.

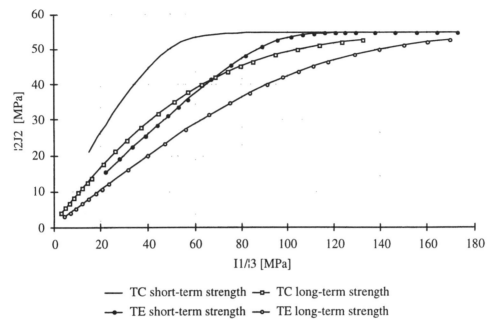

$$\text{— TC short-term strength} \quad \text{—□— TC long-term strength}$$
$$\text{—•— TE short-term strength} \quad \text{—○— TE long-term strength}$$

Figure 4 - Strengths in the invariant plane for rock salt depending on stress state

(3)   Dependencies between the time until creep rupture and permanent stress, the permissible coefficient of utilization and minimum stress

Figure 5 shows the dependencies between the rheological coefficient of utilization η_{rh} related to creep rupture, the time until creep rupture and the minimum stress under TC conditions. Following *Lux (1984)*, the rheological coefficient of utilization η_{rh} is defined as the ratio between the time-dependent strength β_t and the short-term strength β:

$$\eta_{rh} = \frac{\beta_t(\min\sigma,\theta,t_{BK})}{\beta(\min\sigma,\theta)} \tag{21}$$

The reciprocal of the rheological coefficient of utilization η_{rh} is called rheological safety coefficient ξ_{rh} and is:

$$\xi_{rh} = \frac{1}{\eta_{rh}} \tag{22}$$

Figure 5 shows that for the same given time-to-rupture at a lower minimum stress only a smaller rheological coefficient of utilization η_{rh} is permissible as compared to a higher minimum stress, i.e. at lower minimum stresses a higher rheological safety ξ_{rh} against creep rupture must be demanded. Given the same minimum stress – based on the short-term strength with $\eta_{rh} = 1,0$ – the permissible rheological coefficient of utilization η_{rh} can be smaller, the longer the rupture-free time allocated to a salt rock formation.

The gradient, at which the permissible rheological coefficient of utilization η_{rh} decreases over the time, is clearly larger at lower minimum stresses as compared to higher minimum stresses. For example, the permissible rheological coefficient of utilization at $\sigma_3 = 0$ MPa for a time of $t = 1$ year is $\eta_{rh} = 49\%$ and $\eta_{rh} = 27\%$ for a time of $t = 1000$

years. At $\sigma_3 = 15$ MPa, $\eta_{rh} = 76\,\%$ for a time of $t = 1$ year and $\eta_{rh} = 68\,\%$ for $t = 1000$ years.

The tendency to derive from this, that the damage-related strength reduction is much slower and less intensive at higher minimum stresses than at lower minimum stresses, generally agrees with experience.

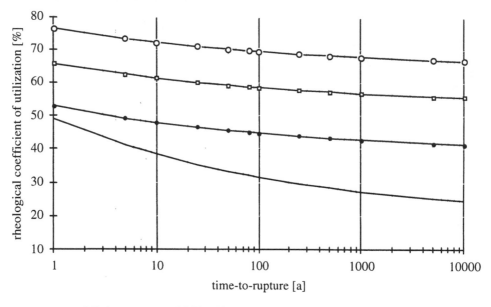

Minimum stress: 0 MPa (TC)   Minimum stress: 10 MPa (TC)
Minimum stress: 5 MPa (TC)   Minimum stress: 15 MPa (TC)

Figure 5 - Relations between the allowed rheological coefficient of utilization, time-to-rupture and minimum stress

(4) Different creep behavior at different minimum stresses at an equal stress of approx. $\sigma_v = 29$ MPa and at a uniform temperature of $T = 323$ K

So far, the worldwide standard for displaying the creep rate has been a representation in dependence of the effective stress and of the temperature. The influences of minimum stress and coefficient of utilization on the creep rate are neglected. In order to investigate these influences, the calculations using the *Hou/Lux* material model are performed at a uniform effective stress of $\sigma_v = 29$ MPa and a uniform temperature of $T = 323$ K but at different minimum stresses of $\sigma_3 = 1, 2, 3, 4, 5, 6$ and 10 MPa.

Qualitatively, the *Lubby2* material model in this case provides the same results as all other material models, which do not account for the influences of minimum stress and coefficients of utilization.

Table 2 shows the impact of the minimum stress on the creep rate and thus the creep deformation. In this case, as early as at a minimum stress of $\sigma_{2,3} = 10$ MPa, the vertical strain is independent of the minimum stress. At a minimum stress of $\sigma_{2,3} = 6$ MPa the vertical strain induced by the damage is so small that until the time $t = 280$ hardly any difference can be detected between the vertical strains calculated with the *Lubby2* material model and the *Hou/Lux* material model. This statement is very helpful for the determination of the material parameters for the *Hou/Lux* material model. The parameters for the partial model *Hou/Lux-ODS* must and can therefore be determined at higher minimum stresses (e.g. $\sigma_{2,3} > 15$ MPa).

Table 2 - Compilation of the vertical strains at different points in time under equal von Mises stress at different minimum stresses.

t_{BK} [d]	ε_1 [%]	ε_1 [%]	ε_1 [%]	ε_1 [%]	ε_1 [%]	ε_1 [%]	ε_1 [%]	ε_1 [%]
$\sigma_v = 29$ [MPa]	$\sigma_3 = 1$ [MPa]	$\sigma_3 = 2$ [MPa]	$\sigma_3 = 3$ [MPa]	$\sigma_3 = 4$ [MPa]	$\sigma_3 = 5$ [MPa]	$\sigma_3 = 6$ [MPa]	$\sigma_3 = 10$ [MPa]	**Lubby2**
2,50	41,71	11,62	9,30	8,60	8,52	8,50	8,50	8,50
6,00	-	57,30	16,99	14,24	13,66	13,56	13,56	13,56
17,50	-	-	77,00	21,23	18,29	17,85	17,81	17,81
62,00	-	-	-	87,40	29,05	26,28	26,08	26,08
165,00	-	-	-	-	61,98	46,09	45,16	45,16
280,00	-	-	-	-	-	69,31	67,02	67,02

(5)   *Load bearing behavior of an axially perforated salt sample under triaxial stress*

The above mentioned stress conditions correspond to the conditions at (TC) model tests with axially perforated rock salt samples of location A under a constant axial pressure ($\sigma_1 = 35$ MPa) and a constant axial confining pressure ($\sigma_{2,3} = 6$ MPa) as well as no pressure in the axial drill hole. Such tests are described in *Lux et al. (1997)*.

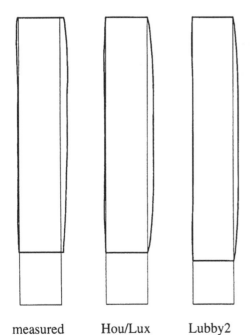

Figure 6 - Comparison of the measured and calculated contour of an axially perforated rock salt sample ($\sigma_1/\sigma_3 = 35/6$ MPa; $t = 69$ d)

measured   Hou/Lux   Lubby2

In *Hou (1998)* the *Hou/Lux* material model was integrated into the existing *MISES3* FEM program, thereby allowing the calculation of complicated load bearing systems. Figure 6 compares the contour of the large sample shown in Figure 2a, measured after completion of the test and after removal of the almost completely destrengthened salt rock at the inner wall, with the contour calculated numerically using the *Hou/Lux* and *Lubby2* material models (by means of parameter variation). The contour area regarded to be completely destrengthened must be estimated to be smaller than 1 mm via a material balance. The good correspondence in tendency and quality between the measured contour and the contour calculated by means of the *Hou/Lux* material model is clearly visible. The measured deformation into the axial bore of approx. 1 mm in the center of the test sample compares to a calculated value of 0.9 – 1.1 mm. As a consequence of the end face friction (-> inhibited transverse expansion) a relatively stronger deformation into the axial bore can be observed near the lower and upper pressure plate, with the model test (approx. 1.5 - 2.4 mm) as well as with the calculations using the *Hou/Lux* material model (approx. 1.7 - 1.9 mm). This deformation directed into the axial bore can only be calculated with the *Hou/Lux* material model or a corresponding material model that takes into account structural damage and dilatancy.

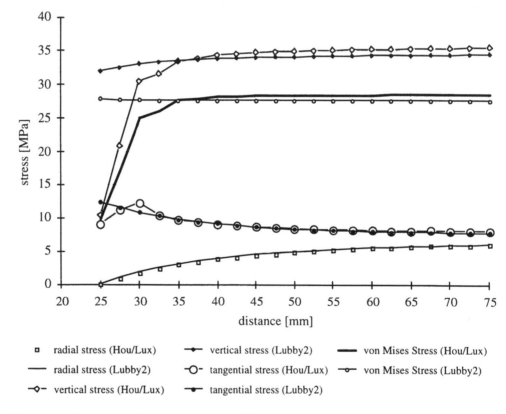

Figure 7 - Stress contributions on the middle horizontal section of a large sample of rock salt with an axial borehole (σ_1/σ_3 = 35/6 MPa; t = 69 d)

Figure 7 shows the stress distributions on the middle horizontal section at a time t = 69 d according to the *Hou/Lux* material model and according to the *Lubby2* material model. Conclusions:

- The stress distributions determined with the *Lubby2* material model after a creep time of 69 days still almost correspond more or less to the elastic stress distribution, *Lux et al. (1997)*. That means the *Lubby2* material model or similarly structured models are incapable of simulating the softening near the axial bore, observed during model tests.

166

- In contrast, the *Hou/Lux* material model taking into account the structural damage, shows that the stresses near the axial bore are continuously rearranged towards the core and the periphery. According to Figure 7, at the time t = 69 d, the vertical stress at the contour of the axial bore is only approx. 9 MPa, a numerical value far below the short-term strength.

- The mechanical process taking place during the model test on axially perforated rock salt sample is very complicated, if the stress exceeds the short-term strength locally. Earlier already, exceeding the dilatancy strength causes material damages and accelerated creep rates, resulting in a decrease of stress, which in turn leads to a slower creep rate and to a relatively longer time until creep rupture. Which processes may dominate and whether there is a local accelerated creep phase, mainly depends on the intensity of exceeding the dilatancy strength and on the bearing system.

- In this special case, the stress intensities are not large enough to make existing or new crack structures unstable and able to spread, so that neither creep rupture nor sudden failure of the axially perforated rock salt sample nor a clearly accelerated creep phase can be observed or calculated.

In summary, the *Hou/Lux* material model is capable of qualitatively and quantitatively determining the major mechanical phenomena observed during model tests on axially perforated rock salt samples. That way it is possible to validate the *Hou/Lux* material model with its parameters and consequently, a basis for new types of mechanical framework analyses, e.g. for salt cavern or underground mine constructions, begins to emerge.

DISCUSSION AND SUMMARY

The mathematical description of the new *Hou/Lux* material model and the exemplary calculations with this material model for rock salt of location A allow the following conclusions regarding its capability as compared to the demands on a material model and the deficits in the field of material models for salt rocks:

- The *Hou/Lux* material model has profited from several material model statements to be found in more recent literature and has been developed based on the *Lubby2* material model according to *Lux (1984)*. Due to additional knowledge the *Hou/Lux* material model has been even further developed regarding the acquisition of salt-specific characteristics, as compared to other material models. To be mentioned here are the consideration of the stress geometry, the representation of the strength in dependence of minimum stress and stress geometry instead of the average stress as e.g. in the *Cristescu & Hunsche* material model or in the *SUVIC-D* and *MDCF* material models as well as the introduction of the fracture strength together with the dilatancy strength into a material model.

- The damage does not just occur with the beginning of tertiary creep but already during the transient creep phase, as the dilatancy strength is exceeded. However, the onset of damage does not immediately lead to tertiary creep. The damage must first accumulate until a certain damage level is reached. Shortly before the occurrence of creep rupture, the damage process quickly accelerates.

- As consequences of the damage, the dilatancy and the additional creep deformations induced by the structural damage can be calculated by means of the *Hou/Lux* material model. Consequences such as a reduced load-bearing cross-sectional area and the consequently increased internal stress $\sigma_v/(1\text{-}D)$ are determined by means of the partial model *Hou/Lux-ODS*. Further consequences of the structural damage such as e.g. increased porosity and permeability can generally be represented in dependence of the dilatancy. First statements in this regard can be found in literature, but do need to be put in more concrete terms, which is currently being done.

- The *Hou/Lux* material model can describe the behavior of tertiary creep and can thus prognostically calculate the creep rupture – a main objective of this material model.

- Because the *Hou/Lux* material model takes into account influences of minimum stress, the coefficient of utilization and the stress geometry, results concerning the ratio between time until creep rupture as well as the coefficient of utilization and minimum stress can be obtained, which are qualitatively plausible and tend to agree with experience. This result also applies to the calculated creep deformations at an equal stress and at different minimum stresses.

- The introduction of the short-term strength and the dilatancy strength under TC and TE conditions and the identification of the dilatancy strength with the fatigue strength establishes a relation between the short-term and the long-term behavior.

- Parameter variations have made it possible to use the *Hou/Lux* material model to qualitatively and quantitatively determine phenomena such as softening, dilatancy, loosening, structural damage, radial deformations into the axial bore as well as spalling, observed during model tests on axially perforated rock salt drill cores. So far it has been impossible to determine and quantify such phenomena using conventional material models such as the *Lubby2* material model.

- Based on these new theories, the authors expect that apart from a deeper insight into the load-bearing behavior e.g. in view of the development of loosening in geological barriers, it will also be possible in the future to design saliniferous mechanical frameworks, e.g. salt cavern constructions, more economically. There shall be a separate article in the future concerning the application in the mechanical framework planning.

References

Aubertin, M., Sgaoula,   J. & Gill, D. E., (1993), "A damage model for rocksalt: Application to tertiary creep," Proc. of 7th Symposium on Salt, Vol. 1, Elsevier Science Publishers B. V., Amsterdam, 117-125.

Aubertin, M., Sgaoula, J., Servant, S., Gill, D. E., Julien, M., and Ladanyi, B., (1996), "A recent version of a constitutive model for rocksalt," 4th Conference on the Mechanical Behavior of Salt (Preprints), Montreal, 1996, 129-134.

Chan, K. S., Bodner, S. R., Fossum, A. F., and Munson, D. E., (1992), "A constitutive model for inelastic flow and damage evolution in solids under triaxial compression," Mechanics of Materials 14, 1-14.

Chan, K. S., Brodsky, N. S., Fossum, A. F., Bodner, S. R., and Munson, D. E., (1994), "Damage-induced nonassociated inelastic flow in rock salt," International Journal of Plasticity, Vol. 10, No. 6, 623-642.

Cristescu, N., and Hunsche, (1993), "A comprehensive constitutive equation for rock salt: Determination and application," Third Conference an the Mechanical Behavior, Ecole Polythechnique, 1993, 177-191.

Glabisch, U., (1997), "Stoffmodell für Grenzzustände im Salzgestein zur Berechnung von Gebirgshohlräumen," Dissertation an der TU Braunschweig.

Hou, Z., (1997), "Untersuchungen zum Nachweis der Standsicherheit für Untertagedeponien im Salzgebirge," Dissertation an der TU Clausthal.

Hou, Z, (1998), "Implementierung des neuen Stoffmodells *Hou/Lux* in das FEM-Programm *MISES3*," Bericht der Professur für Deponietechnik und Geomechanik der TU Clausthal, unveröffentlicht.

Hou, Z., and Lux, K.-H., (1998), "Ein neues Stoffmodell für duktile Salzgesteine mit Einbeziehung von Gefügeschädigung und tertiärem Kriechen auf der Grundlage der Continuum-Damage-Mechanik," Geotechnik 21 (1998) Nr. 3, 259-263.

Hou, Z., Lux, K.-H., and Düsterloh, U., (1998), "Bruchkriterium und Fließmodell für duktile Salzgesteine bei kurzzeitiger Beanspruchung," Glückauf - Forschungshefte, 59 (1998) Nr. 2, 59-67.

Hunsche, U., and Schulze, O., (1994), "Das Kriechverhalten von Steinsalz," Kali und Steinsalz, Band 11, Heft 8/9, 238-255.

Kachanov, L. M., (1986), "Introduction to continuum damage mechanics," Martinus Nijhoff Publishers.

Langer, M., (1991), "General report: Rheological behavior of rock salt," 7th Int. Congress on Rock Mechanics, Aachen 1991, Vol III, 1811-1819.

Lemaitre, J., (1992), "A course on damage mechanics," Springer-Verlag.

Lux, K.-H., (1984), "Gebirgsmechanischer Entwurf und Felderfahrungen im Salzkavernenbau," Ferdinand Enke Verlag Stu tgart.

Lux, K.-H., Düsterloh, U., Bertram, J., and Hou, Z., (1997), "Abschlußbericht zum BMBF-Forschungsvorhaben 02 C 0092 ," Professur für Deponietechnik und Geomechanik der TU Clausthal.

Lux, K.-H., and Hou, Z., (1999), "Gefügeschädigungen als Grundlage zur Formulierung von neuartigen Stoffmodellen für viskoplatische Salinargesteine," Glückauf - Forschungshefte, 60 (1999) Nr. 1, 23-34.

Lux, K.-H., and Yin, J., (2000), "Analytische Lösungen zur Ermittlung des Tragverhaltens von tiefliegenden Tunneln mit Spritzbetonausbau," in Vorbereitung.

Minkley, W., (1997), "Sprödbruchverhalten von Carnallitit und seine Auswirkungen auf die Langzeitsicherheit von Untertagedeponien. Untertägige Entsorgung," wissenschaftliche Berichte FZKA-PTE Nr. 5, 249-275.

Munson, D. E., Devries, K. L., Fossum, A. F., and Callahan, G. D., (1993), "Extension of the M-D model for treating stress drops in salt," 3rd Conference on the Mechanical Behaviour of Salt, Palaiseau, September 14 - 16, 1993, 31-44.

Yin, Ji., (1996), "Untersuchungen zum zeitabhängigen Tragverhalten von tiefliegenden Hohlräumen im Fels mit Spritzbetonausbau," Dissertation an der TU Clausthal.

Basic and Applied Salt Mechanics, Cristescu, Hardy, Jr & Simionescu (eds)
© 2002 Swets & Zeitlinger, Lisse, ISBN 90 5809 383 2

On the third-power creep law for salt in mine conditions

Leo Rothenburg & Maurice B.Dusseault
University of Waterloo, Waterloo, Canada

Dennis Z.Mraz
Mraz Project Consultants, Saskatoon, Canada

Abstract

The paper re-examines published creep tests data on Palo Duro salt and presents parameters of the Munson-Dawson creep law based on regression of steady state creep data. The main outcome of this statistical analysis is that the derived parameters, extrapolated to temperature and stress ranges typical of mining conditions correspond to the power law creep with the exponent 3.0 operating at differential stress below 9.5 MPa. The paper presents a review of several other published cases when the third-power stress dependence of creep was observed for rock salt and discusses possible physical mechanisms leading the third-power creep law.

Introduction

Steady state creep of rock salt is of primary importance for design and long-term performance assessment of conventional underground salt and potash mines, solution cavities in salt rock and repositories for radioactive and toxic wastes. Description of steady state creep is also required for comprehensive viscoplastic models attempting to describe inelastic flow of alkali halides in a wide ranges of transient conditions, stresses and temperatures.

Despite a considerable effort of the past three decades to rationalize physical mechanisms involved in steady state creep, laboratory testing remains the most reliable method of determining the steady state creep parameters. However, while high temperature laboratory tests are usually of relatively short duration, testing at stresses and temperatures typical of underground mining operations is prohibitively long. Typically, it may take a year or longer to accumulate a 1 mm deformation of a 20 cm long salt specimen at room temperature and the differential stress around 10 MPa. As a result, extrapolation of high temperature, high stress tests to lower ranges of both parameters appears to be the only practical alternative for determining steady state creep parameters for typical conditions of underground mines. In this paper a published data set for Palo Duro salt, Senseny et. al. (1985, 1986), Hansen (1988) are re-analyzed and the behavior is extrapolated to low differential stresses ranges and temperatures typical of mining conditions. While performing a multi-parametric regression rarely warrants a separate publication, this analysis gives an important result: a third-power dependence of steady state creep on differential tress for conditions typical of many mining operations. Apart from describing details of data analysis for Palo Duro salt, the paper compiles other experimental studies where the third-power law was noticed and discusses possible microscopic mechanisms and models that could lead to the third-power stress dependence of steady state creep.

Description of steady state creep

Despite considerable differences in the way transient creep is described in various constitutive models for rock salt, from exponential time models, Senseny, 19085, to models based on internal variables. Aubertin, 1991, steady-state creep is almost universally simulated using a single- or multi-mechanism power law or a hyperbolic sine law, as in the Munson-Dawson, 1982, model. More recently a steady-state creep law covering a wide range of stresses and temperatures was proposed by Hampel et. al., 1995, based on physical arguments. This creep law incorporates parameters and empirical functional relationships interpretable in physical terms.

The most straightforward way of determining empirical parameters of steady state creep models is by direct least square fit on steady state rates. In simplest cases it involves a linear regression in log $\dot{\varepsilon}_{ss}$ versus log σ coordinates, where $\dot{\varepsilon}_{ss}$ and σ refer to suitable measures of steady state rate and differential stress respectively. In other cases, a non-linear regression is used in the same coordinates, e.g. Hampel et. al., 1995. One possible disadvantage of this procedure is that determination of steady state rates is not always accurate when laboratory tests are of insufficient duration, Mrugala and Hardy, 1988. In such cases transient states are used to derive both transient and steady state parameters. This approach has frequently been taken in interpretation to tests carried out within the scope of the US high level nuclear waste disposal program, e. g. Senseny, 1985. In ideal conditions of perfect testing and modeling these approaches are expected to give the same results. The reality, however, neither testing no modeling are perfect and different tests stages carry different associated uncertainties. Early stages of creep tests are, generally, more affected by sampling procedures compared to states approaching steady state, Guessous, 1988. Also, models describing transient states are far more certain at approach to steady state when the first order kinetics becomes an accurate approximation. In such circumstances, giving the same weight to all data points in the creep test may affect the accuracy of determining the steady state creep parameters.

Tests on Palo Duro salt

Samples of salt from Palo Duro basin in Texas were tested by RE/SPEC Inc. in 1982-84 as part of a program examining several potential sites for high level nuclear waste repositories in salt. Tests were carried out on salt recovered from Units 4 and 5 of the Lower San Andreas formation. Steady state rates for 12 tests on Unit 4 salt and the same number of tests on samples of Unit 5 salt where summarized by Senseny et. al. in 1985 and 1986 reports. Both reports present parameters for the single-mechanism power creep law as well as parameters of transient states fitted using the exponential time model. Differences between estimated parameters of the steady state creep law for both sets of tests are small and within the estimated standard error, e.g. exponents 5.5±0.4 and 5.3±0.6 for samples from Units 4 and 5 respectively. Differences between parameters of the exponential time model were somewhat greater but still within limits of the standard error.

172

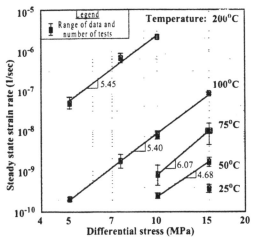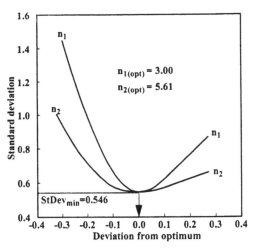

Figure 1: Results of creep tests on Palo Duro salt

Figure 2: Optimum creep law exponents

Parameters of the Munson-Dawson steady state model

These results are consistent with mineral composition of tested samples summarized by Hansen, 1988: average percentages of halite are the same in samples from both units (94%) and percentages of fluid inclusions are very close (0.7% and 0.8%). The same applies to a combined content of clays and anhydrite, with more clays in Unit 4 samples (3.5%) versus more anhydrite in samples from Unit 5 (3.6%). Average grain sizes, 8.2 mm and 6.9 mm for samples from Unit 4 and 5 respectively, are within limits of standard error, estimated to be about 2 mm in both cases. The mentioned similarities of creep test results combined with similarities in mineralogy of samples from both units presents a sufficient rationale to combine test data from Units 4 and 5 for purposes of determining creep parameters of the Munson-Dawson model.

Figure 1 presents a summary of all tests on a conventional log-log plot. The average exponent over groups of test for the same temperature is 5.4, i.e. the average of exponents presented by Senseny et. al. (1985, 1986) for Unit 4 and 5 salts. A very different pattern emerges, however, if this data set is processed using the Munson-Dawson model.

The Munson-Dawson, 1982, model represents steady state creep as consisting of contribution from three mechanisms with the resulting rate computed as follows:

$$\dot{\varepsilon}_{ss} = A_1\left(\frac{\sigma}{\mu}\right)^{n_1}\exp\left(-\frac{Q_1}{RT}\right) + 2\left[B_1\exp\left(-\frac{Q_1}{RT}\right) + B_2\exp\left(-\frac{Q_2}{RT}\right)\right]\sinh\left(q\frac{\sigma-\sigma_0}{\mu}\right) + A_2\left(\frac{\sigma}{\mu}\right)^{n_2}\exp\left(-\frac{Q_2}{RT}\right)$$

where n_1 and n_2 are exponents for the two power law mechanisms above with corresponding activiation energies Q_1, Q_2 and coefficients A_1, A_2 with the dimensionality of strain rate. The mechanism represented by the hyperbolic sine term is considered to be activated for differential stress σ in excess σ_0; its related parameters B_1, B_2 with the

173

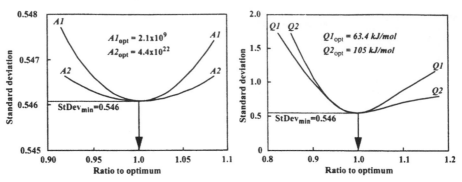

Figure 3: Optimum A-coefficients   Figure 4: Optimum activation energies

dimensionality of strain rate and a non-dimensional parameter q. The shear modulus of Palo Duro salt $\mu = 9620$ MPa is used as a normalizing factor. Temperature, T, in the above expression is in ^0K and $R = 8.314$ J/mol$\cdot^0K$.

There are a total of 10 parameters that were determined using a least square non-linear regression in which the objective function was constructed as the standard deviation between logarithms of test strain rates and corresponding model strain rates.

Figures 2-6 illustrates behavior of the objective function in the vicinity of the optimum. Results are presented as variations of the standard deviation with respect to one variable while others are kept at their optimum values. The plot in Figure 2 presents variations of the standard deviation with respect to exponents n_1 and n_2 varying both around optimum values. To accommodate differences in scales of different variables, deviations from the optimum of other parameters are expressed in terms of the ratio of a particular variable to its optimum value. This type of plots demonstrate existence of an isolated minimum (that was found using the conjugate gradient method). Optimum values are well defined. It should be noted that due to symmetry of the creep rate equation with respect to parameters of the two power law mechanisms there is a another minimum corresponding to a simple interchange of these parameters.

Dominant mechanism at low differential stresses and temperature

Perhaps the most interesting outcome of the described optimization is that one of the two determined exponents in the Munson-Dawson model turned out to be $n_1 = 3$ with two significant figures. As the exponent of another power law mechanism is higher, $n_2 = 5.61$, the exponent 3 mechanism is expected to dominate in the lower stress range. Also, as the corresponding activation energy $Q_1 = 63.4$ kJ/mol is the lowest of the two mechanisms, the exponent 3 mechanism is significant at low temperatures.

Figure 7 illustrates relative contributions of the three mechanisms into the total creep rate. At a relatively high temperature of 150°C all three mechanisms contribute into the total creep rate significantly. The mechanism with the exponent 5.6, most likely a diffusion creep, dominates in the range of stresses around 10 MPa. At temperature of 25°C this mechanism makes almost no contribution into the overall creep rate. In the important range

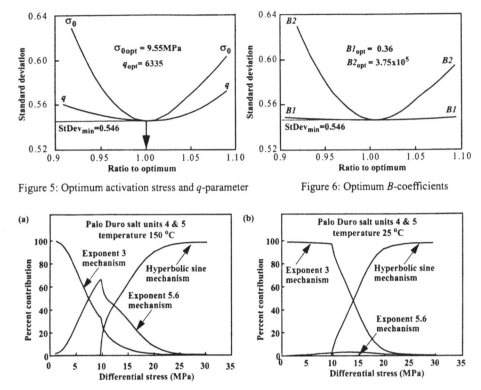

Figure 5: Optimum activation stress and q-parameter   Figure 6: Optimum B-coefficients

Figure 7: Relative contributions of various creep mechanisms at different stress levels
(a) Temperature 150 °C. (b) Temperature 25 °C.

of differential stress below 10 MPa a mechanism with the exponent 3 is the only one responsible for creep at temperatures typical of mining conditions. The mechanism described by the hyperbolic sine and associated with dislocation glide makes almost similar contributions into the overall creek rate in a wide range of temperatures. This is entirely consistent with the mechanism map presented by Munson and Dawson with the "unidentified mechanism" being the third-power law, at least for the Palo Duro salt.

Possible origins of the third-power creep law

References to the third-power creep law for salts are only as rare as creep tests carried out at stresses below 10 MPa and at room temperatures. In a few reported instances e.g. Lajtai and Duncan, 1988, for Rocanville Potash and Mraz et. al., 1995, for tachyhydrite, exponents close 3 were obtained. The same applies to Breese salt in France tested by Pouya, 1993. It should be also noted that Lomenic's, 1968, tests on model pillars fabricated from Lyon's salt, Kansas, point out to the exponent 3 mechanism, even, though these tests were terminated at still transient stages.

In general, the exponent 3 power law corresponds to a process in which the velocity of dislocations is linearly proportional to differential stress. This follows from Orowan's equation for creep rate $\dot{\varepsilon} = b\rho v$, b is the length of the Burger's vector, v is the velocity of dislocations and ρ is the dislocation density, known to be proportional to σ^2, Porier, 1985. The proportionality $v \sim \sigma$, as for ice for example, assures the third-power dependence of creep rate on differential stress. Cannon and Sherby discussed various possible mechanisms that could lead to the exponent 3 power law for non-metals.

More direct experimental studies for conditions typical of mining operations are needed to clarify the nature of the third-power exponent law for salt and its physical origin.

References

Aubertin, Gill, D. E., and B. Lafdanyi, 1991, A unified viscoplastic model for the inelastic flow of alkalihalides, *Mechanics of materials*, Vol. 11, pp. 63-82.

Cannon, W. R., and O. D. Sherby, 1973. Third-Power Stress Dependence in Creep of Polycrystalline Non-metals, *Journal of the American Ceramic Society*, Vol. 56, No. 3, pp. 157-160.

Guessous, Z., Ladanyi, B. and D. E. Gill, 1988, Effects of sampling disturbance on laboratory determined properties of rock salt. *Proceedings, Second conference on mechanical behavior of salt*, Hannover, Sept. 24-28, Trans Tech Publications, pp. 137-158

Hansen, F. D., 1988, Physical and mechanical variability of natural rock salt, *Proceedings, Second conference on mechanical behavior of salt*, Hannover, Sept.24-28, Trans Tech Publications, pp. 23-39.

Lajtai, E. Z., and E. J. S. Duncan, 1988, The mechanism of deformation and fracture in potash rock, *Canadian Geotechnical Journal*, Vol. 25, pp. 262-278.

Lomenick, T. F., 1968, Accelerated deformation of rock salt at elevated temperature and pressure and its implications for high level radioactive waste disposal, *Ph.D. thesis*, University of Tennessee.

Mrugala, M. and H. R. Hardy, 1988, Effects of test duration on the viscoelastic parameters of salt, *Proceedings, Second conference on mechanical behavior of salt*, Hannover, Sept. 24-28, Trans Tech Publications, pp. 244-261.

Munson, D. E. and P. R. Dawson, 1982, A Transient Creep Model for Salt During Loading and Unloading, *SAND82-0962, Sandia National Laboratories*, Albuquerque, NM.

Mraz, D. Z., Rothenburg, L. Valente, A. A. and R. Frank, 1996, Evaluation of a mining panel in Taquari-Vassouras potash mine, in: *Rock Mechanics Tools and Techniques*, M. Aubertin, F. Hasani & H. Mitri (eds.), Balkema.

Pouya, A., 1993, Correlation between mechanical behavior and petrological properties of rock salt, *Proceedings, 32nd US Symposium on Rock Mechanics*, pp. 385-392.

Senseny, P. E., T. W. Pfeifle, K. D. Mellegard, 1985, Constitutive Parameters for Salt and Nonsalt Rocks from the Detten, G. Friemel, and Zeeck Wells in the Palo Duro Basin, Texas, *GWI/ONWI-549*, prepared by RE/SPEC Inc. for Office of Nuclear Waste Isolation, Battelle Memorial Institute, Columbus, OH

Senseny, P. E., T. W. Pfeifle, K. D. Mellegard, 1986, Exponential Time Constitutive Law for Palo Duro Salt from J. Friemel No. 1 Well, *BMI/ONWI-595*, prepared by RE/SPEC Inc. for Office of Nuclear Waste Isolation, Battelle Memorial Institute, Columbus, OH

Basic and Applied Salt Mechanics, Cristescu, Hardy, Jr & Simionescu (eds)
© 2002 Swets & Zeitlinger, Lisse, ISBN 90 5809 383 2

Creep law to describe the transient, stationary and accelerating phases

Klaus Salzer, Wolfgang Schreiner and Ralf-Michael Günther
IfG Institut für Gebirgsmechanik GmbH Leipzig, Leipzig

ABSTRACT

The proposed combined salt creep law describes the deformation-hardening behavior of the transient creep phase, which gives a good description for a relatively short time after the excavation phase and can, by considering the recovery of the hardening, also be extended to the stationary creep phase, which is significant for long times after the creation of openings. The developed material law also includes inverse transient creep, which takes place during load reductions. The accelerating creep phase is included by extending the material law through the accumulation of micro-cracks that come about depending upon the damage work performed. The dilatancy linked to that acts as an inner state variable.

INTRODUCTION

For stability analyses as well as for predictions of the future geomechanical behavior of underground openings in salt rock formations, proper calculation tools are needed. Therefore, efficient numerical calculation methods and material laws have been developed to describe the mechanical behavior of salt.

In principle, the creep curve (i.e. the deformation-time curve under constant load), can be characterised by three phases:

(1) a primary or transient creep phase (Hardening dominates during that part of the creep curve.);

(2) secondary, or stationary, creep phase during which hardening and recovery are in dynamic equilibrium; and

(3) tertiary, or accelerating, creep phase that involves the initiation of a creep fracture. Damage inside the rock grows more quickly than it can be repaired by recovery and heal-up processes.

The observed and measured convergence underground, especially during the phase immediately after the creation of openings, cannot be described with the secondary creep approach alone. Figure 1 shows a typical example of the measured vertical compression of a Sylvinite pillar compared to the calculated values according to the secondary creep approach (IfG, 1992). The dashed bold line shows the calculated pillar compression using the secondary creep approach when the excavation time point corresponds to real time and the secondary creep parameters are fitted to the displacement velocities after 4 years. The deformation parameters for the deformation-hardening approach, which were deduced from laboratory tests in ad-

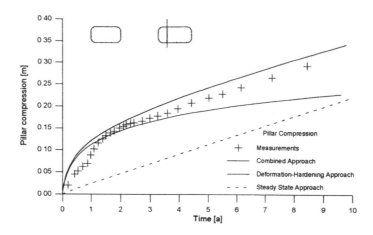

Figure 1: Comparison between in-situ Measured Pillar Compression and Several Calculated Values using Different Creep Approaches

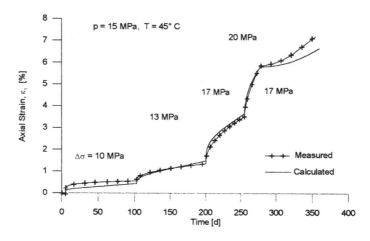

Figure 2: The Inverse Transient Creep demonstrated for Deformation of a Halite Sample during a Karman Test (Hunsche and Schulze 1994)

vance of pillar creation, lead to a good agreement between the in-situ measured deformations and the calculated values (see the thin line in Figure 1). For longer times, the measured values lie above the calculated ones because transition into the stationary phase has already taken place. The deformation-hardening approach with power functions for stress and deformation has been used by several authors (e.g. Lemaitre (1970) or Menzel and Schreiner (1977)). This approach has had good success in describing the creep behavior of salt. However, it does not include the secondary creep phase.

This paper presents a combined creep approach that contains both creep phases by recovery of the hardening, whereby the hardening-part of the creep deformation, ε^{v}, is used as an internal state variable (Salzer and Schreiner 1996). The inverse transient creep is also described with this creep approach. An example of the inverse transient creep that appears during the reduction of the deviatoric stresses is shown in Figure 2 (Hunsche and Schulze 1994), which shows the deformation time curve of a triaxial test on a cylindrical salt sample in the so-called "Karman box" with 5 stress levels. Consideration of these effects is necessary for all geomechanical processes in which the deviatoric stress level decreases (e.g. load changes in a salt storage cavern or the interaction between the salt rock and a sealing construction).

MATHEMATICAL DESCRIPTION OF THE COMBINED LAW

The strain rate tensor, $\dot{\varepsilon}_{ij}$, is given by the following equation:

$$\dot{\varepsilon}_{ij} = \dot{\varepsilon}_{ij}^{el} + \dot{\varepsilon}_{ij}^{cr} \tag{1}$$

where ε_{ij}^{el} and ε_{ij}^{cr} are the elastic and creep portions of the strain tensor, respectively. The corresponding strain rate components are:

$$\dot{\varepsilon}_{ij}^{el} = \frac{v}{E} \cdot \dot{\sigma}_{kk} \delta_{ij} + \frac{(1+v)}{E} \cdot \dot{\sigma}_{ij} \qquad \dot{\varepsilon}_{ij}^{cr} = \frac{3}{2} \dot{\varepsilon}_{eff}^{cr} \frac{S_{ij}}{\sigma_{eff}} \tag{2 and 3}$$

In the deformation-hardening approach, which results in a good description of the primary creep phase with non-inverse transient creep, $\dot{\varepsilon}_{eff}^{cr}$ is given by the following relation:

$$\dot{\varepsilon}_{eff}^{cr} = A^I \frac{\sigma_{eff}^{\beta^I}}{(\varepsilon_{eff}^{cr})^{\mu}} \tag{4}$$

The secondary creep can be considered as a special case of this approach, with $\mu = 0$.
In the proposed combined creep approach, the total creep deformation, ε_{eff}^{cr}, is divided into a hardening part, ε_{eff}^{V} (which represents the internal state variable), and the remainder, ε_{eff}^{R}:

$$\varepsilon_{eff}^{cr} = \varepsilon_{eff}^{V} + \varepsilon_{eff}^{R} \tag{5}$$

The following equations apply in the combined approach

$$\dot{\varepsilon}_{eff}^{cr} = A^I \frac{\sigma_{eff}^{\beta^I}}{(\varepsilon_{eff}^{V})^{\mu}} \qquad \dot{\varepsilon}_{eff}^{V} = A^I \frac{\sigma_{eff}^{\beta^I}}{(\varepsilon_{eff}^{V})^{\mu}} - \frac{\varepsilon_{eff}^{V}}{t_0} \tag{6 and 7}$$

The second term in equation (7) describes the recovery of that part of the deformation that acts as hardening. If the first term of equation (7), which describes the hardening and which is equal to the growth of the total creep deformation [compare to equation (6)] is ignored ($\sigma = 0$, total unloading of the sample), the integration of equation (7) leads to an exponential decay of ε_{eff}^{V} with time, due to the recovering. For short times after the creation of the openings, and with corresponding small values for ε_{eff}^{cr} and ε_{eff}^{V}, the second term of equation (7) is negligible; hence, the creep behavior can be described by the deformation-hardening approach alone. For larger values of ε_{eff}^{V}, the importance of the second term increases. The state of stationary creep is reached if ε^V_{eff} does not grow further—i.e. $\dot{\varepsilon}^V_{eff} = 0$. Recovery and hardening are then in dynamic equilibrium. For that creep state, the value of ε^V_{eff} is given by:

$$(\varepsilon^V_{eff})^{II} = \left(A^I \cdot \sigma^{\beta^I}_{eff} \cdot t_0\right)^{\frac{1}{1+\mu}} \tag{8}$$

and the following relations apply:

$$(\dot{\varepsilon}^{cr}_{eff})^{II} = A^{II} \cdot \sigma^{\beta^{II}}_{eff} \tag{9}$$

where

$$A^{II} = A^I \left(\frac{1}{A^I \cdot t_0}\right)^{\frac{\mu}{1+\mu}} \qquad \qquad \beta^{II} = \frac{\beta^I}{1+\mu} = n \tag{10} \text{ and } (11)$$

As will be shown later, the adjustment of t_0 to the in-situ measured deformation values for mining temperature conditions (20°-30°C) gives recovery times between 10 and 30 years. During the inverse transient creep, ε^V_{eff} lies above the equilibrium hardening, $(\varepsilon^V_{eff})^{II}$, for the new reduced deviatoric stress level, so that ε^V_{eff} reduces to a new equilibrium combined with a corresponding increase of $\dot{\varepsilon}^{cr}_{eff}$ as a result of the recovery process. Due to the thermal nature of the recovery, it is reasonable to use the ARRHENIUS approach for the temperature dependence on the recovery time, t_0:

$$t_0 \sim e^{\frac{Q_V}{R \cdot T}} \tag{13}$$

Based on equation (10), one can deduce the known temperature dependence of the secondary creep—i.e. factor A^{II}:

$$A^{II} \sim e^{-\frac{Q_V \cdot \mu}{R \cdot T \cdot (1+\mu)}} \tag{14}$$

VALIDATION

We started off by implementing the combined creep approach in the MKEN program system (Salzer and Schreiner 1991) and later in the FLAC (ITASCA, 1998) and FLAC^{3D} (ITASCA, 1997) program systems for making verification and validation calculations as well as for comprehensive practical use. All further explanations are valid for mining temperature conditions

Laboratory Tests

As shown in Figure 2, the combined creep law provides a good description of creep tests (load tests including multi-level creep and inverse transient creep).
The following three figures demonstrate that, with the same parameter set describing a specific salt type, the stress-deformation behavior for three different load regimes can be reproduced. These three load regimes are: a standard creep test (σ_{eff} = constant; Figure 3), a triaxial test with constant deformation rate ($\dot{\varepsilon}_1$ = constant; Figure 4) and a triaxial test with relaxation phases ($\dot{\varepsilon}_1 = 0$; Figure 5). For the halite investigated, the following material parameters were

determined based on creep tests (Young's modulus = 30 GPa; Poisson's ratio = 0,23; $A^L = 10^{-58}$; $\beta^I = 28$; $\mu = 7$ and $t_0 = 10$ years). The system units are: σ [MPa]; t [h] The parallel shifting of the creep curve occurs because primary compression loading was applied in several steps in the laboratory test, and in only one step in the numerical simulation.

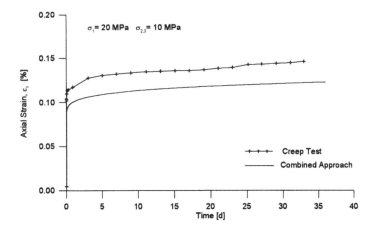

Figure 3: Comparison between Measured and calculated values for a Creep Test of 'Leinesteinsalz'

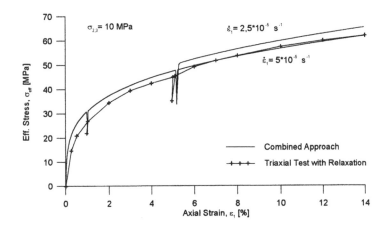

Figure 4: Comparison between Measured and Calculated Values for a Triaxial Test on 'Leinesteinsalz' with two interposed Relaxation Phases

Figure 4 shows the stress-strain curve for a triaxial compression test under constant deformation velocity with relaxation phases at 1% and 5% of the sample compression. Figure 5 shows the corresponding stress-time diagram. Figure 4 also documents the influence of increased deformation velocity on the stress-deformation behavior. After the second relaxation, at a compression of 5%, the deformation velocity was increased from $5 \cdot 10^{-6} s^{-1}$ to $2.5 \cdot 10^{-5} s^{-1}$. An increase of the deformation velocity is connected with a stress increase at the same strain. This is also reproduced quantitatively by the computer model.

Figure 5 shows the result of a back-analysis of both relaxation phases. The relaxation behavior is also reproduced by the computer model with sufficient accuracy. Therefore, it can be assumed that complicated load regimes can also be modelled properly with the material model as long as the deformation has not reached the tertiary creep phase.

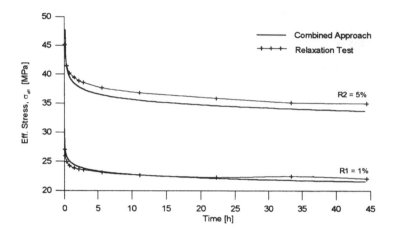

Figure 5: Comparison between Measured and Calculated Stresses during the Relaxation Phase for 'Leinesteinsalz'

Field Measurements

With the help of the three following examples, the assessment of the recovery time, t_0, is demonstrated. The assessment is made on the basis of the adjustment of the calculated deformation values to the in-situ measured values.

Sylvinite Pillar

The measured in-situ pillar compression of a sylvenite pillar was compared with the calculated values (Figure 1). In this example, a quadratic pillar was transformed into a column shape that can be modelled with the axisymmetric calculation option. The zero measurement (reference) was conducted 25 days after creation of the pillar and was taken into consideration in Figure 1. With creep parameters for the deformation-hardening approach determined during the early 70's (IFG 1992): $A^I = 10^{-56}$; $\beta^I = 25$ and $\mu = 7.3$; the later (1978 - 1987) observed pillar compression could be described properly. Only in the last 5 years of the total measuring period did an underestimation of the pillar compression became apparent. Before investigations with the new creep approach were begun, the factor A^{II} was fitted in such a way that the pillar compression rates within the stationary phase were reproduced properly (dashed bold line). On the basis of the so-determined A^{II}, the parameter t_0 can be estimated according to equation (10). For this example, t_0 amounts to 21 years. With these parameters for the combined creep behavior, the bold line was calculated; it shows good agreement between the measured and calculated pillar compression for all considered times.

Rock Salt Pillar ('Werrasteinsalz')

We were able to estimate the "recovery time" t_0 at 30 years with a second example using both a 2-D and a 3-D model in a similar fashion. In the 90's, one were able to measure constant vertical convergence rates in the middle of a chamber of approximately 0.1 mm/year in a room-and-pillar mining system with long-pillars made of Werra Rock Salt that was driven in 1935 (Klügel 1997). The mining system is at a depth of approximately 280 m and pillars were driven with a width of 9 m as well as chambers at a height of 3 m and a width of 11.5 m. From laboratory investigations, we were able to ascertain the following creep parameters for the material of the pillars: $A^I = 5.8 \cdot 10^{-64}$; $\beta^I = 33$ and $\mu = 5.7$.

Rock Salt Pillar ('Leinesteinsalz' Horizon 'Speisesalz')

The third example compares calculated and measured deformation with a 3-D model in a relatively thick salt pillar situated within a room-and-pillar system under work. The pillar width is 35 m, the room width is 20 m, and the mean pillar height is 35 m. After the excavation of a roof chamber 20-m wide and 5-m tall, the total headroom is mined by underhand stoping and haulaged through a system of excavations in the floor (Figure 6). The rock mechanics structural model based upon the FLAC^{3D} program system that is cut up on the level of the roof slice is also shown in Figure 6.

A measuring system was installed with the objective of monitoring the stress-deformation behavior during the room-creation phase. The second excavation phase corresponds to the excavation of the chamber A located Southeast; the third excavation phase corresponds to the chamber B located Northwest; and the fourth phase includes the drivage of a conveyor road in the pillar itself (Figure 6). In a fifth phase, the large-scale chambers further away were mined. The vertical compression of the pillar and its lateral deformation is measured with extensometers in boreholes from the measuring chamber and the conveyor road. Beyond this, both the horizontal and vertical convergences were recorded in the measuring chamber and the conveyor road.

A test calculation showed that the stationary creep approach results in an unsatisfactory agreement between the in situ measured and calculated values. This is because the measurements include the creation process and, therefore, the transient creep part can not be neglected. Subsequently, the new creep approach was used, with recovery time t_0 chosen in such a way that the measured surface settlements above the mining area could be reproduced in a more complex rock mechanics model.

The following deformation parameters were chosen $A^I = 1.7 \cdot 10^{-22}$; $\beta^I = 10$; $\mu = 1.6$ and $t_0 = 10$ years. The hardening process of this very pure halite is not so strong as for the other tested rock salts. This finding results in a lower value μ. In Figure 7, we compare the convergences measured in the conveyor road with the corresponding calculated values. Here, we discover good agreement, although the measured values are somewhat larger than the calculated values. The contour dilatancy observed and measured in the conveyor road supply a plausible explanation, although they should not be taken into consideration in this material model.

183

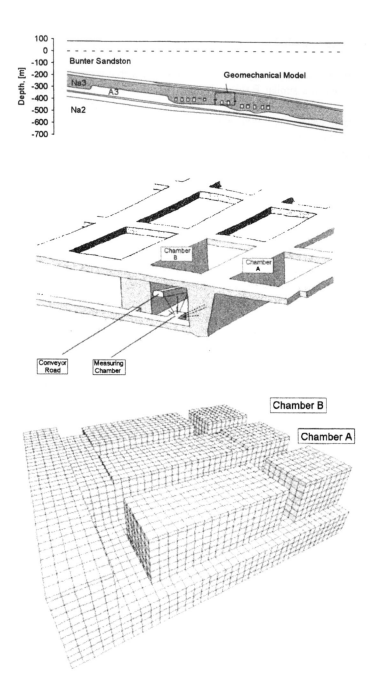

Figure 6: Surroundings of the Measuring Chamber (Plane and Perspective View) and FLAC^{3D}-Model

The minimum stress distribution in the pillar calculated with this rock mechanics model corresponds to the values determined in situ with the hydraulic fracturing method (Figure 8). However, the comparison between the calculated and measured distribution of minimum stress also

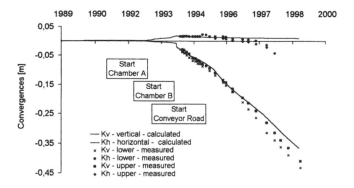

Figure 7: Comparison between Measured and Calculated Convergences of Conveyor Road

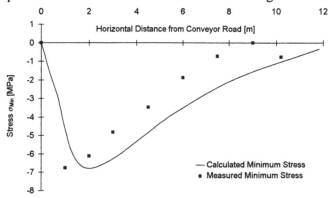

Figure 8: Comparison between Measured and Calculated Minimum Stress between Conveyor Road and Chamber A (in a horizontal borehole)

shows that for improved mapping of the stress-strain-behavior it is necessary to integrate time-dependent softening, i.e. the tertiary creep phase of the salt rocks into the material laws.

ACCELERATING CREEP PHASE

The modern testing technique allows us to record not only the fracture behavior, but also the dilatancy behavior in the phase of initiating the fracture. Figure 9 gives a characteristic example of our own laboratory investigations for the fracture and dilatancy behavior of rock salt. Recording the dilatancy behavior in the pre-failure phase makes it possible to differentiate between a dilatant zone that is afflicted with damage and a zone that is free of dilatancy i.e. free of damage (Figure 10) below the failure surface within the stress space. Based upon these experimental findings, it is possible to develop a solidly founded model for the development of the fracture including accelerated creeping. In the following, we would like to sketch out the basic principles and first results of the development of a material law such as this.

We will introduce a damage parameter D_V according to Kachanov (1986) to take the accelerated creep phase into account that is set equal to the volumetric loosening ε_V^D, i.e. dilatancy.

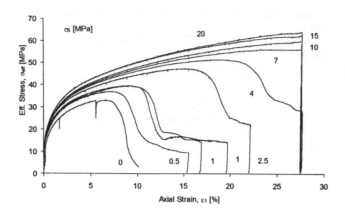

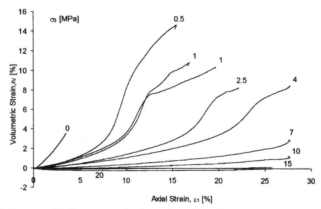

Figure 9: Failure and Dilatancy Behavior of Rock Salt in Triaxial Compression Tests

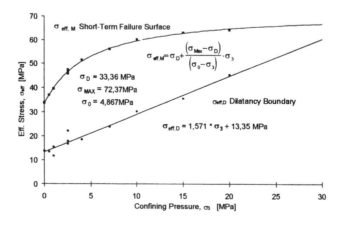

Figure 10: Short-Term Failure Surface and Dilatancy Boundary for Rock Salt

$$\dot{\varepsilon}_{\text{eff}}^{\text{cr}} = \frac{A^{I}}{\left(\varepsilon^{V}\right)^{\mu}} \cdot \left(\frac{\sigma_{\text{eff}}}{1-D}\right)^{\beta^{I}} \tag{14}$$

This means that the creeping rate increases with increasing material damage or dilatancy. In turn, dilatancy is a function of the damage work U^{D} and the minimum stress σ_{3}, which is determined from laboratory experiments.

$$\varepsilon_{V}^{D} = f(\sigma_{3}, U^{D}) \tag{15}$$

By damage work U^{D}, we mean the work that is done above the dilatancy limit and that is associated with the formation of micro-cracks. If the critical damage work U_{B}^{D} that is done in the short-term experiment to the macroscopic fracture has been achieved, the permissible effective stress $\sigma_{\text{eff,M}}$ is reduced by looking upon all further deformations as fracture deformations ε^{B} that lead to a decrease of the permissible effective stress in the post-failure zone in accordance with the experimental observations (Minkley and Menzel 1999).

$$\sigma_{\text{eff,M}} = g(\sigma_{3}, \varepsilon^{B}) \tag{16}$$

By calculation, we achieve this decrease by introducing fictitious fracture damage D_{B}. This produces for the entire damage:

$$D = D_{V} + D_{B} \tag{17}$$

It is necessary to add another term to equation (3) to completely describe creeping behavior including dilatancy:

$$\dot{\varepsilon}_{ij}^{\text{cr}} = \frac{3}{2}\dot{\varepsilon}_{\text{eff}}^{\text{cr}} \frac{S_{ij}}{\sigma_{\text{eff}}} + \dot{\varepsilon}_{V}^{D} \cdot \left[\frac{\delta_{ij}}{3} - \frac{S_{ij}}{\sigma_{\text{eff}}}\right] \tag{18}$$

Beyond this, it is necessary to include the decrease of the elastic parameters depending upon the loosening or dilatancy suffered in order to have a realistic description of softening behavior. Here, we are making the assumption that decrease only has an effect on the Young's-modulus, and not upon Poisson's ratio.

$$E^{D} = \frac{E}{1 - \dfrac{\varepsilon_{V}^{D}}{\varepsilon_{V}^{\text{el}}}} \tag{19}$$

In Figure 11, triaxial short-term tests are modelled on the basis of the set of parameters used for the re-calculations of laboratory experiments with the combined approach (Figures 3, 4 and 5) taking the dilatancy behavior and the post-failure phase into account. This is to compare with the experimental values in Figure 9. Figure 12 shows the failure behavior in triaxial compression tests for different strain rates. In Figure 13, a creep test is calculated with the same set of parameters that includes the accelerating creep phase.

As the example presented shows, the material law reproduces the principle failure behavior of salt rock and also the accelerating creep phase; this means that it may be used for forecasting and evaluating time-dependent softening processes. However, it is necessary to do further

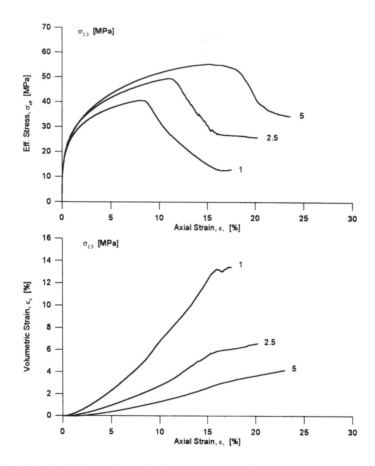

Figure 11: Calculated Stress-Strain Behavior for Triaxial Compression Test of Rock Salt
including Post-Failure Phase

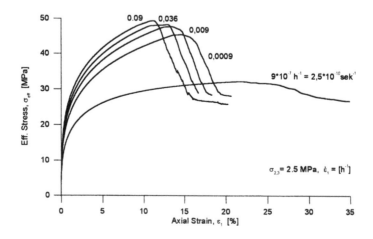

Figure 12: Calculated Stress-Strain Behavior for Triaxial Compression Tests of Rock Salt for
Decreasing Strain Rates

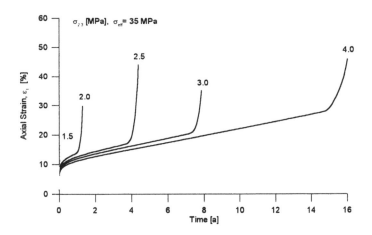

Figure 13 Calculated Strain-Time Curves for Creep Tests of Rock Salt including Accelerating Creep Phase

work especially to better adapt the function $\varepsilon_v^D = f(\sigma_3, U^D)$ to the experimental results. Beyond this, there have to be validation calculations on in-situ measurements and observations.

CONCLUSIONS

The proposed combined creep law, which is implemented into the codes FLAC and FLAC^{3D}, allows a proper description of the secondary, transient and inverse transient creep phases. The proposed combined creep law corresponds to the addition of a new component in the deformation-hardening model formulation to account for a better description of the transient creep phase by consideration of hardening recovery and allow modelling of secondary creep phase and inverse creep. The examples presented show good agreement between the measured and calculated values according to the combined creep approach, although the deduced creep parameters, A^I, β^I and μ were already determined on the basis of creep tests before the excavation had taken place and only recovery time t_0 was adjusted to the in-situ measurements. A first assessment of t_0 under mining temperature conditions (20°-30° C) gives values between 10 and 30 years. The extension of the material law presented here includes the accelerating creep phase through the accumulation of micro-cracks depending upon the damage work performed. This reflects the failure behavior in both the short-term and the creep test. We are planning upon better adjustment and validation calculations of field measurements for further improvement.

ACKNOWLEDGMENT

The presented work was granted by the Federal Ministry of Research and Education under the contracts 02 C0264 and 02 C0274. The authors are responsible for the content of this publication.

REFERENCES

Hunsche, U. and Schulze, O. (1994): Das Kriechverhalten von Steinsalz, Kali und Steinsalz; 11, H8/9, 238-255.

IfG, (1992): Überprüfung vorhandener Modellvorstellungen zum Konvergenzverhalten und zur Standsicherheit komplexer Grubengebäude im Salzgestein. Forschungsvorhaben BMFT-FB/02 E 8161 6, IfG GmbH Leipzig, Januar 1992

ITASCA Consulting Group, Inc. (1997): FLAC^{3D}, Version 2.0, Minneapolis, Minnesota

ITASCA Consulting Group, Inc. (1998): FLAC, Version 3.4, Minneapolis, Minnesota

Kachanov, L. M. (1986): Introduction to continuum damage mechanics. Martinus Nijhoff Publisher

Klügel,T. (1997): Personnel communication

Lemaitre, J. (1970): Sur la détermination des lois de compartement des matériaux. élasto-viscoplastiques, thesis, Public., O.N.E.R.A N° 135 (1970) Paris

Menzel, W. and Schreiner, W. (1977): Zum geomechanischen Verhalten von Steinsalz verschiedener Lagerstätten der DDR. Teil II: Das Verformungsverhalten, Neue Bergbautechnik 5, H.8, 565 - 571

Minkley, W. and Menzel, W. (1999): Pre-calculation of a mine collapse Rock burst at Teutschenthal on 11 September 1996. 9th International Congress on Rock Mechanics, Paris August 25-28,1999

Salzer, K. and Schreiner, W. (1991): Der Rechencode MKEN zur Ermittlung der Zeitabhängigkeit des Spannungs-Verformungs-Zustandes um Hohlräume im Salzgebirge. Kali und Steinsalz; 10, H12, 416-423

Salzer, K. and Schreiner, W. (1996): Long-term safety of salt mines in flat beddings. 4th Conference on the Mechanical Behavior of Salt, Montreal June 17 and 18, 1996

Part IV

Constitutive modeling

Basic and Applied Salt Mechanics, Cristescu, Hardy, Jr & Simionescu (eds)
© 2002 Swets & Zeitlinger, Lisse, ISBN 90 5809 383 2

Extrapolation of creep of rock salt with the composite model

Andreas Hampel
Development & Application of Constitutive Equations, Isernhagen, Germany

Udo Hunsche
Federal Institute for Geosciences and Natural Resources (BGR), Hannover, Germany

ABSTRACT

The composite model describes the transient and steady-state creep of natural rock salt on the basis of the deformation micromechanisms (dislocation glide) and microstructure (dislocation subgrains; impurities in the salt matrix as obstacles against dislocation glide). The model uses physical quantities as fit parameters that can be checked in independent measurements, observations, and in comparisons with deformation theory. Successful checks demonstrate the validity and reliability of predictions of the long-term creep behavior of rock salt that are made with this model. Such calculations are necessary for the safety assessment of a permanent repository for hazardous wastes in salt.

In this paper, the basic ideas of the composite model and an overview about its current modified formulation, used for the application in numerical computer codes, will be presented. Also, comparisons between modeled and measured creep curves will be shown.

INTRODUCTION

Hazardous wastes can be disposed of in repositories in deep geological structures. As an important part of the safety analysis of such a repository, the mechanical behavior of the host rock is to be predicted over thousands of years with high reliability. Therefore, the corresponding numerical calculations should be based as far as possible on constitutive equations which describe the physical mechanisms and resulting microstructure of deformation. This conception has the opportunity that valuable information from different types of experimental investigation including microscopy can be incorporated. Also, theoretical knowledge from materials science about the micro-processes can be applied. This improves the basis of the calculations and, therefore, the validity and reliability of the extrapolations.

Such models use physical quantities as parameters. Their values are usually determined in fits of the model equations to results of deformation tests. However, as a result of the physical nature of these quantities, the parameter values can be checked in other experiments, microscopic observations, and in comparisons with well established results from physical deformation theory. This verification of the fit results yields information about the validity and consistency of the model.

Rock salt is an important host material for the permanent disposal of radioactive wastes. In the development of a physically based constitutive model for salt one can take advantage of the close similarity between the micromechanisms and microstructure in creep deformed rock salt and other crystalline materials like many metals and alloys. Therefore, since the 1980s, the composite model – which was originally developed for the description of the flow stress and deformation behavior of metals – was applied to transient and steady-state creep of natural rock salt [Vogler & Blum (1990), Blum (1991), Weidinger et al. (1998), Hampel et al. (1998), Hunsche & Hampel (1999)].

Recently, the general formulation of the composite model was modified in order 1.) to describe the big differences in creep of various rock salt types; 2.) to model the deformation behavior that comes along with continuously changing stresses; and 3.) to simplify the implementation into numerical computer codes.

In this contribution, the basic ideas of the composite model as well as its current formulation are presented. Also, modeling results are compared with some findings of an extensive testing program.

THE MODIFIED COMPOSITE MODEL

The composite model describes the transient and steady-state creep of natural rock salt, i.e. the shape change of a salt body at constant volume. Dilatancy, fracturing and healing are not yet included in the model. The name "composite model" reflects the heterogeneous dislocation microstructure (Fig. 1) which determines the creep behavior of the material:

1.) the subgrain boundaries with a high dislocation density and, accordingly, with a local internal stress level σ_h which exceeds the externally measured or applied mean stress difference σ ("hard regions"): $\sigma_h > \sigma$;

2.) the subgrain interiors with a much lower dislocation density. Here, a back stress which is generated by the dislocations of the subgrain boundaries, causes a reduced local stress level ("soft regions"): $\sigma_s < \sigma$.

The composite model is based on the description of the thermally activated dislocation glide under the influence of

1.) the subgrain microstructure (Fig. 1) which is already present in natural rock salt because of deformation processes in the geological history. The change of the subgrain structure after a change of deformation conditions (mainly of the differential stress) is modeled by means of evolution equations for the characteristic quantities subgrain size, dislocation density inside of the subgrains, and thickness or volume fraction of the subgrain boundaries. This allows to simulate transient behavior after a stress increase or a stress reduction;

2.) impurities in the salt matrix (Fig. 2) which act as non-dislocation obstacles against

dislocation motion. Observations have shown that the creep behavior of rock salt is heavily influenced by those impurities, when their spacing is small enough [Hunsche et al. (1996)]. This spatial distribution is a result of the salt deposition conditions and are modeled successfully by describing the influence of these obstacles on the the geological history. Thus, the big differences in the creep behavior of various salt dislocation glide.

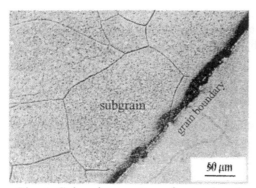

Fig. 1: Subgrain structure of natural rock salt in the as-drilled state (light microscopy, Vogler (1992))

Fig. 2: Impurities in natural rock salt of type z3OSO (electron microscopy, Weidinger (1998))

The relation between the macroscopic creep rate $\dot{\varepsilon}$ and characteristic microscopic processes and quantities is given by the Orowan equation [Frost & Ashby (1982)]:

$$\dot{\varepsilon} = \frac{b}{M} \rho \, v \qquad (1)$$

with b: length of the Burgers vector of a dislocation; M: Taylor factor; ρ: density of moving dislocations within the subgrains; and v: mean glide velocity of these dislocations.

While in the full composite model a similar kinetic expression has to be formulated for the hard as well as for the soft regions, in the modified model the total creep rate is described by an expression for the soft regions alone. A difference between the results of the two model versions can only be observed at the beginning of a transient. The differences are only small, because by far most of the creep deformation is carried by dislocations that move within the subgrains. Therefore, in the modified composite model ρ designates the density of dislocations moving in the soft regions ($\rho = \rho_s$) and v expresses their glide velocity ($v = v_s$).

For the stress dependence of v, a hyperbolic sine function is used [e.g. Frost & Ashby (1982)] which leads to the following creep equation:

$$\dot{\varepsilon} = \frac{b}{M} \rho \, v_0 \, \exp\left\{ -\frac{Q}{RT} \right\} \sinh\left\{ \frac{b \, \Delta a \, \sigma^*}{M \, k \, T} \right\} \qquad (2)$$

with the additional quantities: v_0: velocity constant; Q: microscopic activation energy; R: universal gas constant; T: absolute temperature; Δa: activation area; σ^*: internal effective stress (driving the dislocations within the subgrains); k: Boltzmann constant.

Important quantities are Δa and σ^*. Δa can be regarded as an average area swept by a moving dislocation in one step between two obstacles [Weertman & Weertman (1983)]. The expression $\Delta a \, / \, b$ has the dimension of a length and, therefore, can be related to the average spacing of the dominant obstacles against dislocation motion, namely the other dislocations within a subgrain and non-dislocation obstacles (e.g. impurities) which have a certain density (i.e. an average spacing d_p) and can be sheared by the dislocations. In the modified composite model, this is reflected by the following expression:

$$\Delta a = \frac{b}{\frac{1}{d_p} + \sqrt{\rho}} \qquad (3)$$

The internal effective stress σ^* drives the gliding dislocations. It is the difference between the external (average) stress difference σ and various back stresses which act within a subgrain:

$$\sigma^* = \sigma - \sigma_b - \sigma_{G,\rho} - \sigma_p' \qquad (4)$$

σ_b is the back stress within a subgrain that is caused by dislocations of the subgrain boundaries. In this way, the modified composite model takes into account the hard regions. The difference $\sigma - \sigma_b$ constitutes the stress level in the soft regions σ_s:

$$\sigma_s = \sigma - \sigma_b = \frac{1 - f_h k_h}{1 - f_h} \sigma \qquad (5)$$

with f_h: volume fraction of the hard regions (see below), $k_h = 3.4$ (the value follows from a more detailed analysis by Sedlácek (1995)).

The other two terms, $\sigma_{G,\rho}$ and σ_p' only act on a moving dislocation. For them preliminary expressions are currently used which will be modified in the range of small stress differences (below about 2 MPa). Then, (4) turns into the following equation:

$$\sigma^* = \frac{1 - f_h k_h}{1 - f_h} \sigma - \alpha M G(T) b \sqrt{\rho} - \left(1 - \exp\left\{ -\frac{0.4 \sigma}{\sigma_p} \right\} \right) \sigma_p (6)$$

with α: dislocation interaction constant; $G(T)$: temperature dependent shear modulus [Frost & Ashby (1982)]; σ_p: back stress of hard non-dislocation obstacles (e.g. particles which cannot be sheared by the gliding dislocations, in contrast to those which are described by d_p).

The volume fraction of the hard regions f_h can be estimated by the subgrain size w and the thickness a of the subgrain boundaries (more precisely: the thickness of the hard regions, where $\sigma_h > \sigma$):

$$f_h = \frac{2a}{w} \qquad (7)$$

The quantities w, ρ, and a characterize the deformation microstructure of rock salt. The evolution of the microstructure after a change of the deformation conditions (e.g. a change of the stress difference) is reflected by the following evolution equations (S stands for w, ρ, and a):

$$\frac{\partial S}{\partial t} = \frac{\partial S}{\partial \varepsilon} \frac{\partial \varepsilon}{\partial t} = \frac{\partial S}{\partial \varepsilon} \dot{\varepsilon} \qquad (8)$$

The temporal evolution of the three microstructural quantities w, ρ, and a is put down to their evolution with the creep strain ε, because it takes and produces strain (i.e. the movement of dislocations) to change the internal deformation structure of the material. It is assumed that the magnitude of the change of S with creep strain ε is proportional to the difference between the present value of S and the momentary steady-state value S_∞ (k_S is the corresponding rate constant):

$$\frac{\partial S}{\partial \varepsilon} = \frac{S_\infty - S}{k_S} \tag{9}$$

The combination of (8) and (9) yields

$$\frac{\partial S}{\partial t} = \frac{S_\infty - S}{k_S} \dot{\varepsilon} \tag{10}$$

The occurrence of the creep rate $\dot{\varepsilon}$ in (2) and (10) expresses mathematically the close correlation between the evolution of the macroscopic creep strain and the evolution of the microstructure.

In contrast to the momentary value of S, the steady-state values $S_\infty = w_\infty$, ρ_∞, a_∞ only depend on the stress difference σ and implicitly on temperature T, but not on deformation history. The values for S_∞ are known from investigations in the light and electron microscope (Fig. 3, Hampel et al. (1998), cf. Carter et al. (1982)):

$$w_\infty = 33 \frac{G(T)\, b}{\sigma} \tag{11}$$

$$\frac{1}{\sqrt{\rho_\infty}} = \frac{G(T)\, b}{\sigma_s} = \frac{1 - f_{h,\infty}}{1 - f_{h,\infty}\, k_h} \frac{G(T)\, b}{\sigma} \tag{12}$$

$$a_\infty = \frac{1}{2} f_{h,\infty}\, w_\infty \tag{13}$$

where $f_{h,\infty} = 0.07$ according to a more detailed analysis by Sedlácek (1995).

Observations have shown that the microstructure of natural rock salt in its original (as-drilled) state before deformation in the laboratory is already a steady-state subgrain structure which has developed during geological times under the influence of a natural stress difference (see fig. 1). Therefore, the initial values $S_0 = w_0$, ρ_0, and a_0 which are needed for (10), are physically determined by measuring the microstructure of rock salt in its as-drilled state. By inserting these initial values into equations (11) to (13) a "natural" stress difference of about 0.5 to 4 MPa can be estimated, depending on the salt type and its local geological history [cf. Carter et al. (1982)].

The three equations represented by (10) (with S = w, ρ, and a) and equation (2) form a coupled set of four first order differential equations. It reflects the physical coupling between the macroscopic strain and the evolution of the microstructure during transient and steady-state creep. This coupling is well known from many experimental investigations, also for rock salt.

As an example, in Fig. 4 the result of a uniaxial creep test at room temperature and with two successive stress stages is compared with the result of a simulation with the modified composite model. The calculated subgrain size w is also plotted in the diagram. The figure demonstrates the close relation between the shape of a creep curve and the evolution of the microstructure.

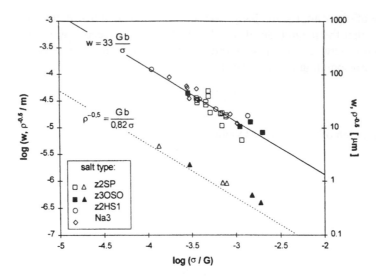

Fig. 3: In steady-state creep the subgrain size w and the dislocation spacing inside of subgrains $\rho^{-0.5}$ (ρ: dislocation density) depend only on stress difference σ and, implicitly, on temperature T (through the temperature dependence of the shear modulus G(T) [Frost & Ashby (1982)]), not on deformation history; data by Weidinger (1998).

In general this system of four differential equations can be solved numerically by using well established methods [see e.g. Press et al. (1989)]. One has to take into account that at the very beginning of a transient and when steady-state creep is approached this system becomes stiff. For the solution of such a problem e.g. the Rosenbrock variation of Runge-Kutta's method and a variation of the Bulirsch-Stoer technique can be applied. However, details cannot be given in this contribution.

For the implementation of the modified composite model into a numerical computer code (e.g. a finite difference or a finite element code), besides an appropriate time integration of the stiff set of equations an extension of equation (2) to three dimensions is necessary. For pure creep deformation at constant volume the following transformation is applied:

$$\dot{\varepsilon}_{ij} \; = \; \frac{3}{2} \; \frac{\dot{\varepsilon}_{eff}}{\sigma_{eff}} \; S_{ij} \qquad\qquad (14)$$

with $\dot{\varepsilon}_{ij}$: component of the strain rate tensor, S_{ij} : corresponding component of the stress deviator, $\dot{\varepsilon}_{eff}$: effective strain rate (= $\dot{\varepsilon}$ in (2)), σ_{eff} : effective stress difference (= σ in (2) with (6)).

DETERMINATION OF THE MODEL PARAMETERS

Figs. 5 to 8 show the dependence of the steady-state creep rate $\dot{\varepsilon}_{\infty}$ on temperature and stress difference for different salt types from various locations. Each creep test of the series was performed up to a high creep strain in order to ensure that steady-state creep was ically achieved. A part of the measurements was carried out at the Institut für Werk-

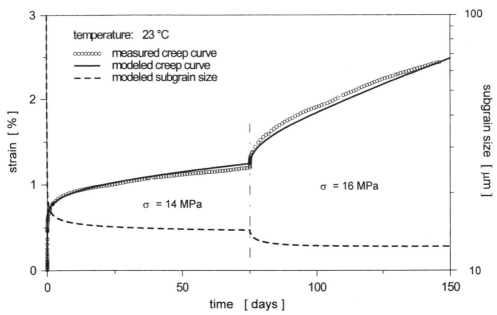

Fig. 4: The uniaxial creep test was performed with rock salt of type z3LS from the Morsleben mine, and modeled with the modified composite model. The figure displays the close relation between the microstructural evolution and the shape of a creep curve.

stoffwissenschaften (Institute for Materials Science) of the University of Erlangen-Nürnberg, Germany. At stress differences exceeding about 15 MPa, the tests were either carried out triaxially with a confining pressure of 15 to 20 MPa, or uniaxial tests were carefully evaluated and corrected for friction at the loading platens and for a dilatancy influence by following well established procedures of Weidinger et al. (1996). Therefore, the experimental data represent really the pure steady-state creep behavior at constant volume of the respective salt type.

The curves in figs. 5, 7, and 8 and the dashed lines in fig. 6 result from a fit of the modified composite model to the experimental data of z2SP in fig. 5. Only three fit parameters v_0, d_p, and σ_p had to be adjusted, the values are given in the diagrams. The full lines in

fig. 6 indicate a fit of the model equation to the data of the z3OSO salt type. Only the value of the parameter d_p had to be altered to transform the model curves of z2SP into those for z3OSO. The fact that the value of d_p was found to be smaller in z3OSO than in z2SP agrees physically with the observation of a smaller spacing of mineral particles in z3OSO (fig. 2) than in z2SP. Inside of subgrains these particles act as obstacles against dislocation motion and, therefore, cause the lower creep rates in z3OSO than in z2SP (compare the dashed and the respective full lines in fig. 6).

In z2SP (from the Asse mine), z2HS1 (Gorleben), and z2HS (Asse) only big particles with a large spacing were found in the microscope. These particles can hardly influence the dislocation motion. Therefore, these salt types show a similar creep behavior with comparatively high rates (see figs. 7 and 8), and are modeled with the same parameter values.

From the careful evaluation of these several 100 creep tests, the **microscopic** activation energy Q in (2) was found to be constant and unique (Q = 180 kJ/mole) for all investigated temperatures (50 ≤ T/°C ≤ 250), stress differences (2 ≤ σ/MPa ≤ 30), and salt types. Therefore, this quantity Q is not regarded as a free fit parameter. However, the **effective** activation energy Q_{eff} which represents the effective temperature dependence of the creep rate ἑ [Frost & Ashby (1982)], can depend on the salt type: In z3OSO a value of Q_{eff} = 135 kJ/mol was determined, while in the other three salt types Q_{eff} = 110 kJ/mol was found. The reason for differences in Q_{eff} is the temperature dependence of the argument of the hyperbolic sine function in (2), where for different salt types a different Δa (through d_p in (3)) or σ* (through σ_p in (6)) may occur. In z3OSO and z2SP the different values of d_p causes the differences in Q_{eff}. The unique value for the microscopic activation energy Q confirms that the modified composite model describes the dominant deformation processes and microstructures of creep deformed rock salt in the right way.

The different values of d_p in z2SP and z3OSO (figs. 5 and 6) which lead to different effective activation energies Q_{eff}, also affect the activation area Δa, see equation (3). As described above, Δa/b can be related to the mean spacing of dislocation obstacles (impurities and intersecting dislocations) inside of the subgrains. This quantity can also be evaluated from the results of stress drop experiments by using Friedel's theory [Vogler (1992)]. This means, a fit result for d_p can be experimentally checked independently from the (creep) tests that were used to determine the parameter value, see fig. 9 for example.

The model curve for Δa/b in fig. 9 demonstrates that for small stress differences impurities in the salt matrix (inside of the grains) are the dominant obstacles against dislocation motion, because the dislocation density is still low, i.e. the dislocation spacing is much

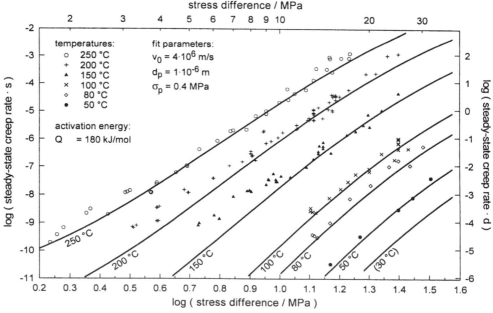

Fig. 5: Results of the reference creep test series with natural rock salt of type z2SP from the Asse mine (800 m level, symbols). The lines represent a fit of the composite model to the experimental data by using the indicated parameter values.

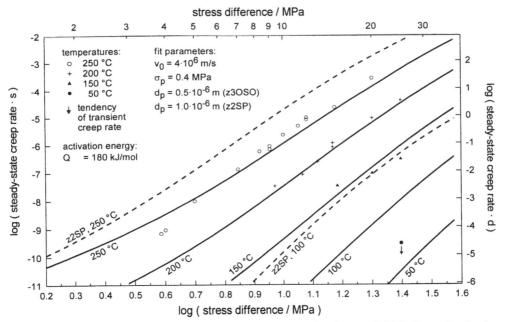

Fig. 6: Results of a creep test series with natural rock salt of type z3OSO from the Gorleben salt dome (depth: 350 m, symbols). Full lines represent a fit of the composite model to the experimental data. Dashed lines indicate model curves for z2SP for comparison (see fig. 5).

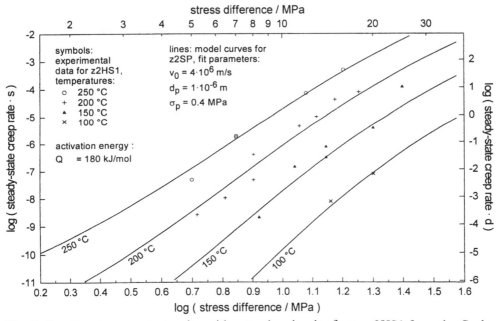

Fig. 7: Results of a creep test series with natural rock salt of type z2HS1 from the Gorleben salt dome (depth: 254 m, symbols). Full lines represent a fit of the composite model to the experimental data of z2SP (see fig. 5), indicating similar creep behavior of both salt types.

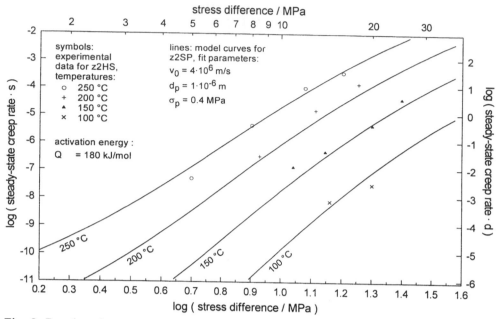

Fig. 8: Results of a creep test series with natural rock salt of type z2HS from the Asse mine (800 m level, TSDE field, symbols). Full lines represent a fit of the composite model to the experimental data of z2SP (see fig. 5), indicating similar creep behavior of both salt types.

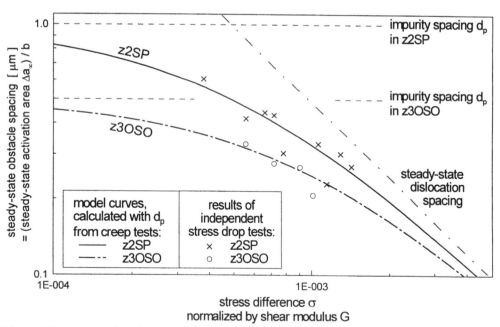

Fig. 9: The modeled activation areas of the salt types z2SP and z3OSO were calculated with the composite model by using values for d_p from the creep test series in fig. 5 and 6. A comparison with the results of independent stress drop tests shows a good agreement and demonstrates that this model consistently describes the deformation processes.

larger than that one of the impurities (cf. equation (3)). On the other hand, at large stress differences and after enough deformation a high dislocation density has developed so that the dislocations are now impeded by each other much more than by the impurities. Fig. 9 shows that these modeled influences of impurities and of intersecting dislocations on a moving dislocation are experimentally confirmed by the results of stress drop tests. This possibility, to check the fit results by independent experiments and observations, is one of the big advantages of the use of physical quantities in the modified composite model.

For a determination of the values of the six model parameters v_0, d_p, σ_p, k_w, k_p, k_a from a single creep curve, a simple and quick explicit algorithm was used instead of the numerical solution of the stiff set of the four differential equations (see above). Here, during transient creep the stepsize Δt was controlled by means of the condition

$$\left|\dot{\varepsilon}_{i-1} - \dot{\varepsilon}_i\right| \Delta t_i \;<\; \Delta\varepsilon_{crit} \qquad (15)$$

where $\dot{\varepsilon}_{i-1}$ is the known creep rate of the previous iteration; $\dot{\varepsilon}_i$ is the creep rate that corresponds to the time $t_i = t_{i-1} + \Delta t_i$ under consideration; and $\Delta\varepsilon_{crit}$ is a strain increment criterion ($\Delta\varepsilon_{crit} = 1\cdot10^{-6}$ was generally a good choice).

An example for such a fit of the composite model to a creep curve is shown in fig. 4. The modeled creep strain is closely correlated with the calculated evolution of the subgrain size w which is also shown in fig. 4. This microstructural evolution is given by equation (10) by using the parameter value for k_w and the relation (11). Similar curves are cal-

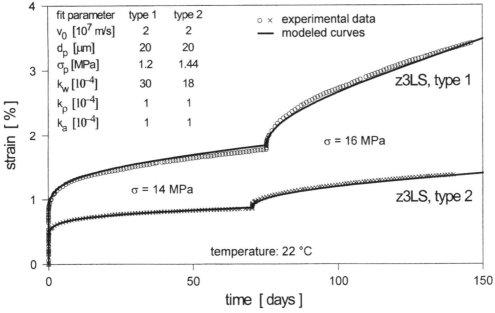

Fig. 10: Fits of the modified composite model to the results of uniaxial creep tests with rock salt from different locations (type 1/2) of a z3LS layer in the Morsleben mine, (depth: 506 m). The differences in the creep behavior are modeled by changing only two parameter values: σ_p affects the steady-state creep rate, k_w alters the length of a creep transient.

culated for the evolution of the dislocation density inside of subgrains ρ and of the thickness of the subgrain walls a (more precisely the thickness of the "hard regions" with σ_h > σ, see above). Always, the values of the rate constants k_ρ and k_a are found to be much smaller than that one of k_w (see fig. 10).

This result agrees with the observation that after a change of the deformation condition (mainly of the stress difference σ) ρ and a approach their new steady-state much faster than w. Weidinger (1998) experimentally confirmed the different rates of the microprocesses by measuring w and ρ on salt specimens of the same type after reaching different strain levels, i.e. at different points along a creep transient. The physical reason for the different evolution rates is that ρ and a change through the rearrangement of single dislocations, while the change of w requires the collective movement of dislocations (formation, movement, or annihilation of subgrain boundaries as a whole) which requires and produces much more strain.

Fig. 10 shows fits of the modified composite model (equation (2)) to the creep curves of natural rock salt from two different locations of the stratigraphic layer z3LS in the Morsleben mine. The different creep behavior of both salt types is explained by a different geological history of the two locations which influences the type or spatial distribution of salt minerals ("impurities", see above): If many small and "hard" impurities are present inside of the grains, they may effectively act as obstacles against dislocation motion and, therefore, cause low creep rates. However, if the salt minerals are aggregated to widely

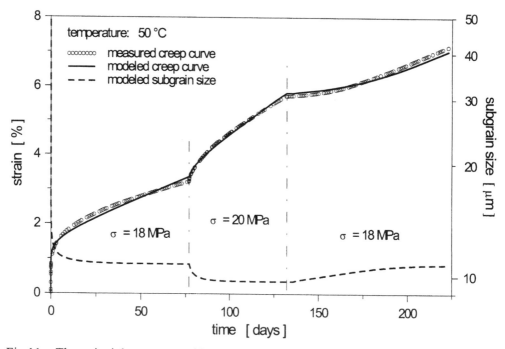

Fig.11 : The uniaxial creep test with a reduced axial stress in the third stage was performed with rock salt of type z2SP from the Asse mine, and simulated with the modified composite model. After a stress decrease the inverse curvature of the creep curve results from a gradual increase of the subgrain size (recovery of the material).

spaced and/or weaker particles or to particles on the grain boundaries, they do hardly affect the dislocation motion. Therefore, such salt types creep comparatively fast. In the composite model the spatial distribution and type of the impurities are described by the parameters d_p and σ_p, respectively, see equations (3) and (6). Indeed, one model curve in fig. 10 can be transformed into the other one by changing only σ_p and k_w. A higher value for the back stress σ_p produces smaller steady-state creep rates. The rate constant k_w for the evolution of the subgrain size w affects the length of a creep transient (see above and fig. 4) and, therefore, takes into account that the faster creeping salt type in fig. 10 has a more pronounced transient phase.

For some constitutive models the description of the inverse curvature of a creep curve after a stress reduction is a problem. The modified composite model naturally produces such an inverse creep transient, because the equations (11) to (13) cause the evolution of an increasing subgrain size and dislocation spacing with a decreasing stress difference and, in this way, models the recovery of the deformation microstructure. This is demonstrated in fig. 11 for an example (see also [Hunsche & Hampel (1999)]).

CONCLUSION

The composite model describes the transient and steady-state creep of natural rock salt on the basis of the deformation mechanisms and the resulting microstructure. This crystalline geomaterial exhibits a fully developed subgrain structure already in its original state in the intact underground. When the local stress difference is altered e.g. because of the excavation of new underground openings (shafts, drifts, galleries), characteristic changes in the subgrain structure appear. In the composite model, the description of this microstructural evolution under different conditions allows to simulate the various features of macroscopic creep, including the inverse transient creep after a stress reduction.

The subgrain structure consists of a network of subgrain walls with high dislocation density (mechanically "hard" regions) which enclose "soft" regions with a much lower dislocation density. This structural heterogeneity results in a modification of the local stress field. Therefore, an internal effective stress is introduced that drives a moving dislocation through a subgrain. In addition, impurities (salt minerals) may impede a moving dislocation. Both, the subgrain structure and the type and spatial distribution of non-dislocation obstacles, have an important influence on creep. Therefore, the description of these influences allows to model the big differences in the creep behavior of various salt types.

The development of the model is based on and supported by a large number (several 100's) of carefully performed and evaluated creep tests with different salt types from various locations. The steady-state creep rate at various temperatures $22 \leq T/°C \leq 250$ and stress differences $2 \leq \sigma/MPa \leq 30$ could only be measured reliably and reproducibly, when in each test stage the specimens were deformed up to a sufficient strain. Otherwise, only higher stages of creep transients are investigated, which may result in incorrect conclusions about steady-state creep (e.g. a too small stress exponent of the steady-state creep rate in a power law).

The modified composite model is developed to simplify the implementation of the constitutive equation into numerical computer codes. Here, the total creep rate is expressed by that one of the soft regions alone. The model parameters are physical quantities that are determined by fits of the constitutive equation to experimental data, and – this is a big advantage of a physically based model – can be checked with independent measurements and observations.

In all cases a good agreement was found between the model curves and the experimental data. Differences in the steady-state creep behavior of different salt types could be modeled by only changing the parameter for the type or for the spatial distribution of impurities inside of the subgrains. This modeling result is consistent with microscopic observations. Moreover, the calculated activation area shows a good agreement with results of independent stress drop experiments. While the effective activation energy may depend on the salt type, the microscopic activation energy, which is used in the Arrhenius term of the model equation, was found to be constant and unique for all investigated salt types. In addition, the rate constant for the evolution of the subgrain size after a stress change was found to be higher than that one for the evolution of the mean dislocation spacing. This modeling result could also be confirmed experimentally.

A good agreement between the results (e.g. parameter values) from fitting and the findings of independent investigations demonstrate that the important microprocesses and structures are described by the composite model in the right way. Therefore, with this model reliable predictions of the long-term creep behavior of natural rock salt can be performed. These are necessary for e.g. the safety assessment of a permanent repository for hazardous wastes in salt.

First test simulations with an implementation of the modified composite model into the computer code FLAC were successful and indicated the applicability of this model for practical engineering purposes.

REFERENCES

W. Blum (1991): High-temperature deformation and creep of crystalline solids.- In: Plastic Deformation and Fracture of Materials, H. Mughrabi (ed.), vol. 6 in series: Materials Science and Technology, R.W. Cahn, P. Haasen & E.J. Kramer (eds. of series), 339-405, VCH Verlagsgesellschaft, Weinheim.

N.L. Carter, F.D. Hansen & P.E. Senseny (1982): Stress magnitudes in natural rock salt.- J. Geophys. Res., vol. 87, 9289-9300.

H.J. Frost & M.F. Ashby (1982): Deformation mechanism maps: the plasticity and creep of metals and ceramics.- 166 pp., Pergamon Press, Oxford.

A. Hampel, U. Hunsche, P. Weidinger & W. Blum (1998): Description of the creep of rock salt with the composite model - II. steady state creep.- In: The Mechanical Behavior of Salt IV, Proc. 4th Conf. (MECASALT IV), Montreal 1996, M. Aubertin & H.R. Hardy, Jr. (eds), 287-299, Trans Tech Publications, Clausthal.

U. Hunsche & A. Hampel (1999): Rock salt – the mechanical properties of the host rock material for a radioactive waste repository.- Engng Geol., vol. 52, 271-291.

U. Hunsche, G. Mingerzahn & O. Schulze (1996): The influence of textural parameters and mineralogical composition on the creep behavior of rock salt.- In: The Mechanical Behavior of Salt III, Proc. 3rd Conf. (MECASALT III), Palaiseau 1993, M. Ghoreychi, P. Berest, H.R. Hardy, Jr. & M. Langer (eds), 143-151, Trans Tech Publications, Clausthal.

W.H. Press, B.P. Flannery, S.A. Teukolsky & W.T. Vetterling (1989): Numerical recipes in Pascal – the art of scientific computing.- Cambridge University Press.

R. Sedláček (1995): Glide dislocation shapes and long-range internal stresses in dislocation wall structures.- Phys. stat. sol. (a), vol. 149, 85-93.

S. Vogler (1992): Kinetik der plastischen Verformung von natürlichem Steinsalz und ihre quantitative Beschreibung mit dem Verbundmodell.- Doctoral thesis, Universität Erlangen-Nürnberg, 148 pp.

S. Vogler & W. Blum (1990): Micromechanical modelling of creep in terms of the composite model.- In: Creep and Fracture of Engineering Materials and Structures, Proc. 4th Internat. Conf., Swansea 1990, B. Wilshire & R.W. Evans (eds.), 65-79, The Institute of Metals, London.

J. Weertman & J.R. Weertman (1983): Mechanical properties, mildly temperature-dependent.- In: Physical Metallurgy, Part II, 3rd ed., RW. Cahn & P. Haasen (eds), 1259-1307, North-Holland Physics Publishing, Amsterdam.

P. Weidinger (1998): Verformungsverhalten natürlicher Steinsalze: Experimentelle Ermittlung und mikrostrukturell begründete Modellierung.- Doctoral thesis, Universität Erlangen-Nürnberg, 164 pp., Shaker Verlag, Aachen.

P. Weidinger, W. Blum, A. Hampel & U. Hunsche (1998): Description of the creep of rock salt with the composite model - I. transient creep. - In: The Mechanical Behavior of Salt IV, Proc. 4th Conf. (MECASALT IV), Montreal 1996, M. Aubertin & H.R. Hardy, Jr. (eds), 277-285, Trans Tech Publications, Clausthal.

P. Weidinger, W. Blum, U. Hunsche & A. Hampel (1996): The influence of friction on plastic deformation in compression tests.- Phys. stat. sol. (a), vol. 156, 305-315.

Basic and Applied Salt Mechanics, Cristescu, Hardy, Jr & Simionescu (eds)
© 2002 Swets & Zeitlinger, Lisse, ISBN 90 5809 383 2

Mechanical behavior around spherical cavities

Doina Massier
Faculty of Mathematics
University of Bucharest, Bucharest, Romania

ABSTRACT

The aim of this paper is the determination of the stress state and of the displacement field around spherical cavities in rock salt. The rheological behavior of the salt is described by a linear viscoelastic model. For the spherical cavity the problem is solved using Neuber's representation [Neuber (1934)] for the Laplace transform of the displacement field as well as the Maxwell's representation for the stress tensor.

The results are illustrated through diagrams for various primary stresses and at different moments of time.

INTRODUCTION

More often than not the problem of the rheological behavior of a rock around a closed cavity was solved with simplifying assumptions: the cavity surface a spherical one; the primary stress state a hydrostatic one; frequently the material is considered isotropic linear elastic [Brethauer (1974)] or isotropic incompressible nonlinear elastic. The domain, for which the mechanical problem has to be solved, is not well known and is usually assimilated to a half or a whole space with a closed hole. There is not possible to give an analytical solution of the fundamental problem even in the case of a linear elastic material. Therefore simplified problems are associated to the real situation. The assumptions done for these problems are presented further down. The problem is solved for deep closed cavities, but the way in which the solving is performed suggests the approach for shallow closed cavities, in a similar manner to this the author has used for horizontal cylindrical cavities. Because of the linear viscoelastic model, through the Laplace transform, there exists a perfect correspondence with the problem defined in the linear elasticity [similarly to Massier (1995a, 1995b)].

STATEMENT OF THE PROBLEMS

Assumptions concerning the geometry of the domain

The depth h of the center of the cavity is much larger than its diameter, $h \gg 2a$ (so that the influence of the free surface of the massif may be neglected). Therefore the domain for which the problem will be solved is the whole space with a cavity.

There will be used spherical coordinates

$$\begin{cases} x^1 = r \sin\theta \cos\varphi, \\ x2 = r \sin\theta \sin\varphi, \\ x^3 = r \cos\theta, \end{cases} \quad r \in [0,+\infty), \theta \in [0,\pi], \varphi \in [0,2\pi]. \tag{1}$$

(x^3 along the vertical axis of the cavity).

Assumptions concerning the primary stress state

Even before the cutting of an underground opening, in the virgin massif there exists a stress field. The reference configuration is considered that characterized by the stresses existing in the massif without mining workings, the primary stress state at the level of the cavity center, the same in the whole domain. Otherwise, the primary stress field is a general one

$$\sigma^P = \sigma^P_{jk} \iota_j \otimes \iota_k .$$

In spherical coordinates there are

$$\sigma^P_{rr} = \left(\sigma^P_{11} \cos^2\varphi + \sigma^P_{22} \sin^2\varphi + 2\sigma^P_{12} \sin\varphi \cos\varphi\right) \sin^2\theta + \sigma^P_{33} \cos^2\theta +$$
$$+ 2\left(\sigma^P_{13} \cos^2\varphi + \sigma^P_{23} \sin\varphi\right) \sin\theta \cos\theta,$$

$$\sigma^P_{\theta\theta} = \left(\sigma^P_{11} \cos^2\varphi + \sigma^P_{22} \sin^2\varphi + 2\sigma^P_{12} \sin\varphi \cos\varphi\right) \cos^2\theta + \sigma^P_{33} \sin^2\theta +$$
$$- 2\left(\sigma^P_{13} \cos^2\varphi + \sigma^P_{23} \sin\varphi\right) \sin\theta \cos\theta,$$

$$\sigma^P_{r\theta} = \left(\sigma^P_{11} \cos^2\varphi + \sigma^P_{22} \sin^2\varphi + 2\sigma^P_{12} \sin\varphi \cos\varphi\right) \sin\theta \cos\theta - \sigma^P_{33} \sin\theta \cos\theta +$$
$$+ \left(\sigma^P_{13} \cos^2\varphi + \sigma^P_{23} \sin\varphi\right)\left(\cos^2\theta - \sin^2\theta\right), \tag{2}$$

$$\sigma^P_{\varphi\varphi} = \sigma^P_{11} \sin^2\varphi + \sigma^P_{22} \cos^2\varphi - 2\sigma^P_{12} \sin\varphi \cos\varphi$$

$$\sigma^P_{r\varphi} = \left[-\left(\sigma^P_{11} - \sigma^P_{22}\right) \sin\varphi \cos\varphi + \sigma^P_{12}\left(\cos^2\varphi - \sin^2\varphi\right)\right] \sin\theta -$$
$$- \left(\sigma^P_{13} \sin\varphi - \sigma^P_{23} \cos\varphi\right) \cos\theta,$$

$$\sigma^P_{\theta\varphi} = \left[-\left(\sigma^P_{11} - \sigma^P_{22}\right) \sin\varphi \cos\varphi + \sigma^P_{12}\left(\cos^2\varphi - \sin^2\varphi\right)\right] \cos\theta +$$
$$+ \left(\sigma^P_{13} \sin\varphi - \sigma^P_{23} \cos\varphi\right) \sin\theta,$$

Assumptions concerning the constitutive model

In order to describe the rheological properties there is considered that
- the rock is homogeneous and isotropic;
- only a small displacement gradient is possible; it follows the use of the infinitesimal strain tensor and the objectivity of the time derivatives of the stress and strain tensors;
- the constitutive model is a linear viscoelastic one, relative to the reference configuration discussed above

210

$$\begin{cases} \dot{\varepsilon}' = -k\left[\varepsilon' - \dfrac{1}{2G_o}\sigma^{R\,\prime}\right] + \dfrac{1}{2G}\dot{\sigma}^{R\,\prime}, \\[4mm] \dot{\varepsilon} = -k_v\left[\varepsilon - \dfrac{1}{3K_o}\sigma^{R}\right] + \dfrac{1}{3K}\dot{\sigma}^{R}, \end{cases} \tag{3}$$

$$\sigma^{R} = \sigma - \sigma^{P},\ \sigma^{R} = \frac{1}{3}\operatorname{tr}\sigma^{R},\ \sigma^{R\,\prime} = \sigma^{R} - \sigma^{R}\mathbf{1},\ \varepsilon = \frac{1}{3}\operatorname{tr}\varepsilon,\ \varepsilon' = \varepsilon - \varepsilon\mathbf{1},$$

σ^{P} – primary stress tensor, σ – secondary current stress tensor, σ^{R} – relative stress tensor, ε – infinitesimal strain tensor, " $\cdot$ " denotes time derivatives. The instantaneous response and the relaxation state are assumed linear elastic

$$\varepsilon' = \frac{1}{2G}\sigma^{R\,\prime},\ \varepsilon = \frac{1}{3K}\sigma^{R}, \tag{i.r.}$$

$$\varepsilon' = \frac{1}{2G_o}\sigma^{R\,\prime},\ \varepsilon = \frac{1}{3K_o}\sigma^{\prime R}. \tag{r.s.}$$

Constitutive restrictions imposed to the material constants, are [Massier (1997a)]

$$0 < 2G < 3K, 0 < k \le k_v,\ k\left(\frac{1}{2G} - \frac{1}{2G_o}\right) \le k_v\left(\frac{1}{3K} - \frac{1}{3K_o}\right) < 0. \tag{4}$$

Assumptions concerning the loading
    There is supposed that the body forces acting before and after the excavation of the cavity are the same. This assumption implies the fact that relative stresses satisfy equilibrium equations without body forces

$$\operatorname{div}\sigma^{R} = \mathbf{O}.$$

On the cavity surface is acting a pressure,

$$\sigma\mathbf{n}\big|_{\Sigma} = -p_o(t)\mathbf{n}\big|_{\Sigma} \tag{5}$$

or in terms of relative stresses

$$\sigma^{R}\mathbf{n}\big|_{\Sigma} = \left(p_o(t)\mathbf{n} - \sigma^{P}\mathbf{n}\right)\big|_{\Sigma}. \tag{6}$$

The perturbation of the primary state is damping with the distance

$$\lim_{r \to \infty}\sigma^{R} = \mathbf{O}, \quad \lim_{r \to \infty}\mathbf{u} = \mathbf{o}. \tag{7}$$

Assumption concerning the time evolution
    From the point of view of the order of magnitude of the time intervals in which an opening is excavated and exploited, two intervals are to be distinguished [Cristescu (1989)]: a relatively short one in which the excavation is done and a much larger one in which the opening is operated. To make things simple, it is assumed that in the first time interval, the rock deforms "instantaneously" while in the second one the deformation continues slowly in time.
    This assumption may be replaced by any other suggested by practice, providing the initial conditions. If such an assumption regards the advance of the face of working, even in the case of the tunnel, it will be solved a problem for plane generalized relative stresses.

Further, it will be used the Laplace transform with respect to the time

$$\tilde{f}(\cdot,\lambda) = \int_0^{\infty} f(\cdot,t)\exp(-\lambda t')\,dt', \qquad t' = \frac{t}{t_*} \tag{8}$$

(t_* an arbitrary chosen time for adimensionalisation). This transform will preserve the aspect of all operations with respect to the spatial coordinates.

Taking into account the instantaneous response, the constitutive law becomes similar to the Hooke's law [Massier (1997b)]

$$\tilde{\varepsilon}' = \frac{1+\tilde{v}}{\tilde{E}}\tilde{\sigma}^R{}', \tilde{\varepsilon} = \frac{1-2\tilde{v}}{\tilde{E}}\tilde{\sigma}^R,$$

$$\tilde{E} = \frac{S(\lambda)}{Q(\lambda)}, \tilde{v} = \frac{R(\lambda)}{Q(\lambda)},$$

$$Q(\lambda) = \left(k + \frac{\lambda}{t_*}\right)\left(\frac{k_v}{3K_o} + \frac{1}{3K}\frac{\lambda}{t_*}\right) + 2\left(k_v + \frac{\lambda}{t_*}\right)\left(\frac{k}{2G_o} + \frac{1}{2G}\frac{\lambda}{t_*}\right), \tag{9}$$

$$R(\lambda) = -\left(k + \frac{\lambda}{t_*}\right)\left(\frac{k_v}{3K_o} + \frac{1}{3K}\frac{\lambda}{t_*}\right) + \left(k_v + \frac{\lambda}{t_*}\right)\left(\frac{k}{2G_o} + \frac{1}{2G}\frac{\lambda}{t_*}\right),$$

$$S(\lambda) = 3\left(k + \frac{\lambda}{t_*}\right)\left(k_v + \frac{\lambda}{t_*}\right).$$

The constitutive restrictions ensure for $\lambda > 0$ that $\tilde{E} > 0, 0 < \tilde{v} < 0.5$. This transform preserves the aspect of all operations with respect to the spatial coordinates.

3. NEUBER'S REPRESENTATION

Some results concerning the three dimensional problems in the linear elasticity [Neuber (1934)] may be extended to linear viscoelasticity, using the Laplace transform of the constitutive law.

Proposition 1. The equilibrium of a homogeneous, isotropic, linear viscoelastic body, in the absence of body forces, is described through the Lamé equation

$$\Delta\tilde{u} + \frac{1}{1-2\tilde{v}}\operatorname{grad}\operatorname{div}\tilde{u} = o, \tag{10}$$

with the boundary conditions (8) [Solomon (1968)].

Proposition 2. (Neuber's representation) Any vector

$$\tilde{u} = \tilde{B} - \frac{1}{4(1-\tilde{v})}\operatorname{grad}\left(x \cdot \tilde{B} + \tilde{B}_o\right), \qquad \Delta\tilde{B} = o, \ \Delta\tilde{B}_o = 0, \tag{11}$$

verifies the equilibrium equation (12).

The problem solution will be searched under the following form

$$u = \sum_{k=0}^{\infty} u_k, \ B = \sum_{k=0}^{\infty} B_k, \ B_o = \sum_{k=0}^{\infty} B_k^o, \ \tilde{u}_k = \tilde{B}_k - \frac{1}{4(1-\tilde{v})}\left(x \cdot \tilde{B}_k + \tilde{B}_{k-1}^o\right).$$

Proposition 3. The stress vector on the spherical boundary $r = a$ has the expression

$$\widetilde{\sigma}^R \mathbf{n}(\widetilde{\mathbf{u}}) = \frac{\widetilde{E}}{1 + \widetilde{v}} \frac{1}{r} \left\{ \frac{\widetilde{v}}{1 - \widetilde{v}} \mathbf{x} \operatorname{div} \widetilde{\mathbf{u}} + (\mathbf{x} \cdot \operatorname{grad}) \widetilde{\mathbf{u}} + \frac{1}{2} \mathbf{x} \times \operatorname{grad} \widetilde{\mathbf{u}} \right\}. \tag{12}$$

Proposition 4. On the spherical boundary $r = a$, there is

$$\widetilde{\sigma}^R \mathbf{n}(\widetilde{\mathbf{u}}_k) = \frac{\widetilde{E}}{1 + \widetilde{v}} \frac{1}{r} \left\{ \widetilde{\mathbf{H}}_k - \frac{k - 2}{4(1 - \widetilde{v})} \left[\frac{r^2}{2k - 1} \operatorname{grad} \operatorname{div} \widetilde{\mathbf{B}}_k + \operatorname{grad} \widetilde{B}^o_{k-1} \right] \right\},$$

$$\widetilde{\mathbf{H}}_k = \frac{k - 1}{2} \widetilde{\mathbf{B}}_k + \frac{k - 2 + 2\widetilde{v}}{4(1 - \widetilde{v})(2k + 1)} \mathbf{K}(\widetilde{\mathbf{B}}_k) + \frac{k(1 - 2\widetilde{v}) - 2 + \widetilde{v}}{2(1 - \widetilde{v})(4k^2 - 1)} \mathbf{T}(\widetilde{\mathbf{B}}_k). \tag{13}$$

$$\mathbf{K}(\mathbf{B}_k) = \operatorname{grad} \left[r^{2k+2} \operatorname{div}(r^{-2k-1} \mathbf{B}_k) \right], \quad \mathbf{T}(\mathbf{B}_k) = r^{2k+1} \operatorname{grad}(r^{-2k+1} \operatorname{div} \mathbf{B}_k)$$

Proposition 5. If the boundary conditions on the spherical boundary $r = a$ have the form

$$\widetilde{\sigma}^R \mathbf{n}(\mathbf{u}) \big|_{r=a} = \widetilde{\mathbf{f}}(\theta, \varphi, \lambda), \quad \widetilde{\mathbf{f}}(\theta, \varphi, \lambda) = \frac{\widetilde{E}}{1 + \widetilde{v}} \frac{1}{a} \sum_{k=0}^{\infty} \widetilde{\mathbf{Y}}_k(\theta, \varphi, \lambda),$$

then for the outside domain

$$\widetilde{\sigma}^R \mathbf{n}(\mathbf{u}) = \frac{\widetilde{E}}{1 + \widetilde{v}} \frac{1}{a} \sum_{k=0}^{\infty} \left(\frac{a}{r} \right)^{k+1} \widetilde{\mathbf{Y}}_k(\theta, \varphi, \lambda) -$$

$$- \frac{\widetilde{E}}{1 + \widetilde{v}} \frac{r^2 - a^2}{2} \sum_{k=0}^{\infty} \frac{k - 2}{k^2 - (1 - 2\widetilde{v})k + 1 - \widetilde{v}} \operatorname{grad} \left[\left(\frac{a}{r} \right)^{k+1} \widetilde{\mathbf{Y}}_k(\theta, \varphi, \lambda) \right]. \tag{13}$$

Proposition 6. The displacement field corresponding to (12), (13) is for the outside domain

$$\widetilde{\mathbf{u}}_k = \frac{1}{k} \widetilde{\mathbf{H}}_k - \frac{1}{k(k - 1)} \mathbf{x} \times \operatorname{rot} \widetilde{\mathbf{H}}_k + \frac{(1 - 4\widetilde{v})k - 2(1 - \widetilde{v})}{k\left[k^2 - (1 - 2\widetilde{v})k + 1 - \widetilde{v} \right]} \mathbf{x} \operatorname{div} \widetilde{\mathbf{H}}_k +$$

$$+ \frac{(2k\widetilde{v} - \widetilde{v} + 1)a^2}{k(k - 1)\left[k^2 - (1 - 2\widetilde{v})k + 1 - \widetilde{v} \right]} \operatorname{grad} \operatorname{div} \widetilde{\mathbf{H}}_k - \frac{r^2 - a^2}{2\left[k^2 - (1 - 2\widetilde{v})k + 1 - \widetilde{v} \right]} \operatorname{grad} \operatorname{div} \widetilde{\mathbf{H}}_k,$$

$$\mathbf{H}_0 = \mathbf{o}, \quad \widetilde{\mathbf{H}}_1 = \operatorname{grad} \left(\frac{1}{2} \sum_{m,n=1}^{3} \widetilde{h}_{mn} x^m x^n \right).$$

The boundary conditions for the spherical cavity, expressed in terms of relative stresses are

$$\sigma^R \mathbf{n}(\mathbf{u}) \big|_{r=a} = \left(p_o \mathbf{n} - \sigma^P \mathbf{n} \right) \big|_{r=a}, \quad \lim_{r \to \infty} \sigma^R = \mathbf{O}.$$

Explicitly, there are

$$\sigma^R \mathbf{n}(\mathbf{u}) \big|_{r=a} = \frac{1}{r} \left(p_o \mathbf{x} - \sigma^P \mathbf{x} \right) \big|_{r=a}.$$

Proposition 7. The harmonical polynomial function corresponding to the boundary condition is

$$\widetilde{\mathbf{H}}_{-2} = \frac{1+\widetilde{v}}{\lambda\widetilde{E}}\left(\frac{a}{r}\right)^3\left(p_o\mathbf{x} - \sigma^P\mathbf{x}\right).$$

Proposition 8. The displacement field is

$$\widetilde{\mathbf{u}} = \frac{1+\widetilde{v}}{\widetilde{E}}\frac{a^3}{r^3}\left\{-\frac{1}{2}\widetilde{p}_o\mathbf{x} + \frac{1}{\lambda}\left[5\frac{1-2\widetilde{v}}{7-5\widetilde{v}} + \frac{3}{7-5\widetilde{v}}\frac{a^2}{r^2}\right]\sigma^P\mathbf{x} + \right.$$

$$\left. + \frac{1}{\lambda}\frac{15}{2(7-5\widetilde{v})}\left(1-\frac{a^2}{r^2}\right)\frac{\mathbf{x}\cdot\sigma^P\mathbf{x}}{r^2}\mathbf{x} - \frac{1}{2\lambda}\left[\frac{6-5\widetilde{v}}{7-5\widetilde{v}} - \frac{3}{7-5\widetilde{v}}\frac{a^2}{r^2}\right](\mathrm{tr}\sigma^P)\mathbf{x}\right\}. \tag{14}$$

Proposition 9. The relative stress tensor is

$$\sigma^R = \frac{1}{2}\widetilde{p}_o\frac{a^3}{r^3}\left(3\frac{\mathbf{x}\otimes\mathbf{x}}{r^2}-\mathbf{1}\right) + \frac{1}{\lambda}\left(5\frac{1-2\widetilde{v}}{7-5\widetilde{v}}\frac{a^3}{r^3} + \frac{3}{7-5\widetilde{v}}\frac{a^5}{r^5}\right)\sigma^P +$$

$$+ \frac{1}{\lambda}\left[(\sigma^P\mathbf{x})\otimes\mathbf{x} + \mathbf{x}\otimes(\sigma^P\mathbf{x})\right]\left(\frac{15\widetilde{v}}{7-5\widetilde{v}}\frac{a^3}{r^5} - \frac{15}{7-5\widetilde{v}}\frac{a^5}{r^7}\right) +$$

$$\frac{1}{2\lambda}\frac{15}{7-5\widetilde{v}}\left(\frac{a^3}{r^5}-\frac{a^5}{r^7}\right)(\mathbf{x}\cdot\sigma^P\mathbf{x})\mathbf{1} + \frac{1}{2\lambda}\frac{15}{7-5\widetilde{v}}\left(-5\frac{a^3}{r^7}+7\frac{a^5}{r^9}\right)(\mathbf{x}\cdot\sigma^P\mathbf{x})\mathbf{x}\otimes\mathbf{x} + \tag{15}$$

$$+ \frac{1}{2\lambda}\left(3\frac{6-5\widetilde{v}}{7-5\widetilde{v}}\frac{a^3}{r^5} - \frac{15}{7-5\widetilde{v}}\frac{a^5}{r^7}\right)(\mathrm{tr}\sigma^P)\mathbf{x}\otimes\mathbf{x} +$$

$$+ \frac{1}{2\lambda}\left(-\frac{6-5\widetilde{v}}{7-5\widetilde{v}}\frac{a^3}{r^3} + \frac{3}{7-5\widetilde{v}}\frac{a^5}{r^5}\right)(\mathrm{tr}\sigma^P)\mathbf{1} + \frac{1}{\lambda}\frac{5\widetilde{v}}{7-5\widetilde{v}}\frac{a^3}{r^3}\left(\mathrm{tr}\sigma^P - 3\frac{\mathbf{x}\cdot\sigma^P\mathbf{x}}{r^2}\right)\mathbf{1}.$$

Proposition 10. The displacements in spherical coordinates are

$$\widetilde{u}_r = \frac{1+\widetilde{v}}{\widetilde{E}}\frac{a^3}{r^2}\left\{-\frac{1}{2}\widetilde{p}_o + \frac{1}{\lambda}\left[5\frac{1-2\widetilde{v}}{7-5\widetilde{v}} + \frac{3}{7-5\widetilde{v}}\frac{a^2}{r^2}\right]\sigma_{rr}^P + \right.$$

$$\left. + \frac{1}{\lambda}\frac{15}{2(7-5\widetilde{v})}\left(1-\frac{a^2}{r^2}\right)\sigma_{rr}^P - \frac{1}{2\lambda}\left[\frac{6-5\widetilde{v}}{7-5\widetilde{v}} - \frac{3}{7-5\widetilde{v}}\frac{a^2}{r^2}\right]\left(\sigma_{rr}^P + \sigma_{\theta\theta}^P + \sigma_{\varphi\varphi}^P\right)\right\}, \tag{16}$$

$$\widetilde{u}_\theta = \frac{1+\widetilde{v}}{\lambda\widetilde{E}}\frac{a^3}{r^2}\left[5\frac{1-2\widetilde{v}}{7-5\widetilde{v}} + \frac{3}{7-5\widetilde{v}}\frac{a^2}{r^2}\right]\sigma_{r\theta}^P,$$

$$\widetilde{u}_\varphi = \frac{1+\widetilde{v}}{\lambda\widetilde{E}}\frac{a^3}{r^2}\left[5\frac{1-2\widetilde{v}}{7-5\widetilde{v}} + \frac{3}{7-5\widetilde{v}}\frac{a^2}{r^2}\right]\sigma_{r\varphi}^P.$$

Proposition 11. The relative stresses in spherical coordinates are

$$\tilde{\sigma}_{rr}^{R} = \tilde{p}_o \frac{a^3}{r^3} - \frac{1}{\lambda}\left(5\frac{5-\tilde{v}}{7-5\tilde{v}}\frac{a^3}{r^3} - \frac{18}{7-5\tilde{v}}\frac{a^5}{r^5}\right)\sigma_{rr}^{P} + \frac{1}{\lambda}\frac{6}{7-5\tilde{v}}\left(\frac{a^3}{r^3} - \frac{a^5}{r^5}\right)\left(\sigma_{rr}^{P} + \sigma_{\theta\theta}^{P} + \sigma_{\varphi\varphi}^{P}\right),$$

$$\tilde{\sigma}_{\theta\theta}^{R} = -\frac{1}{2}\tilde{p}_o\frac{a^3}{r^3} + \frac{1}{\lambda}\left(5\frac{1-2\tilde{v}}{7-5\tilde{v}}\frac{a^3}{r^3} + \frac{3}{7-5\tilde{v}}\frac{a^5}{r^5}\right)\sigma_{\theta\theta}^{P} + \frac{15}{2\lambda}\left(\frac{1-2\tilde{v}}{7-5\tilde{v}}\frac{a^3}{r^3} - \frac{1}{7-5\tilde{v}}\frac{a^5}{r^5}\right)\sigma_{rr}^{P} -$$

$$- \left(2\frac{3-5\tilde{v}}{7-5\tilde{v}}\frac{a^3}{r^3} - \frac{3}{7-5\tilde{v}}\frac{a^5}{r^5}\right)\left(\sigma_{rr}^{P} + \sigma_{\theta\theta}^{P} + \sigma_{\varphi\varphi}^{P}\right),$$

$$\tilde{\sigma}_{\varphi\varphi}^{R} = -\frac{1}{2}\tilde{p}_o\frac{a^3}{r^3} + \frac{1}{\lambda}\left(5\frac{1-2\tilde{v}}{7-5\tilde{v}}\frac{a^3}{r^3} + \frac{3}{7-5\tilde{v}}\frac{a^5}{r^5}\right)\sigma_{\varphi\varphi}^{P} + \frac{15}{2\lambda}\left(\frac{1-2\tilde{v}}{7-5\tilde{v}}\frac{a^3}{r^3} - \frac{1}{7-5\tilde{v}}\frac{a^5}{r^5}\right)\sigma_{rr}^{P} -$$

$$- \left(2\frac{3-5\tilde{v}}{7-5\tilde{v}}\frac{a^3}{r^3} - \frac{3}{7-5\tilde{v}}\frac{a^5}{r^5}\right)\left(\sigma_{rr}^{P} + \sigma_{\theta\theta}^{P} + \sigma_{\varphi\varphi}^{P}\right),$$

$$\tilde{\sigma}_{r\theta}^{R} = \frac{1}{\lambda}\left(5\frac{1+\tilde{v}}{7-5\tilde{v}}\frac{a^3}{r^3} - \frac{12}{7-5\tilde{v}}\frac{a^5}{r^5}\right)\sigma_{r\theta}^{P}, \quad \tilde{\sigma}_{\theta\varphi}^{R} = \frac{1}{\lambda}\left(5\frac{1-2\tilde{v}}{7-5\tilde{v}}\frac{a^3}{r^3} + \frac{3}{7-5\tilde{v}}\frac{a^5}{r^5}\right)\sigma_{\theta\varphi}^{P},$$

$$\tilde{\sigma}_{\varphi r}^{R} = \frac{1}{\lambda}\left(5\frac{1+\tilde{v}}{7-5\tilde{v}}\frac{a^3}{r^3} - \frac{12}{7-5\tilde{v}}\frac{a^5}{r^5}\right)\sigma_{\varphi r}^{P}.$$

Proposition 12. For constant pressure, the time functions involved in the expressions of the displacement vector and the relative stress tensor are combinations of the exponentials $\exp(-kt)$, $\exp\left(\frac{\lambda_1}{t_*}t\right)$, $\exp\left(\frac{\lambda_2}{t_*}t\right)$, where λ_1, λ_2 are the real negative roots of the polynomial $7Q(\lambda) - 5R(\lambda)$.

4. MAXWELL'S REPRESENTATION

Proposition 13. The equilibrium equations are satisfied for any potentials A, B, C of class C^3 if

$$\sigma_{11}^{R} = B_{,33} + C_{,22}, \quad \sigma_{22}^{R} = C_{,11} + A_{,33}, \quad \sigma_{33}^{R} = A_{,22} + B_{,11},$$

$$\sigma_{23}^{R} = -A_{,23}, \quad \sigma_{13}^{R} = -B_{,13}, \quad \sigma_{12}^{R} = -C_{,12}.$$

Proposition 14. For the linear viscoelastic law (5) the Maxwell's potentials are biharmonical functions

$$\Delta\Delta A(x^1, x^2, x^3, \lambda) = 0, \quad \Delta\Delta B(x^1, x^2, x^3, \lambda) = 0, \quad \Delta\Delta C(x^1, x^2, x^3, \lambda) = 0.$$

Proposition 15. In spherical coordinates the Maxwell's representation is

$$\sigma_{rr}^{R} = \sin^2\theta\left[\frac{1}{r^2\sin^2\theta}\frac{\partial^2 C}{\partial\varphi^2} + \frac{1}{r}\frac{\partial C}{\partial r} + \frac{1}{r^2}\frac{\cos\theta}{\sin\theta}\frac{\partial C}{\partial\theta}\right] + \frac{1}{r^2}\frac{\partial^2 f}{\partial\theta^2} + \frac{1}{r}\frac{\partial f}{\partial r} +$$

$$+ \cos^2\theta\left[\frac{1}{r^2\sin^2\theta}\frac{\partial^2(A+B-f)}{\partial\varphi^2} + \frac{1}{r}\frac{\partial(A+B-f)}{\partial r} + \frac{1}{r^2}\frac{\cos\theta}{\sin\theta}\frac{\partial(A+B-f)}{\partial\theta}\right] +$$

$$+ 2\frac{\sin\varphi\cos\varphi}{r^2}\frac{\cos\theta}{\sin\theta}\left[\frac{\cos\theta}{\sin\theta}\frac{\partial(A-B)}{\partial\varphi} + \frac{\partial(A-B)}{\partial\theta\partial\varphi}\right] + 2\frac{\cos^2\varphi - \sin^2\varphi}{r^2}\frac{\cos^2\theta}{\sin^2\theta}(A-B).$$

$$\sigma_{\theta\theta}^{R} = \cos^2\theta\left[\frac{1}{r^2\sin^2\theta}\frac{\partial^2 C}{\partial\varphi^2} + \frac{1}{r}\frac{\partial C}{\partial r} + \frac{1}{r^2}\frac{\cos\theta}{\sin\theta}\frac{\partial C'}{\partial\theta}\right] + \frac{\partial^2 f}{\partial r^2} +$$

$$+ \sin^2\theta\left[\frac{1}{r^2\sin^2\theta}\frac{\partial^2(A+B-f)}{\partial\varphi^2} + \frac{1}{r}\frac{\partial(A+B-f)}{\partial r} + \frac{1}{r^2}\frac{\cos\theta}{\sin\theta}\frac{\partial(A+B-f)}{\partial\theta}\right] +$$

$$+ 2\frac{\sin\varphi\cos\varphi}{r}\left[\frac{1}{r}\frac{\partial(A-B)}{\partial\varphi} + \frac{\partial(A-B)}{\partial r\partial\varphi}\right] + 2\frac{\cos^2\varphi - \sin^2\varphi}{r^2}(A-B),$$

$$\sigma_{\varphi\varphi}^{R} = \sin\theta\frac{\partial}{\partial r}\left[\sin\theta\frac{\partial C}{\partial r} + \frac{\cos\theta}{r}\frac{\partial C'}{\partial\theta}\right] + \frac{\cos\theta}{r}\frac{\partial}{\partial\theta}\left[\sin\theta\frac{\partial C}{\partial r} + \frac{\cos\theta}{r}\frac{\partial C}{\partial\theta}\right] +$$

$$+ \cos\theta\frac{\partial}{\partial r}\left[\cos\theta\frac{\partial(A+B-f)}{\partial r} - \frac{\sin\theta}{r}\frac{\partial(A+B-f)}{\partial\theta}\right] -$$

$$- \frac{\sin\theta}{r}\frac{\partial}{\partial\theta}\left[\cos\theta\frac{\partial(A+B-f)}{\partial r} - \frac{\sin\theta}{r}\frac{\partial(A+B-f)}{\partial\theta}\right],$$

$$\sigma_{r\theta}^{R} = \sin\theta\cos\theta\left[\frac{1}{r^2\sin^2\theta}\frac{\partial^2 C}{\partial\varphi^2} + \frac{1}{r}\frac{\partial C}{\partial r} + \frac{1}{r^2}\frac{\sin\theta}{\cos\theta}\frac{\partial C}{\partial\theta}\right] - \frac{1}{r}\frac{\partial^2 f}{\partial r\partial\theta} + \frac{1}{r^2}\frac{\partial f}{\partial\theta} -$$

$$- \sin\theta\cos\theta\left[\frac{1}{r^2\sin^2\theta}\frac{\partial^2(A+B-f)}{\partial\varphi^2} + \frac{1}{r}\frac{\partial(A+B-f)}{\partial r} + \frac{1}{r^2}\frac{\cos\theta}{\sin\theta}\frac{\partial(A+B-f)}{\partial\theta}\right] -$$

$$- \sin\varphi\cos\varphi\left[\frac{1}{r}\frac{\cos\theta}{\sin\theta}\frac{\partial^2(A-B)}{\partial r\partial\varphi} + \frac{1}{r^2}\frac{\partial^2(A-B)}{\varsigma\partial\partial\varphi} + \frac{1}{r^2}\frac{\cos\theta}{\sin\theta}\frac{\partial(A-B)}{\partial\varphi}\right] -$$

$$- 2\frac{\cos^2\varphi - \sin^2\varphi}{r^2}\frac{\cos\theta}{\sin\theta}(A-B)$$

$$\sigma_{r\varphi}^{R} = -\sin\theta\left[\frac{1}{r}\frac{\partial^2 C}{\partial r\partial\varphi} + \frac{1}{r^2}\frac{\cos\theta}{\sin\theta}\frac{\partial^2 C'}{\partial\theta\partial\varphi} - \frac{1}{r^2\sin^2\theta}\frac{\partial C'}{\partial\varphi}\right] -$$

$$- \frac{\cos\theta}{r}\left[\frac{\cos\theta}{\sin\theta}\frac{\partial^2(A+B-f)}{\partial r\partial\varphi} - \frac{1}{r}\frac{\partial^2(A+B-f)}{\partial\theta\partial\varphi}\right] + \sin\varphi\cos\varphi\left[\frac{\sin\theta}{r^2}\frac{\partial^2(A-B)}{\partial\theta^2} -\right.$$

$$\left.- \frac{\cos\theta}{r}\frac{\partial^2(A-B)}{\partial r\partial\theta} + \frac{1}{r}\left(\sin^2\theta - 2\frac{\cos^2\theta}{\sin\theta}\right)\frac{\partial(A-B)}{\partial r} + 3\frac{\cos\theta}{r^2}\frac{\partial(A-B)}{\partial\theta}\right],$$

$$\sigma_{\theta\varphi}^{R} = -\cos\theta\left[\frac{1}{r}\frac{\partial^2 C}{\partial r\partial\varphi} + \frac{1}{r^2}\frac{\cos\theta}{\sin\theta}\frac{\partial^2 C'}{\partial\theta\partial\varphi} - \frac{1}{r^2\sin^2\theta}\frac{\partial C'}{\partial\varphi}\right] +$$

$$+ \frac{\sin\theta}{r}\left[\frac{\cos\theta}{\sin\theta}\frac{\partial^2(A+B-f)}{\partial r\partial\varphi} - \frac{1}{r}\frac{\partial^2(A+B-f)}{\partial\theta\partial\varphi}\right] + \sin\varphi\cos\varphi\left[\cos\theta\frac{\partial^2(A-B)}{\partial r^2} -\right.$$

$$\left.- \frac{\sin\theta}{r}\frac{\partial^2(A-B)}{\partial r\partial\theta} + 2\frac{\cos\theta}{r}\frac{\partial(A-B)}{\partial r} - \frac{\sin\theta}{r^2}\frac{\partial(A-B)}{\partial\theta}\right],$$

$$f = A\sin^2\varphi + C'\cos^2\varphi.$$

Proposition 16. For the primary stress state with axial symmetry $\sigma^P = \sigma_h\mathbf{1} + (\sigma_v - \sigma_h)\iota_3 \otimes \iota_3$, the Maxwell's representation becomes

$$\sigma_{rr}^R = \sin^2\theta\left(\frac{1}{r}\frac{\partial C}{\partial r} + \frac{1}{r^2}\frac{\cos\theta}{\sin\theta}\frac{\partial C}{\partial\theta}\right) + \frac{1}{r^2}\frac{\partial^2 A}{\partial\theta^2} + \frac{1}{r}\frac{\partial A}{\partial r} + \cos^2\theta\left(\frac{1}{r}\frac{\partial A}{\partial r} + \frac{1}{r^2}\frac{\cos\theta}{\sin\theta}\frac{\partial A}{\partial\theta}\right),$$

$$\sigma_{\theta\theta}^R = \cos^2\theta\left(\frac{1}{r}\frac{\partial C}{\partial r} + \frac{1}{r^2}\frac{\cos\theta}{\sin\theta}\frac{\partial C}{\partial\theta}\right) + \frac{\partial^2 A}{\partial r^2} + \sin^2\theta\left(\frac{1}{r}\frac{\partial A}{\partial r} + \frac{1}{r^2}\frac{\cos\theta}{\sin\theta}\frac{\partial A}{\partial\theta}\right),$$

$$\sigma_{\varphi\varphi}^R = \sin\theta\frac{\partial}{\partial r}\left[\sin\theta\frac{\partial C}{\partial r} + \frac{\cos\theta}{r}\frac{\partial C}{\partial\theta}\right] + \frac{\cos\theta}{r}\frac{\partial}{\partial\theta}\left[\sin\theta\frac{\partial C}{\partial r} + \frac{\cos\theta}{r}\frac{\partial C}{\partial\theta}\right] +$$

$$+ \cos\theta\frac{\partial}{\partial r}\left[\cos\theta\frac{\partial A}{\partial r} - \frac{\sin\theta}{r}\frac{\partial A}{\partial\theta}\right] - \frac{\sin\theta}{r}\frac{\partial}{\partial\theta}\left[\cos\theta\frac{\partial A}{\partial r} - \frac{\sin\theta}{r}\frac{\partial A}{\partial\theta}\right],$$

$$\sigma_{r\theta}^R = \sin\theta\cos\theta\left(\frac{1}{r}\frac{\partial C}{\partial r} + \frac{1}{r^2}\frac{\sin\theta}{\cos\theta}\frac{\partial C}{\partial\theta}\right) - \frac{1}{r}\frac{\partial^2 A}{\partial r\partial\theta} + \frac{1}{r^2}\frac{\partial A}{\partial\theta} -$$

$$- \sin\theta\cos\theta\left[\frac{1}{r}\frac{\partial A}{\partial r} + \frac{1}{r^2}\frac{\cos\theta}{\sin\theta}\frac{\partial A}{\partial\theta}\right],$$

$$\sigma_{r\varphi}^R = \sigma_{\theta\varphi}^R = 0.$$

Proposition 17. The function

$$F(r,\theta,t) = F_0(r,t) + F_{c2}(r,t)\cos 2\theta \tag{17}$$

is biharmonical if and only if

$$F_0(r,t) = \frac{a_{01}(t)}{r} + a_{02}(t) + a_{03}(t)r + a_{04}(t)r^2 + \frac{7}{15}\left[\frac{a_{21}(t)}{r^3} + a_{04}(t)r^4\right],$$

$$F_{c2}(r,t) = \frac{a_{21}(t)}{r^3} + \frac{a_{22}(t)}{r} + a_{23}(t)r^2 + a_{24}(t)r^4. \tag{18}$$

Because the considered domain is a multyconnected one, the supplementary compatibility conditions, on curves $r = r_0, \theta = \theta_0$ (parallel circles on the sphere of radius r_0) or $\alpha = \alpha_0, \beta = \beta_0$ (parallel circles on surface $\alpha = \alpha_0$) must be taken into account the supplementary compatibility conditions

$$\int_0^{2\pi}\left\{(1+\tilde{v})\left(\tilde{\sigma}_{11}^R\frac{\partial x^1}{\partial\varphi} + \tilde{\sigma}_{12}^R\frac{\partial x^2}{\partial\varphi}\right) - \tilde{v}\tilde{\sigma}_{mm}^R\frac{\partial x^1}{\partial\varphi} - \right.$$

$$\left. - (x^2 - x_I^2)\left[(1+\tilde{v})\left((\tilde{\sigma}_{11,2}^R - \tilde{\sigma}_{12,1}^R)\frac{\partial x^1}{\partial\varphi} - (\tilde{\sigma}_{22,1}^R - \tilde{\sigma}_{21,2}^R)\frac{\partial x^2}{\partial\varphi}\right) - \tilde{v}\left(\tilde{\sigma}_{mm,2}^R\frac{\partial x^1}{\partial\varphi} - \tilde{\sigma}_{mm,1}^R\frac{\partial x^2}{\partial\varphi}\right)\right]\right\}d\varphi = 0,$$

$$\int_0^{2\pi}\left\{(1+\tilde{v})\left(\tilde{\sigma}_{21}^R\frac{\partial x^1}{\partial\varphi} + \tilde{\sigma}_{22}^R\frac{\partial x^2}{\partial\varphi}\right) - \tilde{v}\tilde{\sigma}_{mm}^R\frac{\partial x^2}{\partial\varphi} + \right.$$

$$\left. + (x^1 - x_I^1)\left[(1+\tilde{v})\left((\tilde{\sigma}_{11,2}^R - \tilde{\sigma}_{12,1}^R)\frac{\partial x^1}{\partial\varphi} - (\tilde{\sigma}_{22,1}^R - \tilde{\sigma}_{21,2}^R)\frac{\partial x^2}{\partial\varphi}\right) - \tilde{v}\left(\tilde{\sigma}_{mm,2}^R\frac{\partial x^1}{\partial\varphi} - \tilde{\sigma}_{mm,1}^R\frac{\partial x^2}{\partial\varphi}\right)\right]\right\}d\varphi = 0,$$

$$\int_0^{2\pi}\left\{(1+\tilde{v})\left(\tilde{\sigma}_{31}^R\frac{\partial x^1}{\partial\varphi} + \tilde{\sigma}_{32}^R\frac{\partial x^2}{\partial\varphi}\right) - \right.$$

$$- (x^1 - x_I^1)\left[(1+\tilde{v})\left((\tilde{\sigma}_{13,1}^R - \tilde{\sigma}_{11,3}^R)\frac{\partial x^1}{\partial\varphi} + (\tilde{\sigma}_{23,1}^R - \tilde{\sigma}_{21,3}^R)\frac{\partial x^2}{\partial\varphi}\right) + \tilde{v}\tilde{\sigma}_{mm,3}^R\frac{\partial x^1}{\partial\varphi}\right] -$$

$$\left. - (x^2 - x_I^2)\left[(1+\tilde{v})\left((\tilde{\sigma}_{13,2}^R - \tilde{\sigma}_{12,3}^R)\frac{\partial x^1}{\partial\varphi} + (\tilde{\sigma}_{23,2}^R - \tilde{\sigma}_{22,3}^R)\frac{\partial x^2}{\partial\varphi}\right) + \tilde{v}\tilde{\sigma}_{mm,3}^R\frac{\partial x^2}{\partial\varphi}\right]\right\}d\varphi = 0.$$

Proposition 18. From the supplementary compatibility conditions, for any boundary conditions, in the case of axial symmetry, it results that among the coefficients of the potentials $A = B$ and C of the form (18)-(19) must exist the following relations

$$a_{21} = c_{21} = 0, \quad a_{22} = c_{22}, \quad \tilde{a}_{01} - \tilde{c}_{01} = 4(1 - \tilde{v})\tilde{a}_{22}. \tag{19}$$

Proposition 19. The boundary conditions $(9)_1$ determine the coefficients

$$a_{24} = c_{24} = 0, \quad a_{04} = a_{23}, c_{04} = c_{23} - 2a_{23}. \tag{20}$$

Proposition 20. The boundary conditions (8) determine the coefficients

$$\frac{\tilde{a}_{21}}{a^5} = \frac{3}{4}\frac{1}{7 - 5\tilde{v}}\frac{\sigma_h - \sigma_v}{\lambda}, \quad \frac{\tilde{a}_{22}}{a^3} = -\frac{5}{4}\frac{1}{7 - 5\tilde{v}}\frac{\sigma_h - \sigma_v}{\lambda},$$

$$2\frac{\tilde{a}_{01}}{a^3} = -\left(\tilde{p}_o - \frac{\sigma_h + \sigma_v}{2\lambda}\right) + \frac{6 + 5\tilde{v}}{7 - 5\tilde{v}}\frac{\sigma_h - \sigma_v}{2\lambda}, \tag{21}$$

$$2\frac{\tilde{a}_{02}}{a^3} = -\left(\tilde{p}_o - \frac{\sigma_h + \sigma_v}{2\lambda}\right) + \frac{26 - 15\tilde{v}}{7 - 5\tilde{v}}\frac{\sigma_h - \sigma_v}{2\lambda}.$$

Proposition 21. The relative stress state is

$$\tilde{\sigma}_{rr}^R = \tilde{p}_o\frac{a^3}{r^3} - \frac{1}{\lambda}\left(5\frac{5 - \tilde{v}}{7 - 5\tilde{v}}\frac{a^3}{r^3} - \frac{18}{7 - 5\tilde{v}}\frac{a^5}{r^5}\right)\left(\frac{\sigma_h + \sigma_v}{2} - \frac{\sigma_h - \sigma_v}{2}\cos 2\theta\right) +$$

$$+ \frac{1}{\lambda}\frac{6}{7 - 5\tilde{v}}\left(\frac{a^3}{r^3} - \frac{a^5}{r^5}\right)(2\sigma_h + \sigma_v),$$

$$\tilde{\sigma}_{\theta\theta}^R = -\frac{1}{2}\tilde{p}_o\frac{a^3}{r^3} - \frac{1}{\lambda}\left(5\frac{1 - 2\tilde{v}}{7 - 5\tilde{v}}\frac{a^3}{r^3} + \frac{3}{7 - 5\tilde{v}}\frac{a^5}{r^5}\right)\left(\frac{\sigma_h + \sigma_v}{2} + \frac{\sigma_h - \sigma_v}{2}\cos 2\theta\right) +$$

$$+ \frac{15}{2\lambda}\left(\frac{1 - 2\tilde{v}}{7 - 5\tilde{v}}\frac{a^3}{r^3} - \frac{1}{7 - 5\tilde{v}}\frac{a^5}{r^5}\right)\left(\frac{\sigma_h + \sigma_v}{2} - \frac{\sigma_h - \sigma_v}{2}\cos 2\theta\right)$$

$$+ \frac{1}{\lambda}\frac{6}{7 - 5\tilde{v}}\left(\frac{a^3}{r^3} - \frac{a^5}{r^5}\right)(2\sigma_h + \sigma_v), \tag{22}$$

$$\tilde{\sigma}_{\varphi\varphi}^R = -\frac{1}{2}\tilde{p}_o\frac{a^3}{r^3} + \frac{1}{\lambda}\left(5\frac{1 - 2\tilde{v}}{7 - 5\tilde{v}}\frac{a^3}{r^3} + \frac{3}{7 - 5\tilde{v}}\frac{a^5}{r^5}\right)\sigma_h +$$

$$+ \frac{15}{2\lambda}\left(\frac{1 - 2\tilde{v}}{7 - 5\tilde{v}}\frac{a^3}{r^3} - \frac{1}{7 - 5\tilde{v}}\frac{a^5}{r^5}\right)\left(\frac{\sigma_h + \sigma_v}{2} - \frac{\sigma_h - \sigma_v}{2}\cos 2\theta\right) -$$

$$- \left(2\frac{3 - 5\tilde{v}}{7 - 5\tilde{v}}\frac{a^3}{r^3} - \frac{3}{7 - 5\tilde{v}}\frac{a^5}{r^5}\right)(2\sigma_h + \sigma_v)$$

$$\tilde{\sigma}_{r\theta}^R = \frac{1}{\lambda}\left(5\frac{1 + \tilde{v}}{7 - 5\tilde{v}}\frac{a^3}{r^3} - \frac{12}{7 - 5\tilde{v}}\frac{a^5}{r^5}\right)\frac{\sigma_h - \sigma_v}{2}\sin 2\theta,$$

$$\tilde{\sigma}_{r\varphi}^R = \tilde{\sigma}_{\theta\varphi}^R = 0.$$

Proposition 22. The displacement field resulting from the geometrical equations and the conditions $(9)_2$

$$\tilde{u}_r = \frac{1 + \tilde{v}}{\tilde{E}}\frac{a^3}{r^2}\left\{\frac{1}{\lambda}\left[\frac{5}{2}\frac{5 - 4\tilde{v}}{7 - 5\tilde{v}} - \frac{9}{2}\frac{1}{7 - 5\tilde{v}}\frac{a^2}{r^2}\right]\left(\frac{\sigma_h + \sigma_v}{2} - \frac{\sigma_h - \sigma_v}{2}\cos 2\theta\right) - \right.$$

$$\left. - \frac{1}{2\lambda}\left[\frac{6 - 5\tilde{v}}{7 - 5\tilde{v}} - \frac{3}{7 - 5\tilde{v}}\frac{a^2}{r^2}\right](2\sigma_h + \sigma_v) - \frac{1}{2}\tilde{p}_o\right\}, \tag{23}$$

$$\tilde{u}_\theta = \frac{1 + \tilde{v}}{\lambda\tilde{E}}\frac{a^3}{r^2}\left[5\frac{1 - 2\tilde{v}}{7 - 5\tilde{v}} + \frac{3}{7 - 5\tilde{v}}\frac{a^2}{r^2}\right]\frac{\sigma_h - \sigma_v}{2}\sin 2\theta, , \tilde{u}_\varphi = 0,$$

Proposition 23. The boundary of the axial section is at every moment an ellipse having the axes

$$a \sin \theta \left\{ 1 + \frac{1 + \tilde{\nu}}{2\tilde{E}} \left[-\tilde{p}_o + \frac{2\sigma_h + \sigma_v}{3\lambda} + 4 \frac{4 - 5\tilde{\nu}}{7 - 5\nu} \frac{\sigma_h - \sigma_v}{3\lambda} \right] \right\},$$

$$a \sin \theta \left\{ 1 + \frac{1 + \tilde{\nu}}{2\tilde{E}} \left[-\tilde{p}_o + \frac{2\sigma_h + \sigma_v}{3\lambda} - 8 \frac{4 - 5\tilde{\nu}}{7 - 5\nu} \frac{\sigma_h - \sigma_v}{3\lambda} \right] \right\}.$$

(24)

5. ANALYSIS OF THE ROCK SALT BEHAVIOR

The behavior of the rock salt from Slanic-Prahova, characterized by the material constants

$G = 2950$ MPa, $K = 6100$ MPa, $G_o = 930$ MPa, $K_o = 1900$ MPa, $k = k_v = 8 \cdot 10^{-6}$ sec^{-1},

for axial symmetric primary stress state, is illustrated through some diagrams.

In Figure 1 it is presented the variation of the nonzero displacements in an axial section, for $r \in [a, 4a]$, for the ratio $m = \frac{\sigma_h}{\sigma_v} = 0.5$, at $t = 0$ and for $t \to \infty$ (relaxation state). In Figure 2, in the same conditions as before, it is presented the variation of the displacements on the boundary. It must be noticed the concentration for the radial displacement at the top and the bottom of the spherical cavern respectively for $\theta = \frac{\pi}{4}$ and $\theta = \frac{3\pi}{4}$. This concentration at $t = 0$ for some values of the ratio m ($m = 0.25$, 0.5, 0.75, 1, 1.5) is illustrated in Figure 3.

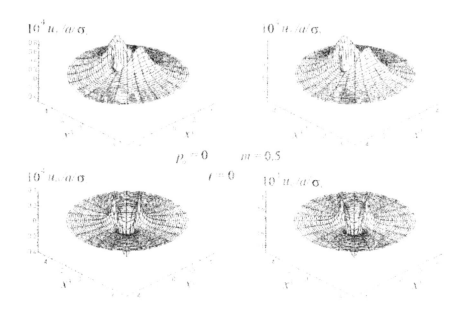

Figure 1. Displacement field around the sperical cavern for $m = 0.5$,
for instantaneous response and in the relaxation state

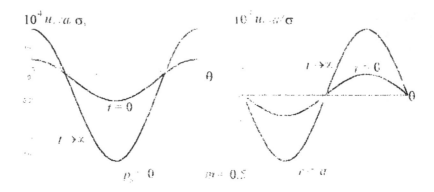

Figure 2. Concentration of displacements on the boundary of the spherical cavern
for $m = 0.5$, for instantaneous response and in the relaxation state

As for the stress state, his variation around the spherical cavern in the same
conditions as for the displacement field ($m = 0.5$), through instantaneous response is
presented in Figure 4. For the used material constants there does not exist a significant
variation in time. If for cylindrical cavities, for null pressure, the stresses in the plane of

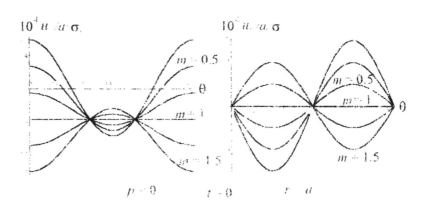

Figure 3. Boundary displacement concentration depending on the ratio m

the cross section were constant in time, for the spherical cavity all the components of the
stress tensor are depending of time even if not very strong. In Figure 5 is considered the
boundary concentration of nonconstant stresses with respect to the ratio m ($m = 0.25$,
0.5, 0.75, 1, 1.5). For small values of m, in the top and the bottom of the cavern may
come out stretching stresses. For greater values of m may come out important
compression stresses.

6. CONCLUSIONS

This way of approach may be extend for closed cavities with axial symmetry,
using a Neuber's displacement representation through four/three harmonical potentials

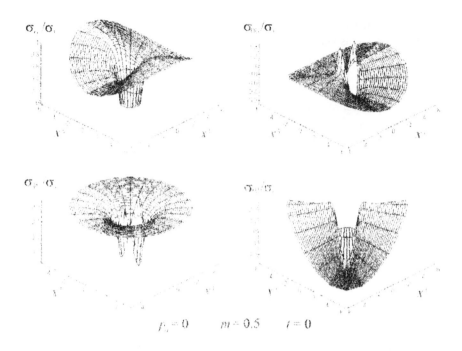

i·igure 4. Stress state around the spherical cavern for $m = 0.5$, by instantaneous response

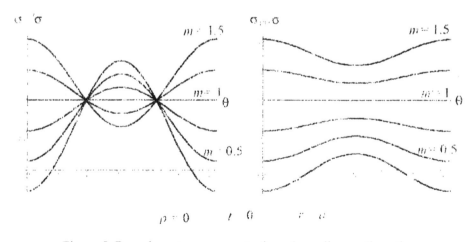

Figure 5. Boundary stress concentrations depending on the ratio m

or a Maxwell's stress representation through two/three biharmonical potentials. As well as for the spherical coordinates, for generalized spatial Greenspan coordinates, the form of the potentials are suggested by the boundary conditions. The peculiar form of the Neuber's displacement representation or Maxwell's stress representation is valid for any considered cavity, but the form of the potentials are very dependent of the used curvilinear coordinates.

The same problem may be tackled considering a Morera's representation for the relative stresses, through three harmonical potentials; the same observation as before

221

may be done for these potentials: the Morera's representation is a general one, but the form of the potentials is very dependent of the used curvilinear coordinates.

The complete determination of the potentials for displacement representation is obtained using the boundary conditions (pressure on the cavity surface, the decrease to null of the relative stresses with the distance); the complete determination of the potentials for stress representations is obtained using the boundary conditions but also supplementary compatibility conditions for multiple connected domains. Further on, the determination of the displacement field is done using a harmonical displacement potential and the corresponding diminution of the displacements with the distance.

The linear viscoelastic model may be used in several steps, changing the material constants (considering the material constants as function of the reference configuration).

The obtained results may be used as a first step in the approach of the problem for more complex constitutive models (instantaneous response) and give a comparison term for results obtained only through numerical methods.

REFERENCES

Brethauer, G.E., [1974], "Stress Around Pressurized Spherical Cavities in triaxial Stress Fields", Int.J.Rock Mech.Min.Sci.&Geomech.Abstr., Vol.11, 91-96.

Cristescu, N.D., [1989], "Rock Rheology", Kluwer Academic Publ., Dordrecht.

Massier, D., [1995a], "In situ Creep of a Linear Viscoelastic Rock around Some Noncircular Tunnels. Theoretical Solution", Rev.Roum.Sci.Tech., Ser.Mec.Appl., Tome 40, 2-3, 385-395.

Massier, D., [1995b], "In situ creep of a linear viscoelastic rock around some noncircular tunnels. Boundary stress and displacement concentrations", Rev.Roum.Sci.Tech., Ser.Mec.Appl., Tome 40, 4-5-6. 485-502.

Massier, D., [1997a], "The Behaviour of a Linear Viscoelastic Rock around a Horizontal Tunnel with Vertical Axis of Symmetry", Rev.Roum.Sci.Tech., Ser.Mec.Appl., Tome 42, 5-6, 411-420.

Massier, D., [1997b], "Rock Mechanics", Bucharest University Press.

Neuber, H., [1934], "Ein neuer Ansatz zur Lösung Räumlicher Problem der Elastizitätstheorie. Der Hohlkugel unter Einzellast als Beispiel", ZAMM, 14, 4.

Solomon, L., [1968], "Elasticité lineaire", Masson. Paris.

Basic and Applied Salt Mechanics, Cristescu, Hardy, Jr & Simionescu (eds)
© 2002 Swets & Zeitlinger, Lisse, ISBN 90 5809 383 2

Non-elastic instantaneous response in rock rheology

Sanda Cleja-Tigoiu
Faculty of Mathematics, University of Bucharest, Bucharest, Romania

Ervin Medves
S.C. IPROMIN S.A., Bucharest, Romania

ABSTRACT

In the paper two equivalent constitutive equations for rock salt (of creep and relaxation type) with loading dependent instantaneous response are derived, for experiments carried out either with stress control or with the deformation control. Our new constitutive equations describe the behavior of rock type materials relative to a configuration which was previously deformed under possible unknown large deformations, a typical feature for rocks. The rock salt compressibility and dilatancy are also described by the model. To high loading rate the rock behavior is quite different from creep (or relaxation) tests and the irreversible deformations appear in our model even for small stresses. From the propagation equation we emphasize the existence of the elastic waves as well as the non-elastic waves. The stability of the elasto-plastic process and the uniqueness of the solution of the variational inequality related to the quasi-static boundary problem are also proved.

INTRODUCTION

In the paper the constitutive equations for rock salt are proposed, based on experimental data from Cristescu [1989], Cristescu and Hunsche [1998], as well as on the laboratory test performed with rock salt specimen from Slanic Prahova Salt Mines, and on the theoretical considerations relative to the existing constitutive models for rock like materials. We remark that plasticity like constitutive equations (see for instance (7.3.1.) in Cristescu [1989], or our relations (17), (18) written for $F \equiv 0$) does not describe the creep, since if the stresses are kept constant during a certain time-interval the rate of strain becomes zero. On the other hand they are rate independent and consequently only the stress path in the stress space (but not the loading rate in running this curve) defines the material response. If the constitutive equations are reduced to elasto-viscoplastic ones (see for instance (8.4.1) in the same book, or our relations (17), (18) written for $g \equiv 0$), for high loading rate the rock behavior is purely elastic in loading and unloading process. More over in the last case the acceleration waves are propagating through the rock with the elastic velocities only. These features contradict the experimental evidences (see also Figures 1 - 3 from our paper), which emphasize the complex coupling between elastic, plastic and viscoplastic properties, as well as a strong influence of the loading rate on the stress and strain state developed in rock salt. The experimental conclusions are used in the formulation of the mathematical models for rock salt.

We fit together elasto-plastic and elasto-viscoplastic models (see chapters 7 and 8 in Cristescu [1989]) with the constitutive equations introduced in Cleja-Ţigoiu [1991], to describe the behavior of rock type materials relative to a configuration which was previously deformed under possible unknown large deformations, a typical feature for rocks. We also pay attention to the initial conditions (different in situ or in the laboratory test) that are to be added to the differential equations governing the material response in the rock salt. In our approach to rock type materials, like rock salt, the behavior of the material will be described at each material particle by rate constitutive equation with loading dependent instantaneous response.

Further we shall use the following notations: $Sym-$ the set of all second order symmetric tensor; $\mathbf{I}$ - the identity tensor ; $\mathbf{A} \cdot \mathbf{B} = \text{tr}(\mathbf{AB}^T)$ - the inner product on tensor space; $\mathbf{a} \otimes \mathbf{b}$ the tensor product of the vectors $\mathbf{a}, \mathbf{b}$; $\dot{h}$ the material time derivative of the function h; $< z >$ the positive part of the real number $z \in \mathbf{R}$; $\mathcal{H}-$ Heaviside function, i.e. $\mathcal{H}(z) = 0$ for $z < 0$ and $\mathcal{H}(z) = 1$ for $z \geq 0$.

CONSTITUTIVE ASSUMPTIONS

We assume that:

H1. The material undergoes the small relative deformations with respect to a fixed configuration, say at time t_0, obtained by a certain, possible finite deformation process. We denote by $\mathbf{u}-$ the relative displacement field, and by

$$\epsilon = \frac{1}{2}(\nabla \mathbf{u} + \nabla^T \mathbf{u}) \equiv \epsilon(\mathbf{u}) \tag{1}$$

the small relative strain tensor and by $\mathbf{T}$ the secondary stress tensor.

H2. The rate of small strain is additively decomposed into its symmetric components

$$\dot{\epsilon} = \dot{\epsilon}^e + \dot{\epsilon}^i + \dot{\epsilon}^{vp} \tag{2}$$

ϵ^e- the elastic deformation tensor, $\epsilon^{vp}-$ the viscoplastic deformation tensor, ϵ^i- the irreversible instantaneous deformation tensor will be defined below.

H3. The rate of the reversible, elastic deformation is related to the rate of Cauchy stress tensor, $\dot{\mathbf{T}}$, via the elastic type constitutive equation

$$\dot{\mathbf{T}} = \mathcal{E} \, \dot{\epsilon}^e \tag{3}$$

Here $\mathcal{E}-$ the fourth order tensor is symmetric and positive definite, i.e.

$$\mathcal{E} : Sym \longrightarrow Sym \quad \text{linear}, \quad \mathcal{E} \, \mathbf{A} \cdot \mathbf{B} = \mathcal{E} \, \mathbf{B} \cdot \mathbf{A} \quad \forall \, \mathbf{A}, \mathbf{B} \in Sym$$
$$\mathcal{E} \, \mathbf{A} \cdot \mathbf{A} \geq 0 \quad \forall \, \mathbf{A} \in Sym \quad \text{and} \, \mathcal{E} \, \mathbf{A} \cdot \mathbf{A} = 0 \Longleftrightarrow \mathbf{A} = 0 \tag{4}$$

Hence the inverse $\mathcal{E}^{-1}$ and the fourh order tensor $\mathcal{E}^{1/2}$ defined by

$$\mathcal{E}^{1/2} \, \mathbf{A} \cdot \mathcal{E}^{1/2} \mathbf{B} = \mathcal{E} \, \mathbf{A} \cdot \mathbf{B} \quad \forall \, \mathbf{A}, \mathbf{B} \in Sym \tag{5}$$

224

exist and have the same properties.

Two kinds of irreversible deformations are involved in our approach to rock salt, one occuring in the instantaneous response, ϵ^i, and the other one, ϵ^{vp} — the viscoplastic deformation, describing the creep or (relaxation) behavior.

H4. There are two internal variables, χ and w, which represent the irreversible stress power per unit volume at a given time t. They are related to the irreversible deformations by the following relationships:

$$\dot{\chi} = \mathbf{T} \cdot \dot{\epsilon}^i, \quad \dot{w} = \mathbf{T} \cdot \dot{\epsilon}^{vp} \tag{6}$$

H5. The irreversible instantaneous deformation is described via the (instantaneous) plastic potential (called also the yield function) $g(\cdot,\cdot) : \mathcal{D}_g \in Sym \times \mathbf{R} \longrightarrow \mathbf{R}_{\leq 0}$, i.e. $g(\mathbf{T},\chi) \leq 0$, by

$$\dot{\epsilon}^i = \lambda \, \partial_\mathbf{T} \, g(\mathbf{T},\chi), \quad \lambda \geq 0 \tag{7}$$

$$\lambda \, g = 0, \quad \lambda \, \dot{g} = 0 \quad \text{consistency condition} \tag{8}$$

Proposition 1. $\lambda \geq 0$ can be expressed either by the inequality

$$(\mu - \lambda) \, \dot{g} \leq 0, \quad \text{for all } \mu \geq 0, \quad \text{together with} \quad \lambda \, g = 0, \tag{9}$$

or under its explicit dependence on the rate of stress:

$$\lambda = \frac{1}{h_c} < \partial_\mathbf{T} \, g(\mathbf{T},\chi) \cdot \dot{\mathbf{T}} > \quad \text{on} \quad g(\mathbf{T},\chi) = 0, \tag{10}$$

where h_c — the hardening (creep) parameter is given by

$$h_c = -g'_\chi(\mathbf{T},\chi)(\partial_\mathbf{T} \, g(\mathbf{T},\chi) \cdot \mathbf{T}) > 0. \tag{11}$$

H6. The viscoplastic response is defined by the function H (describing the creep behavior, put into evidence when $\dot{\mathbf{T}} = 0$ on a certain time interval), together with the viscoplastic potential F,

$$\dot{\epsilon}^{vp} = k < 1 - \frac{w}{H(\mathbf{T})} > \partial_\mathbf{T} \, F(\mathbf{T}) \mathcal{H}(-g), \quad k > 0 \tag{12}$$

k is the viscosity coefficient. The condition $H(\mathbf{T}) = w$ characterizes the stress state laying on the stabilization boundary. Hence $\dot{\epsilon}^{vp} = 0$ when the stress is in such a way that $g(\mathbf{T},\chi) > 0$, i.e. the viscoplastic strain can be developed only inside of current yield domain: $\{\mathbf{T} \mid g(\mathbf{T},\chi) \leq 0\}$.

H7. For isotropic rock type material, all constitutive functions are isotropic with respect to their arguments, χ and w being scalar invariants.

For instance when $g(\mathbf{T},\chi) = \bar{g}(\mathbf{T}) - \chi$ it results

$$g(\mathbf{T},\chi) = \bar{g}(\sigma,\tau,\Delta) - \chi, \quad \text{where } \sigma := \frac{1}{3}\text{tr } \mathbf{T} , \Delta := \det \mathbf{T} , \tau := (\frac{1}{3}\mathbf{T}' \cdot \mathbf{T}')^{1/2}, \tag{13}$$

are stress invariants, and $\mathbf{T}' := \mathbf{T} - \sigma \mathbf{I}$ — the stress deviator. When we neglect the influence of the third invariant, the constitutive functions are defined inside of the stress domain characterized by

$$\sigma > 0 \quad \text{and} \quad \frac{1}{2}[\text{tr}\,(\mathbf{T}')^2 - (\text{tr}\,\mathbf{T})^2] \equiv \frac{3}{2}(\tau^2 - 3\,\sigma^2) \tag{14}$$

H8. The "initial conditions" have to be added in formulating quasi- static boundary value problem for rock like material, or at the different stages of the loading or of the deformation precess (in stress or strain controlled tests), in order to obtain a mathematically coherent and physically motivated description.

For instance: the undisturbed *in situ* rock is already under a certain σ^p, primary stress, in the irreversible deformed configuration (at time t_0) taken as a reference configuration:

$$\begin{aligned}
&\mathbf{T}(t_0) = \sigma^p, \quad \epsilon(t_0) = 0, \quad \chi(t_0) = \chi^p, \quad w(t_0) = w^p, \quad \text{with} \\
&\chi^p = \bar{g}(\sigma^p), \quad w^p = H(\sigma^p),
\end{aligned} \tag{15}$$

χ and w being secondary internal variables.

In the laboratory test we assume that the rock specimen is stress free, due to the unloading process (for instance after excavation), but again in an irreversible deformed configuration, with χ^p and w^p defined in the last relationship (15). Hence

$$\mathbf{T}(t_0) = 0, \quad \epsilon(t_0) = 0, \quad \chi(t_0) = \chi^p, \quad w(t_0) = w^p. \tag{16}$$

CONSTITUTIVE EQUATIONS OF CREEP AND RELAXATION TYPE

Two equivalent constitutive equations with loading dependent instantaneous response are derived, under the hypothesis that the hardening (creep) parameter defined in (11) is positive, i.e. $h_c > 0$.

As a direct consequences of the assumptions H2.- H6. it can be proved that: For the given stress the *constitutive equation of creep type* is defined by

$$\dot{\epsilon} = \mathcal{E}^{-1}\,\dot{\mathbf{T}} + \lambda\,\partial_{\mathbf{T}}\,g(\mathbf{T},\chi) + k \; < 1 - \frac{w}{H} > \; \partial_{\mathbf{T}}\,F(\mathbf{T})\mathcal{H}(-g),$$

$$\dot{\chi} = \lambda\,\partial_{\mathbf{T}}\,g(\mathbf{T},\chi)\cdot\mathbf{T}, \quad \text{with } \lambda = \frac{1}{h_c} < \partial_{\mathbf{T}}\,g(\mathbf{T},\chi)\cdot\dot{\mathbf{T}} > \quad \text{on } g(\mathbf{T},\chi) = 0, \tag{17}$$

$$\dot{w} = k \; < 1 - \frac{w}{H} > \; \partial_{\mathbf{T}}\,F(\mathbf{T})\cdot\mathbf{T}\,\mathcal{H}(-g),$$

For the given strain the *constitutive equation of relaxation type* is defined by

$$\dot{\mathbf{T}} = \mathcal{E}\,\dot{\epsilon} - \frac{\beta}{h_r}\mathcal{E}[\,\partial_{\mathbf{T}}\,g(\mathbf{T},\chi)] - k \; < 1 - \frac{w}{H} > \; \mathcal{E}[\,\partial_{\mathbf{T}}\,F(\mathbf{T})]\,\mathcal{H}(-g),$$

$$\dot{\chi} = \frac{\beta}{h_r}\,\partial_{\mathbf{T}}\,g(\mathbf{T},\chi)\cdot\mathbf{T}, \quad \text{with } \beta = < \mathcal{E}\partial_{\mathbf{T}}\,g(\mathbf{T},\chi)\cdot\dot{\epsilon} > \; \mathcal{H}(g), \tag{18}$$

$$h_r = \mathcal{E}\,\partial_{\mathbf{T}}\,g(\mathbf{T},\chi)\cdot\partial_{\mathbf{T}}\,g(\mathbf{T},\chi) + h_c,$$

and $\dot{w}$ given in the last relationship from (17).

Loading Dependent Instantaneous Response

Theorem 1. The behavior of rock salt subjected to *high loading rate* is emphasized by the loading dependent instantaneous response associated with (17), for a given stress process

$$\dot{\epsilon} = \mathcal{E}^{-1}\,\dot{\mathbf{T}} + \lambda\,\partial_{\mathbf{T}}\,g(\mathbf{T},\chi), \quad \dot{\chi} = \lambda\,\partial_{\mathbf{T}}\,g(\mathbf{T},\chi)\cdot\mathbf{T}, \quad \dot{w} = 0. \tag{19}$$

Proof. By the procedure proposed by Suliciu [1989] we change the time scale by $t = cs$, with $c > 0$, $c-$ a fixed dimentionless parameter, and we introduce the appropriate function corresponding to time rescalling:

$$\tilde{\mathbf{T}}(s) := \mathbf{T}(cs) \equiv \mathbf{T}(t), \quad \text{and} \quad \frac{d\mathbf{T}}{dt}(t) = \frac{1}{c}\frac{d}{ds}\tilde{\mathbf{T}}(s). \tag{20}$$

Taking into account (20) we find the constitutive representation (17) in terms of the new time variable s, under the form

$$\frac{1}{c}[\frac{d\tilde{\epsilon}}{ds} - \mathcal{E}^{-1}\frac{d\tilde{\mathbf{T}}}{ds} - \tilde{\lambda}\,\partial_{\mathbf{T}}\,g(\tilde{\mathbf{T}}(s),\tilde{\chi}(s))] = k \ <1 - \frac{\tilde{w}(s)}{H(\tilde{\mathbf{T}}(s))}> \partial_{\mathbf{T}}\,F(\tilde{\mathbf{T}}(s))\mathcal{H}(-g),$$

$$\frac{1}{c}[\frac{d\tilde{\chi}(s)}{ds} - \tilde{\lambda}\,\partial_{\mathbf{T}}\,g(\tilde{\mathbf{T}}(s),\tilde{\chi}(s))\cdot\tilde{\mathbf{T}}(s)] = 0, \tag{21}$$

$$\frac{1}{c}\frac{d\tilde{w}(s)}{ds} = k \ <1 - \frac{\tilde{w}(s)}{H(\tilde{\mathbf{T}}(s))}> \partial_{\mathbf{T}}\,F(\tilde{\mathbf{T}}(s))\cdot\tilde{\mathbf{T}}(s)\mathcal{H}(-g),$$

Multiplying by $c > 0$, when we pass c to zero we get just the relationships (19).

Remark 1. The *physical meaning* of such kind of rescalling with very small $c > 0$ is put into evidence in (20): $|\frac{d\mathbf{T}}{dt}(t)| \to \infty$ for $c \to 0$, under the hypothesis that $|\frac{d}{ds}\tilde{\mathbf{T}}(s)| < \infty$. Hence the rock is subjected to very high loading rate, or the change in stress is performed in a very short time interval.

An equivalent form of $(19)_1$ is given by

$$\dot{\mathbf{T}} = \mathcal{E}\,\dot{\epsilon} - \lambda\,\mathcal{E}\,\partial_{\mathbf{T}}\,g(\mathbf{T},\chi), \tag{22}$$

Using (22) in the consistency condition (8), for $\lambda > 0$, on the yield surface $g(\mathbf{T},\chi) = 0$, we get

$$0 = \dot{g}(\mathbf{T},\chi) \equiv g'_\chi(\mathbf{T},\chi)\dot{\chi} + \partial_{\mathbf{T}}\,g(\mathbf{T},\chi)\cdot\mathcal{E}\,\dot{\epsilon} - \lambda\,\partial_{\mathbf{T}}\,g(\mathbf{T},\chi)\cdot\mathcal{E}\,\partial_{\mathbf{T}}\,g(\mathbf{T},\chi) \tag{23}$$

We substitute in (23) the evolution of χ predicted by $(17)_2$ and the new hardening parameter defined in $(18)_4$. It follows

$$\lambda\,h_r = \partial_{\mathbf{T}}\,g(\mathbf{T},\chi)\cdot\mathcal{E}\,\dot{\epsilon} \equiv \beta, \quad \text{or} \quad \lambda = \frac{\beta}{h_r} \quad \text{and} \quad \lambda \geq 0 \quad \Longleftrightarrow \quad \beta \geq 0, \tag{24}$$

since when $h_c > 0$, due to the property H3. of $\mathcal{E}$ the hardening parameter h_r defined in $(18)_4$ is also positive. Consequently, the following assertion was proved

Theorem 2. The behavior of rock salt subjected to *high loading rate* can be equivalently described by the loading dependent instantaneous response for a given strain process under the form

$$\dot{\mathbf{T}} = \mathcal{E}\,\dot{\epsilon} - \frac{\beta}{h_r}\mathcal{E}[\,\partial_{\mathbf{T}}\,g(\mathbf{T},\chi)], \quad \dot{\chi} = \frac{\beta}{h_r}\,\partial_{\mathbf{T}}\,g(\mathbf{T},\chi)\cdot\mathbf{T}, \quad \dot{w} = 0 \qquad (25)$$

Remark 2. The constitutive equation (18) is derived from (17) by applying $\mathcal{E}$, and substituting the expression (24) for λ. The constitutive equation of the relaxation type is more general than the creep one, since h_r remains still positive even when $h_c \leq 0$, i.e. when the softening occurs in rock. The instantaneous response associated with (18) is just given by (25), as it can be proved using the procedure of the time-rescalling.

Two type of *unloading processes* can be put into evidence in our model:
a) The instantaneous unloading process occurs at a fixed time t, when $\mathbf{T}(t)$ satisfies $g(\mathbf{T},\chi) = 0$ and $\partial_{\mathbf{T}}\,g(\mathbf{T},\chi)\cdot\dot{\mathbf{T}}^+ < 0$.
b) The viscoplastic unloading related to creep behavior is defined at time t by the condition $H(\mathbf{T}(t)) < w(t)$.

BEHAVIOR OF ROCK SALT PREDICTED BY MODELS

We analyze the behavior of isotropic material, like rock salt, predicted by the model in confined and triaxial tests.

Hydrostatic Tests

We put into evidence the compressibility of the volume and the *initial conditions* associated to the problem in confined test.

When the hydrostatic stress $\mathbf{T} = \sigma\mathbf{I}$ is applied incrementally in short time from 0 to σ^H, the volume compressibility can be obtained from (19) rewritten under the form

$$\frac{d\epsilon_v}{d\sigma} = \frac{1}{K} + \frac{1}{\sigma}\bar{g}_{,1}(\sigma,0) \equiv \frac{1}{K(\sigma)}, \quad \frac{d\chi}{d\sigma} = \bar{g}_{,1}(\sigma,0), \quad \frac{dw}{d\sigma} = 0,$$

$$\epsilon_v(0) = 0, \quad \chi(0) = \chi^p, \quad w(0) = w^p. \qquad (26)$$

Here the function $\bar{g}$, introduced in (13) satisfies the restrictions

$$\lim_{\sigma\searrow0}\frac{\bar{g}_{,1}(\sigma,0)}{\sigma} = (\frac{1}{K_1} - \frac{1}{K}) > 0, \quad \bar{g}_{,1}(\sigma,0) > 0 \quad \forall\,0 < \sigma < \sigma_0$$

$$\bar{g}_{,1}(\sigma,0) = 0 \quad \forall\,\sigma > \sigma_0 \qquad (27)$$

σ_0 is the hydrostatic pressure that corresponds to the closure of pores and K is the bulk elastic modulus.

Let us denote by $\hat{\epsilon}_v(\sigma,0)$, $\hat{\chi}(\sigma,0) = \chi^p + \bar{g}(\sigma,0)$ and $w(\sigma) = w^p$ the solution of the problem (26).

The deformation by creep is described by (17) in which the stress is kept constant, $\mathbf{T} = \sigma^H\mathbf{I}$ during a certain interval $[t_0,t_1)$:

228

$$\dot{\epsilon}_v = k < 1 - \frac{w}{H(\sigma^H \mathbf{I})} > F_{,1}(\sigma^H \mathbf{I}), \quad \dot{w} = k < 1 - \frac{w}{H(\sigma^H \mathbf{I})} > \sigma^H F_{,1}(\sigma^H \mathbf{I}), \quad (28)$$

the *initial conditions* are chosen to be just values reached in the previous instantaneous hydrostatic response, i.e. $\epsilon_v(t_0) = \hat{\epsilon}_v(\sigma^H, 0)$ and $w(t_0) = w^p$.

Only under the viscoplastic loading condition $H(\sigma^H \mathbf{I}) > w^p$ the deformation by creep occurs in rock specimen and the stabilization boundary $H(\sigma^H \mathbf{I}) = w(t_*)$ is reached when $t_* \to \infty$.

Triaxial Tests

We consider the clasical (Kármán) triaxial tests which reveal the properties of rock specimen subjected to a deviatoric stress superposed on a hydrostatic one. First of all we analyze the material response for the stress characterized by

$$\mathbf{T} = \sigma_1(t)\mathbf{i}_1 \otimes \mathbf{i}_1 + \sigma_2(t)(\mathbf{i}_2 \otimes \mathbf{i}_2 + \mathbf{i}_3 \otimes \mathbf{i}_3) \quad \text{where}$$
$$\sigma_1(t) = \sigma_2(t) = \sigma(t) \quad t_0 \le t \le t_1, \quad \sigma_1(t_1) = \sigma_2(t_1) = \sigma(t_1) = \sigma^H, \quad (29)$$
$$\sigma_2(t) = \sigma^H, \dot{\sigma}_1 = const > 0, \quad t_1 \le t < t_2$$

On the interval $[t_0, t_1]$ the behavior of rock is predicted by the *hydrostatic test* and the final value at time t_1 are addopted as the *initial conditions* on the second stage of the process. In the test the stress invariants are calculated by

$$\sigma = \frac{1}{3}(\sigma_1 + 2\sigma^H) \equiv \frac{1}{3}\sigma_1^R + \sigma^H, \quad \tau = \frac{\sqrt{2}}{3}(\sigma_1 - \sigma^H) \equiv \frac{\sqrt{2}}{3}\sigma_1^R \quad (30)$$

i.e. the stress variation takes place along the line $\sqrt{2}\sigma - \tau = const = \sqrt{2}\sigma^H$.

Proposition 2. Under the hypothesis that the function $\bar{g}$ introduced in (13) satisfies the restriction

$$\lim_{\tau \searrow 0} \frac{\bar{g}_{,2}(\sigma,\tau)}{\sigma} = 0, \quad \forall \sigma > 0, \quad (31)$$

the axial strain and the volumetric strain up to time t_1 are predicted by the solution of the differential system

$$\frac{d\epsilon_{11}}{d\sigma_1} = (\frac{1}{9K} + \frac{1}{3G}) + \frac{[\bar{g}_{,1}(\sigma,\tau) + \sqrt{2}\bar{g}_{,2}(\sigma,\tau)]^2}{9[\sigma\bar{g}_{,1}(\sigma,\tau) + \tau\bar{g}_{,2}(\sigma,\tau)]} +$$

$$+\frac{k}{3\dot{\sigma}_1} < 1 - \frac{w}{\bar{H}(\sigma,\tau)} > [\bar{F}_{,1}(\sigma,\tau) + \sqrt{2}\bar{F}_{,2}(\sigma,\tau)]\mathcal{H}(-g),$$

$$\frac{d\epsilon_v}{d\sigma_1} = \frac{1}{3K} + \frac{\bar{g}_{,1}(\sigma,\tau) + \sqrt{2}\bar{g}_{,2}(\sigma,\tau)}{3[\sigma\bar{g}_{,1}(\sigma,\tau) + \tau\bar{g}_{,2}(\sigma,\tau)]}\bar{g}_{,1}(\sigma,\tau) +$$

$$\hspace{10cm} (32)$$

$$+\frac{k}{\dot{\sigma}_1} < 1 - \frac{w}{\bar{H}(\sigma,\tau)} > \bar{F}_{,1}(\sigma,\tau)\mathcal{H}(-g),$$

$$\frac{dw}{d\sigma_1} = \frac{k}{\dot{\sigma}_1} < 1 - \frac{w}{\bar{H}(\sigma,\tau)} > [\sigma\bar{F}_{,1}(\sigma,\tau) + \tau\bar{F}_{,2}(\sigma,\tau)]\mathcal{H}(-g),$$

$$\frac{d\chi}{d\sigma_1} = \frac{1}{3}\left[\bar{g}_{,1}(\sigma,\tau) + \sqrt{2}\bar{g}_{,2}(\sigma,\tau)\right] \iff \chi = \bar{g}(\sigma,\tau) + \text{const}$$

The last equality defines a prime integral of the differential system. Here σ and τ are related to σ_1 by (30).

On the considered stress process

the loading condition $\lambda > 0$ $\iff$ $\dot{\sigma}_1[\bar{g}_{,1}(\sigma,\tau) + \sqrt{2}\bar{g}_{,2}(\sigma,\tau)] > 0$;

the hardening (creep) condition $h_c > 0$ $\iff$ $\sigma\,\bar{g}_{,1}(\sigma,\tau) + \tau\bar{g}_{,2}(\sigma,\tau) > 0$.

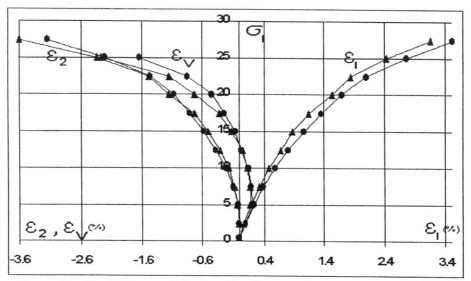

Figure 1 - The Influence of the Loading Rate on Stress- Strain Curves (ploted for $\dot{\sigma}_1 = 0.8$ MPa/min (o) and 2.5 MPa/min (Δ)) and on Dilatancy.

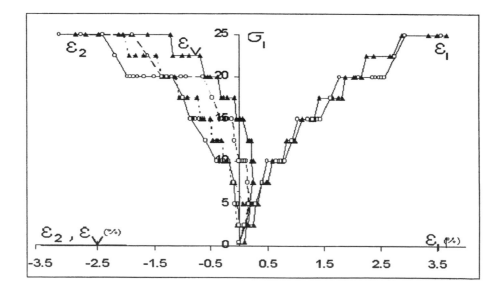

Figure 2 - Incremental Creep Curves for Rock Salt

230

In conclusion we put into evidence the influence of the loading rate on stress-strain curves, revealed by the experimental data, and the rock plasticity even for small stresses (see Figure 1). For high loading rate (i.e. for sufficiently large $\dot{\sigma}_1$) the influence of the last term from (32) disappeares.

Similarly we can study the influence of the loading rate on the material response, when succesive stress increments are applied and after each increase of the axial stress, the stresses are kept constant during a long or short time interval. Hence our model describes the experimental results, ploted in Figure 2, which are performed for short time creep (with stationary time interval 30 min (o) and 15 min (Δ)).

Compressibility/dilatancy boundary

A typical feature of rock is the volumetric behaviour: compressibility for small octahedral shear stress and dilatant for higher values of shear stress. The compressibility/ dilatancy boundary is obtained in classical triaxial tests performed with $\dot{\sigma}_1 =$ const, for various confining pressure $\sigma_2 =$ const, by finding for what stress state the rate of the irreversible volumetric strain is zero, i.e. $\dot{\epsilon}_v^i + \dot{\epsilon}_v^{vp} = 0$ in our model. Consequently, the compressibility/dilatancy boundary is defined from (32) by

$$\frac{\bar{g}_{,1}(\sigma, \tau) + \sqrt{2}\bar{g}_{,2}(\sigma, \tau)}{3[\sigma\bar{g}_{,1}(\sigma, \tau) + \tau\bar{g}_{,2}(\sigma, \tau)]}\bar{g}_{,1}(\sigma, \tau) = -\frac{k}{\dot{\sigma}_1} < 1 - \frac{w}{\bar{H}(\sigma, \tau)} > \bar{F}_{,1}(\sigma, \tau) \qquad (33)$$

Remark 3. 1. The threshold compressibility/dilatancy is dependent on the loading rate, that is in ageement with experiments (see Figure 1 and Figure 2).

2. On the other hand from the equation of the c/d boundary (33) we conclude
$\bar{g}_{,1}(\sigma, \tau) = 0$ defines the c/d boundary for hight loading rate;
$\bar{F}_{,1}(\sigma, \tau) = 0$ defines the c/d boundary during the creep;
$\bar{g}_{,1}(\sigma, \tau) \equiv \bar{F}_{,1}(\sigma, \tau) = 0$ defines the c/d boundary independent on loading rate.

The <u>irreversible work</u> along a loading process in which at time t_k the stress reaches the value $\mathbf{T}_k$, can be calculated from

$$W^{ir}(t_k) = \int_0^{t_k} \mathbf{T} \cdot \dot{\epsilon}^{vp} dt + \int_0^{t_k} \mathbf{T} \cdot \dot{\epsilon}^i dt = w(t_k) + \chi(t_k) \quad \text{and}$$

$$W^{ir}(t_k) = \int_0^{t_k} \mathbf{T} \cdot \dot{\epsilon} dt - \int_0^{t_k} \mathbf{T} \cdot \mathcal{E}(\dot{\mathbf{T}}) dt \equiv \mathcal{P}(t_k) - \frac{1}{2}\mathbf{T} \cdot \mathcal{E}(\mathbf{T}) \mid_0^{t_k}$$

$$(34)$$

Consequently the viscoplastic work can be calculated by

$$w(t_k) = \mathcal{P}(t_k) - \frac{1}{2}\mathbf{T} \cdot \mathcal{E}(\mathbf{T}) \mid_0^{t_k} -\chi(t_k), \qquad (35)$$

where $\mathcal{P}(t_k)$ is the total work done by the stress along the considered process, the second term represent the elastic work while the last term is the instantaneous irreversible work.

231

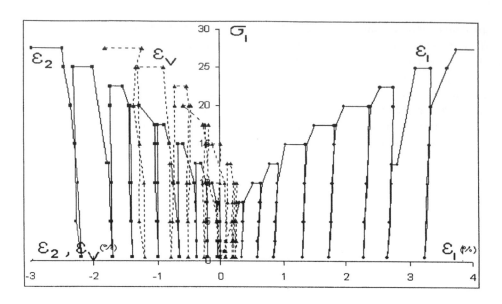

Figure 3 - Loading- Unloading Test with Stress Kept Constant
for 4 min. During the Creep.

Using the above results and an appropriate procedure developed in Cristescu
[1994] the constitutive functions from our models can be completely determined.

VARIATIONAL INEQUALITY

We derive the variational inequality, related to the rate quasi-static boundary
value problem (written in the rock domain $\Omega-$ bounded and open set in $\mathbf{R}^3$) and
associated with a generic stage of the process, using an appropriate procedure as in
Nguyen [1994] and Cleja-Ţigoiu [1996].

The rate quasi-static boundary value problem involves the time differentiation
of the equilibrium equations ($\forall \mathbf{x} \in \Omega$) and of the boundary condition :

$$\text{Div } \mathbf{T} + \mathbf{b} = 0 , \quad \mathbf{T}\mathbf{n} = \mathbf{S} \quad \text{on } S_T, \quad \mathbf{u} = \mathbf{u}_0 \quad \text{on } S_U; \quad S_T \cup S_U = \partial\Omega, \quad (36)$$

Here $\mathbf{b}-$ the volume forces are supposed to be constant in time, $\mathbf{n}-$ is the unit
external normal to the boundary $\partial\Omega$ of the domain Ω, $\mathbf{S}$ and $\mathbf{u}_0$, the surface loading
and the displacemnt vector are prescribed time dependent functions. At a generic
stage of the process the current values, i.e. at the time t, of $\mathbf{u}, \epsilon(\mathbf{u}), \mathbf{T}$ are known
for all $\mathbf{x} \in \Omega$, and so it is the set of all material particles in which the stress reached
the current yield surface

$$\Omega_i \equiv \{\mathbf{x} \in \Omega \mid g(\mathbf{T}(\mathbf{x}, t), \chi(\mathbf{x}, t)) = 0\} \quad (37)$$

The set of kinematically admissible (at time t) velocity (or the displacement
rate) fields is denoted by

$$\mathcal{V}_{ad} \equiv \{\mathbf{v} : \Omega \longrightarrow \mathbf{R}^3 \mid \mathbf{v}\mid_{S_U} = \dot{\mathbf{u}}_0\}. \quad (38)$$

232

Theorem 3. At every time t the displacement rate field, $\dot{\mathbf{u}}$, and the equivalent plastic factor β satisfy the following relationships

$$\int_\Omega \mathcal{E}\,\epsilon(\dot{\mathbf{u}}) \cdot \epsilon(\mathbf{v})dx - \int_\Omega \frac{\beta}{h_r}\mathcal{E}\,\partial_{\mathbf{T}}\,g \cdot \epsilon(\mathbf{v})\mathcal{H}(g)dx - \int_\Omega k \; < 1 - \frac{w}{H} > \mathcal{E}\,\partial_{\mathbf{T}}\,F\cdot$$

$$\cdot\epsilon(\mathbf{v})\mathcal{H}(-g)dx = \int_{S_T} \dot{\mathbf{S}} \cdot \mathbf{v}da + \int_{S_U} \dot{\mathbf{T}}\mathbf{n} \cdot \dot{\mathbf{u}}_0 da, \tag{39}$$

$$\int_\Omega (\gamma - \beta)\frac{1}{h_r}[\beta - \partial_{\mathbf{T}}\,g \cdot \mathcal{E}\epsilon(\dot{\mathbf{u}})]\mathcal{H}(g)dx \geq 0$$

which hold for every admissible vector field $\mathbf{v} \in \mathcal{V}_{ad}$, and for all $\gamma : \Omega \longrightarrow \mathbf{R}_{\geq 0}$.
Proof. In the theorem of virtual power, derived from the rate quasi-static equilibrium equation:

$$\int_{\partial\Omega} \dot{\mathbf{T}}\mathbf{n} \cdot \mathbf{v}da = \int_\Omega \dot{\mathbf{T}} \cdot \epsilon(\mathbf{v})dx, \quad \forall \mathbf{v} \in \mathcal{V}_{ad} \tag{40}$$

we substitute the constitutive equation of the relaxation type (18) and we derive the first identity from (39). The inequality defining the equivalent plastic factor β, given in (9) with (23) and (24) $\forall \mathbf{x} \in \Omega$, can be written in its integral form $(39)_2$.

We introduce the bilinear forms:

$$K[\mathbf{v}, \mathbf{w}] = \int_\Omega \mathcal{E}\epsilon(\mathbf{v}) \cdot \epsilon(\mathbf{w})dx\,, \quad A[\mu, \gamma] = \int_\Omega \frac{\mu\,\gamma}{h_r}\mathcal{H}(g)dx$$

$$C[\mu, \mathbf{v}] = \int_\Omega \frac{\mu}{h_r}\mathcal{E}[\partial_{\mathbf{T}}g] \cdot \epsilon(\mathbf{v})\mathcal{H}(g)dx \tag{41}$$

which hold for every admissible vector fields $\mathbf{v}, \mathbf{w} \in \mathcal{V}_{ad}$, and for all $\mu, \gamma : \Omega \longrightarrow \mathbf{R}_{\geq 0}$.
We define on $\mathcal{V}_{ad}$ the linear functional

$$f[\mathbf{v}] := \int_{S_T} \dot{\mathbf{S}} \cdot \mathbf{v}da + \int_\Omega k \; < 1 - \frac{w}{H} > \mathcal{E}\,\partial_{\mathbf{T}}\,F \cdot \epsilon(\mathbf{v})\mathcal{H}(-g)dx \tag{42}$$

which represents the virtual power produced by the variation in time of the forces $\mathbf{S}$ acting on S_T and by creep occuring in the material at time t.
The displacement rate $\dot{\mathbf{u}}$ is also a kinematically admissible velocity field. As a consequence of $(39)_1$ we obtain an equivalent form of the rate quasi-static quilibrium equation

$$K[\dot{\mathbf{u}}, \mathbf{v} - \dot{\mathbf{u}}] - C[\beta, \mathbf{v} - \dot{\mathbf{u}}] = f[\mathbf{v} - \dot{\mathbf{u}}]\,, \quad \forall \mathbf{v} \in \mathcal{V}_{ad} \tag{43}$$

The inequality $(39)_2$ defining the equivalent plastic factor becomes

$$- C[\gamma - \beta, \dot{\mathbf{u}}] + A[\beta, \gamma - \beta] \geq 0 \tag{44}$$

Let us define the convex set $\tilde{\mathcal{K}}$ in the appropriate functional space of the solution $\mathbf{H}_{ad}$, by

$$\tilde{\mathcal{K}} := \{(\mathbf{v}, \mu) \mid \mathbf{v} \in \mathcal{V}_{ad}, \quad \mu : \Omega \longrightarrow \mathbf{R}_{\geq 0}\}, \tag{45}$$

and the bilinear and symmetric form

$$a[\cdot,\cdot] : \mathbf{H}_{ad} \times \mathbf{H}_{ad} \longrightarrow \mathbf{R} \quad \text{such that} \quad \forall \mathbf{V} = (\mathbf{v},\mu), \mathbf{W} = (\mathbf{w},\gamma)$$

$$a[\mathbf{V},\mathbf{W}] := K[\mathbf{v},\mathbf{w}] - C[\mu,\mathbf{w}] - C[\gamma,\mathbf{v}] + A[\mu,\gamma] \tag{46}$$

As a consequence of (43), (44) and (46) the statement bellow is proved:
Theorem 4. The rate quasi- static boundary value problem requires the following *variational inequality* to be solved: Find $\mathbf{U} = (\dot{\mathbf{u}},\beta) \in \tilde{\mathcal{K}}$, such that

$$a[\mathbf{U},\mathbf{V}-\mathbf{U}] \geq f[\mathbf{V}-\mathbf{U}] \quad \forall \mathbf{V} \in \tilde{\mathcal{K}} \tag{47}$$

holds.

Stability and Uniqueness

Now we analyse the behavior of the solutions of the variational inequality. We prove that no bifurcations occur in rocks, durring the processes in which the hardening (creep) parameter remains positive, based on Hadamard's stability concept appropriate to our constitutive framework:
Definition. The elasto-plastic process is stable at time t if the biliniar form, $a[\cdot,\cdot]$, attached to the rate quasi-static problem is symmetric and positive definite.
Lemma. Under the hypothesis that $h_c > 0$, in Ω_i it follows that

$$\mathcal{E}\,\epsilon(\mathbf{v})\cdot\epsilon(\mathbf{v}) - \frac{2\,\mu}{h_r}\mathcal{E}\,\partial_{\mathbf{T}}\,g\cdot\epsilon(\mathbf{v}) + \frac{\mu^2}{h_r} \geq 0. \tag{48}$$

Proof. As a consequence of (18) and (4) it follows $h_r > 0$ and $\mid \mathcal{E}^{1/2}\partial_{\mathbf{T}}\,g(\mathbf{T},\chi)\mid <\sqrt{h_r}$. We rewrite the left hand side into an equivalent form and using Schwartz-Cauchy inequality we get

$$\mathcal{E}^{1/2}\,\epsilon(\mathbf{v})\cdot\mathcal{E}^{1/2}\epsilon(\mathbf{v}) - \frac{2\,\mu}{h_r}\mathcal{E}^{1/2}\,\partial_{\mathbf{T}}\,g\cdot\mathcal{E}^{1/2}\epsilon(\mathbf{v}) + \frac{\mu^2}{h_r} >$$
$$(\mid \mathcal{E}^{1/2}\epsilon(\mathbf{v})\mid - \frac{\mu}{\sqrt{h_r}}\mid \mathcal{E}^{1/2}\,\partial_{\mathbf{T}}\,g\mid)^2 \geq 0 \tag{49}$$

Based on the definition of the the biliniar forms (46), (41) and on the fact that $\epsilon(\mathbf{v}) = 0$ if and only if the velocity field characterizes a rigid motion, as a consequence of the *Lemma* we can prove that
Theorem 5. The biliniar form is symmetric and positive definite on $\mathbf{H}_{ab}$ apart from the element of $\mathcal{R} := (\mathbf{v},\mu) \equiv (\mathbf{v}_0 + \omega \times (\mathbf{x}-\mathbf{x}_0),0)$:

$$a[\mathbf{V},\mathbf{V}] \geq 0, \quad \text{and} \quad a[\mathbf{V},\mathbf{V}] = 0 \quad \text{if and only if } \mathbf{V} \in \mathcal{R} \tag{50}$$

In conclusion, the equality with zero in (50) takes place if and only if the velocity field characterizes a rigid motion and there is no evolution of irreversible instantaneous deformation $\dot{\epsilon}^i = 0$. As a direct consequence of *Theorems 4.and 5.* we can prove:
Theorem 6. When $h_c > 0$, the solution of the rate quasi-static boundary problem, i.e. the rate of the displacement field $,\dot{\mathbf{u}}$, and the equivalent plastic factor, β, is uniquelly defined if the measure of the part $S_U \subset \partial\Omega$ is positive. Moreover the deformation process is stable at every time t.

PROPAGATION CONDITIONS

We adapt for our approach to elasto-viscoplastic rock having an instantaneous loading dependent response given by (25), the accoustic tensor derived within the framework of relative deformation in Cleja-Țigoiu [1991]. Consequently we can prove:

Theorem 7. The propagation condition for the accelaration waves through the rock, travelling in the direction $\mathbf{n}$ with the intrinsec speed U and the amplitude vector $\mathbf{w}$, is characterized by

$$[\rho_0 U^2 \mathbf{I} - \mathbf{Q}(\mathbf{n})]\mathbf{w} = 0 , \quad \text{with}$$
$$\mathbf{Q}(\mathbf{n})\mathbf{w} = \mathbf{T} \cdot (\mathbf{n} \otimes \mathbf{n}) \, \mathbf{w} + \frac{1}{2}\mathcal{L}(\mathbf{T}, \chi)[\mathbf{n} \otimes \mathbf{w} + \mathbf{w} \otimes \mathbf{n}]\mathbf{n}, \tag{51}$$

$$\mathcal{L}(\mathbf{T}, \chi) := \mathcal{E} - \frac{sgn(\beta)\mathcal{H}(g)}{h_r}[\mathcal{E}\partial_{\mathbf{T}} \, g(\mathbf{T}, \chi) \otimes \mathcal{E}\partial_{\mathbf{T}} \, g(\mathbf{T}, \chi)]$$

We use the formula $(\mathbf{A} \otimes \mathbf{B})(\mathbf{n} \otimes \mathbf{w}) = \mathbf{A}(\mathbf{n} \cdot \mathbf{Bw}) \; \forall \mathbf{A}, \mathbf{B} \in Sym$, and $\mathbf{a}, \mathbf{b}$ vectors and we get for isotropic rock the following expression of the *accoustic tensor:*

$$\mathbf{Q}(\mathbf{n})\mathbf{w} = (\mathbf{Tn} \cdot \mathbf{n}) \, \mathbf{w} + (\lambda_0 + \mu_0) \, (\mathbf{n} \cdot \mathbf{w})\mathbf{n} + \mu_0 \mathbf{w} -$$
$$- \frac{sgn(\beta)\mathcal{H}(g)}{h_r}[\lambda_0 \, (\mathrm{tr} \, \partial_{\mathbf{T}} \, g(\mathbf{T}, \chi))(\mathbf{w} \cdot \mathbf{n}) + \tag{52}$$

$$+ 2\,\mu_0 \partial_{\mathbf{T}} \, g(\mathbf{T}, \chi)\mathbf{w} \cdot \mathbf{n}][\lambda_0 \, (\mathrm{tr} \, \partial_{\mathbf{T}} \, g(\mathbf{T}, \chi))\mathbf{n} + 2\,\mu_0 \partial_{\mathbf{T}} \, g(\mathbf{T}, \chi)\mathbf{n}]$$

where Lamé coefficients are expressed by elastic constants $\lambda_0 = K - \frac{2}{3} G$, $\mu_0 = G$.

We exemplify the above formulae for the undisturbed in situ rock which is under the primary stress state

$$\mathbf{T} = \sigma_v(t)\mathbf{i}_1 \otimes \mathbf{i}_1 + \sigma_h(t)(\mathbf{i}_2 \otimes \mathbf{i}_2 + \mathbf{i}_3 \otimes \mathbf{i}_3) \quad \text{where} \tag{53}$$

$\mathbf{i}_1$ is the vertical direction and $\mathbf{i}_2, \mathbf{i}_3$ in the horizontal plane. By direct calculus from (51)- (53) we get: when $\mathbf{n} = \mathbf{i}_1$ there is a longitudinal wave which is propagating with the velocity $U_L = (\frac{\sigma_v + \lambda_0 + 2\,\mu_0}{\rho_0})^{1/2}$ and there are two transversal waves with the speed $U_T = (\frac{\sigma_v + \mu_0}{\rho_0})^{1/2}$; when $\mathbf{n}$ belongs to the horizontal plane we have the similar results but σ_h replaces σ_v in the formulae of the propagation velocities.

Remark 4. The propagation speeds are determined by the stress state existing in the material at the considered time t, by the elastic properties of the rock as well as by the plastic properties, when the stress corresponds to a loading state. Generally there are also waves which are propagating with non- elastic speeds.

REFERENCES

Cleja-Țigoiu, S., [1991], "Elasto-Viscoplastic Constitutive Equations for Rock-type Materials (Finite Deformations)," Int. J. Engng. Sci., 29, 1531-1544.

Cleja-Țigoiu, S., [1996], "Bifurcations of Homogeneous Deformations of the Bar in Finite Elasto-Plasticity," Eur. J. Mech., A/Solids, 15, 761-786.

Cristescu, N., [1989], "Rock rheology", <u>Kluwer</u>, Dordrecht.

Cristescu, N., [1994], "A Procedure to Determine Nonassociated Constitutive Equations for Geomaterials," <u>Int. J. Plasticity, 10</u>, 103-131.

Cristescu, N.D., and Hunsche, U., [1998], "Time Effects in Rock Mechanics," <u>John Wiley & Sons</u>.

Suliciu, I., [1989], "Some Remarks on the Instantaneous Response in Rate- type Viscoplasticity,"<u>Int. J. Plasticity 5</u>, 173- 181.

Nguyen, Q.S., [1994], "Some Remarks on Plastic Bifurcation," <u>Eur. J. Mech. A/Solids. 13</u> , 485- 500.

Part V

Crushed salt behavior

Basic and Applied Salt Mechanics, Cristescu, Hardy, Jr & Simionescu (eds)
© 2002 Swets & Zeitlinger, Lisse, ISBN 90 5809 383 2

Crushed-salt constitutive model

Gary D.Callahan
RESPEC, P.O. Box 725, Rapid City, SD, USA 57709

Frank D.Hansen
Sandia National Laboratories, MS 1395, 115 N. Main, Carlsbad, NM, USA 88220

ABSTRACT

The constitutive model used to describe the deformation of crushed salt is presented. Two mechanisms — dislocation creep and grain boundary diffusional pressure solution— are combined to form the basis for the constitutive model governing the deformation. The constitutive model is generalized to represent three-dimensional states of stress. Upon complete consolidation, the crushed-salt model reproduces the multimechanism deformation (M-D) model typically used for the Waste Isolation Pilot Plant (WIPP) host geological formation salt. Parameter values are determined through nonlinear least-squares model fitting to an experimental database. Using the fitted parameter values, the constitutive model is validated against constant strain-rate tests. Shaft seal problems are analyzed to demonstrate model-predicted consolidation of the shaft seal crushed-salt component. Based on the fitting statistics, the ability of the model to predict the test data, and the ability of the model to predict load paths and test data outside of the fitted database, the model appears to capture the creep consolidation behavior of crushed salt reasonably well.

INTRODUCTION

Crushed salt is a key component of the shaft seal system for the Waste Isolation Pilot Plant (WIPP). Construction of the WIPP was initiated by the U.S. Department of Energy (DOE) in 1981 to develop the technology for the safe management, storage, and disposal of transuranic (TRU) radioactive wastes generated by the DOE defense programs. Recently, the U.S. Environmental Protection Agency (EPA) determined that the WIPP can safely contain transuranic waste and that it will comply with EPA's radioactive waste disposal standards. The WIPP received its first shipment of waste on March 26, 1999. Safe containment of the waste will rely on a seal system that includes crushed salt dynamically compacted to an initial fractional density of about 0.9. Continual improvement in understanding the consolidation processes in crushed salt is fundamental

---

[1] Work Supported by U. S. Department of Energy (DOE) Contract DE-AC04-94AL85000.

[2] A multiprogram laboratory operated by Sandia Corporation, a Lockheed Martin Company, for the U. S. DOE.

to the eventual construction of a credible seal system that will provide for safe containment of the waste.

Because crushed salt is an important seal material, its constitutive model must satisfy the needs of performance assessment and regulatory compliance. To gain an understanding of the crushed-salt consolidation processes, laboratory testing and theoretical model development have been conducted over the last several years. The WIPP crushed-salt material model or constitutive model, which relates stress and strain, predicts deformations and stresses in the underground sealing system. Laboratory testing has been used to develop the constitutive model from a phenomenological viewpoint and to evaluate the material constants associated with the model. The present work finalizes studies dedicated to the development of a constitutive model for crushed salt. The development program included theoretical considerations to arrive at the ultimate form of the constitutive model and laboratory testing to provide data for parameter estimation and verification of the model.

EARLIER WORK

Our first study (Callahan et al., 1995) relied on existing laboratory data for model formulation. The existing data at the time consisted primarily of results from hydrostatic consolidation tests augmented by one-fourth as many shear consolidation tests. All tests were conducted at initial fractional densities in the range of 0.65 to 0.75. The constitutive model derived in this study was found to be inadequate at high fractional densities. At higher fractional densities, three possible deformation modes (depicted in Figures 1 and 2) are associated with a shear consolidation test where the axial strain rate is negative: (1) lateral strain rate is negative, (2) lateral strain rate is positive, and (3) lateral strain rate is initially negative and later becomes positive. Since all shear consolidation tests available for the first study exhibited the first mode of deformation, the derived constants associated with the flow potential made it impossible for the simulated crushed salt to move toward a volume-preserving type of flow (as is exhibited by intact salt) as the fractional density approached one. In other words, the lack of mode two and mode three types of deformation in the experimental database was a restricting data bias that limited the predictive range of the constitutive model.

Because of these shortcomings, a follow-on study (Callahan et al., 1996) included two additional shear consolidation tests with initial fractional densities near 0.8. These two tests exhibited the second type of deformation mode. The functional form of the flow potential was modified, and the crushed-salt constitutive model was refit to the updated database. The second study resulted in satisfactory model predictions over a wide range of fractional densities and stress triaxialities (ratio of mean stress to effective or deviatoric stress). However, the governing deformation mechanisms for crushed salt with varying degrees of added moisture were contained independently in two separate models, and model predictions for fractional densities greater than 0.9 were much improved but still inadequate. Thus, the next study (Callahan et al., 1998) incorporated the governing mechanisms into a single, comprehensive model, and the flow potential was modified. Two additional shear consolidation laboratory tests were conducted. One test exhibited the second deformation mode while the other test exhibited the third deformation mode. The updated constitutive model was fit to the updated database to obtain new parameter values. Numerical predictions with the updated model showed that

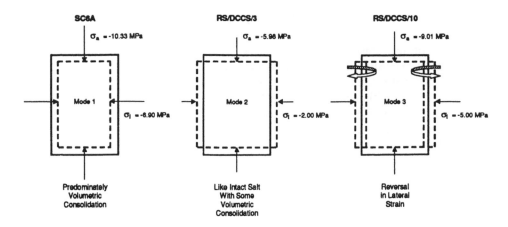

Figure 1. Three Potential Deformation Modes for Crushed-Salt Specimens During Shear Consolidation Tests.

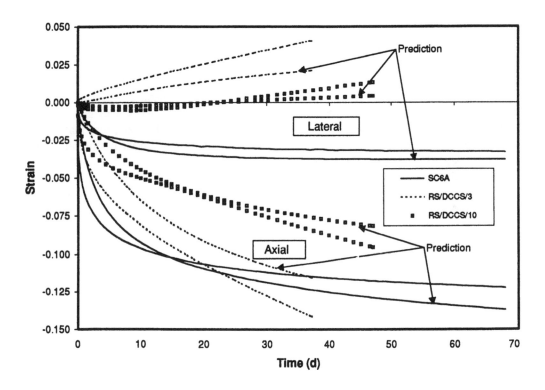

Figure 2. Three Crushed-Salt Shear Consolidation Tests Compared With the Constitutive Model Predictions.

the constitutive model predicts all three deformation modes, which is essential for evolution to the intact salt constitutive model as the fractional density approaches one.

LABORATORY DATA

Creation of the crushed-salt experimental database used for parameter estimation began with an extensive library search and compilation of potentially useful test results. The test data were compiled from the original studies conducted by Holcomb and Hannum (1982), Pfeifle and Senseny (1985), Holcomb and Shields (1987), Zeuch et al. (1991), and Brodsky (1994). The database compilation was reviewed to document those tests deemed inappropriate for parameter estimation, and the surviving tests formed the database for parameter estimation work. The laboratory test results were assembled, analyzed, and reported in previous work (Callahan et al., 1995). This work initiated the process of formulating the functional form of a constitutive model for crushed salt and assembled existing test data into a database to obtain quantitative estimates for the material parameters appearing in the model. The assembled database consists of a collection of tests conducted for a variety of reasons. Thus, the database contains biases and shortcomings when viewed strictly from the standpoint of developing constitutive models. The crushed-salt constitutive model has deviatoric and mean stress components that drive the strain response, and the database has shear and hydrostatic consolidation tests used for evaluating the parameters within those respective stress components.

Since this initial work was completed, additional tests have been conducted to fill gaps in the database. Two additional shear consolidation tests (Callahan et al., 1996) were conducted and added to the database. These two tests represent the first tests with high initial fractional densities (near 0.8) that emulate intact salt behavior accompanied by volumetric consolidation. Subsequently, additional laboratory work was conducted to add test data to the shear consolidation portion of the extant database because that portion of the database was relatively sparse, particularly in the higher initial fractional density range expected at WIPP. Four additional shear consolidation tests were conducted on dynamically compacted crushed salt, which are summarized by Hansen et al. (1998) and Mellegard et al. (1998). Associated with the current study, two additional shear consolidation tests were conducted to add test data to the shear consolidation portion of the database and verify earlier test results. One of the test's (RS/DCCS/10) results are shown in Figure 2. Therefore, the final database used for model development comprises 58 tests including 40 hydrostatic consolidation tests and 18 shear consolidation tests. The original data files often contained thousands of data points, which presented an impractical size for regression analyses. Each data file was reduced in size to create representative data sets containing only 100 data points with the first and last data points always being retained and the remaining data points uniformly sampled.

In addition to the shear and hydrostatic consolidation tests, three constant strain-rate tests were conducted under a confining pressure of -1 MPa at strain rates of $-0.5 \cdot 10^{-7}$ s^{-1}, $-1.0 \cdot 10^{-7}$ s^{-1}, and $-2.0 \cdot 10^{-7}$ s^{-1}. The constant strain-rate test results were not used as part of the fitted database. Instead, the tests were used to assess the model's predictive capability. The constant strain-rate tests provide experimental results and load paths outside of the fitted database. The ability of the crushed-salt constitutive model to predict these tests provides confidence in the model's predictive ability under general states of stress in situ.

MODEL DESCRIPTION

The final form of the crushed-salt, creep-consolidation constitutive model includes two deformation mechanisms operative in moist crushed salt — dislocation creep and pressure solution. The pressure solution portion of the model is only active when moisture is present. Generalization of the model to arbitrary three-dimensional states of stress is accomplished through a density-dependent flow potential. The general form of the kinetic equation governing creep consolidation inelastic flow of crushed salt can be written as:

$$\dot{\varepsilon}_{ij}^c = \left[\dot{\varepsilon}_{eq}^d\left(\sigma_{eq}^f\right) + \dot{\varepsilon}_{eq}^w\left(\sigma_{eq}^f\right)\right]\frac{\partial\sigma_{eq}}{\partial\sigma_{ij}} \tag{1}$$

where $\dot{\varepsilon}_{ij}^c$ is the inelastic strain rate tensor and σ_{eq}^f is an equivalent stress measure. The equivalent inelastic strain rate measures for dislocation creep ($\dot{\varepsilon}_{eq}^a$) and pressure solution ($\dot{\varepsilon}_{eq}^w$) are written as functions of the equivalent stress measure. For use in the flow potential, another equivalent stress measure, σ_{eq}, is used to provide a nonassociative type of formulation that provides flexibility in governing the magnitude of the volumetric behavior. The equivalent stress measures are:

$$\sigma_{eq}^f = \left[\eta_0\Omega_f^{\eta_1}\sigma_m + \left(\frac{2-D}{D}\right)^{\frac{2n_f}{n_f+1}}(\sigma_1-\sigma_3)^2\right]^{1/2} \tag{2}$$

$$\sigma_{eq} = \left[\kappa_0\Omega^{\kappa_1}\sigma_m + \left(\frac{2-D}{D}\right)^{\frac{2n}{n+1}}(\sigma_1-\sigma_3)^2\right]^{1/2}$$

where:

$$\Omega_f = \left[\frac{(1-D)n_f}{(1-(1-D)^{1/n_f})^{n_f}}\right]^{\frac{2}{n_f+1}} \tag{3}$$

$$\Omega = \left[\frac{(1-D_v)n}{(1-(1-D_v)^{1/n})^n}\right]^{\frac{2}{n+1}}$$

$$D_v = \begin{cases} D_t, & D < D_t \\ D & D \geq D_t \end{cases} \tag{4}$$

$$
\begin{aligned}
D &= \text{current fractional density} \\
\sigma_m &= \text{mean stress} \\
\sigma_1 &= \text{maximum principal stress} \\
\sigma_3 &= \text{minimum principal stress}
\end{aligned}
$$

$\eta_0, \eta_1, \kappa_0, \kappa_1,$
$n_f, n, \text{ and } D_t = \text{material parameters.}$

243

The flow potential enables the reversal in lateral strains, as observed in laboratory experiments (Figure 2), and provides for a smooth evolution to an intact salt constitutive model as the density increases. Thus, under compressive loads, the model evolves from a volumetric-consolidation model to a volume-preserving model as the crushed-salt porosity decreases and its density approaches that of intact salt.

To complete the description of the crushed-salt constitutive model, definitions of the inelastic strain rate measures for dislocation creep ($\dot{\varepsilon}_{eq}^{d}$) and pressure solution ($\dot{\varepsilon}_{eq}^{w}$) need to be supplied. The M-D model is used for the dislocation creep model as described by Munson et al. (1989). The only change made in adapting the M-D model to the crushed-salt constitutive model is that the effective stress measure used in the M-D model is replaced by the effective stress measure (σ_{eq}') given in Equation 2. Spiers and Brzesowsky (1993) model is adapted as the equivalent inelastic strain rate for grain boundary diffusional pressure solution of wet crushed salt. The functional form presented by Spiers and Brzesowsky was modified by changing the stress-dependent term (using the equivalent stress measure defined in Equation 2), writing the equation in terms of true strain, and adding the effect of moisture (w is the percent moisture by weight). The moisture function (simply w raised to a power, a) eliminates any strain-rate contribution to the crushed-salt consolidation model when moisture content is zero. With these modifications, the pressure solution portion of the consolidation model becomes:

$$\dot{\varepsilon}_{eq}^{w} = \frac{r_1^{*}}{\exp(\varepsilon_v)} \frac{\exp\left(\dfrac{Q_s}{RT}\right)}{T} \left[\frac{\exp(r_3 \varepsilon_v)}{\left|\exp(\varepsilon_v)-1\right|^{r_4}}\right] \Gamma \sigma_{eq}^{f} \tag{5}$$

where:

$$r_1^{*} = \frac{r_1 w^{a}}{d^{p}} \tag{6}$$

and r_1, r_3, r_4, a, p, and Q_s are material constants; ε_v is the volumetric strain, w is the moisture fraction by weight; d is the average grain size; T is absolute temperature; and R is the universal gas constant. The model consists of two functional forms — one for small strain and one for large strain, which are invoked depending on the prescribed value for Γ given as:

$$\Gamma = \begin{cases} 1 & \text{small strain } (\exp(\varepsilon_v)-1 > -15\%) \\[2ex] \left[\dfrac{\exp(\varepsilon_v)+\phi_0-1}{\phi_0 \exp(\varepsilon_v)}\right]^{n_s} & \text{large strain } (\exp(\varepsilon_v)-1 < -15\%) \end{cases} \tag{7}$$

where ϕ_0 is the initial porosity and n_s is a material parameter. Spiers and Brzesowsky developed the function Γ to account for increasing surface contact (increasing area and decreasing stress) as the strains become large. This geometrically interpreted variable serves to decrease the magnitude of the consolidation driving force and approaches zero as materials approach full consolidation. A problem is evident in Equation 5 at time zero if the initial volumetric strain is zero. To eliminate this problem during computations, some initial value must be assumed for the volumetric strain. Typically, this initial strain value will be computed based on an original fractional density of 0.64 (random dense packing of spherical particles) and the fractional density at the beginning of creep consolidation. The pressure solution model has seven material constants — a, p, Q_s, r_1, r_3, r_4, and n_s, excluding those that appear in the effective stress measure. Callahan et al. (1998) and Callahan (1999) give further mathematical details of the model.

PARAMETER VALUE DETERMINATION

The crushed-salt constitutive model contains 31 material parameters. Seventeen of these material parameters are contained in the dislocation creep (M-D) portion of the model. The M-D model parameters were fixed at the values provided by Munson et al. (1989) for clean salt. The pressure solution portion of the model contains seven material parameters, and the remaining seven material parameters comprise four flow potential material parameters and three equivalent stress measure parameters. These 14 remaining parameters were determined by fitting the equations that define the crushed-salt constitutive model to the laboratory data such that the weighted squared difference between the measured and calculated response was minimized. First, the flow potential parameters (κ_0, κ_1, n, and D_s) were determined by fitting the lateral-to-axial strain-rate ratio to the data measured in the 18 shear consolidation tests. Second, these flow rate parameters were fixed and the 10 creep consolidation parameters (η_0, η_1, n_f, a, p, n_s, r_1, r_3, r_4, and Q_s/R) were determined by fitting the integrated rate equations that define the axial and lateral strains to the data measured in the 18 shear consolidation and 40 hydrostatic consolidation tests. Two different fits were made: (1) shear consolidation tests and (2) both the shear and hydrostatic consolidation tests. Thus, two different parameter value sets were determined, which are named the shear and combined parameter values. The nonlinear regressions were performed using the personal-computer platform BioMedical Data Processing (BMDP, Version 7.01) statistical software package (Frane et al., 1985). Table 1 lists the parameter values determined for the flow potential and Table 2 lists the remaining creep consolidation parameter values. Table 2 lists two sets of values — one set for the fit to the shear consolidation tests and one set for the fit to the combined database.

CONSTANT STRAIN RATE TEST COMPARISONS

After a material model has been developed, the question remains as to how well it represents real-world problems. Constitutive model verification is best accomplished through the prediction of carefully conducted in situ experiments. When these in situ experiments do not exist, confidence in the model and partial verification can be accomplished by predicting bench-scale and other laboratory tests outside of the database upon which the model development relied. The constant strain-rate tests provide an excellent verification test of the constitutive model since these tests provide a constant strain rate and varying stress rate; whereas, the constitutive model was developed based on tests that included a constant stress and a varying strain rate. Since the constant strain-rate tests were not included in the model fitting database and represent a totally different load path than the creep consolidation tests, they are ideal for testing the predictive ability of the constitutive model.

Table 1. Flow Potential Parameters

Parameter	Units	Magnitude
κ_0	—	10.119
κ_1	—	1.005
n	—	1.331
D_t	—	0.896

Table 2. Creep Consolidation Parameters

Parameter	Units	Shear Tests	Combined Tests
η_0	—	0.1029	0.0140
η_1	—	3.9387	3.3881
n_f	—	3.5122	3.9660
a	—	0.3147	$1.0(10^{-5})$
p	—	1.6332	1.0341
n_s	—	0.5576	1.5318
r_1	$m^p \cdot K/(MPa \cdot s)$	$1.041(10^{-6})$	$2.663(10^{-5})$
$r_1^{\cdot}$	$K/(MPa \cdot s)$	0.0194	0.3369
r_3	—	15.1281	0.1286
r_4	—	0.1678	0.1680
Q_s/R	K	1,077.46	2,897.08

$r_1^{\cdot}$ assumes grain size $(d) = 1$ mm and moisture content $(w) = 1$ percent.

Three constant strain-rate tests were performed on laboratory-scale specimens. The specimens were dynamically compacted using the same technique and raw crushed-salt materials that were employed for fabrication of six of the shear consolidation specimens. All three tests were performed at a confining pressure of -1 MPa, a temperature of $20°C$, and a constant axial strain rate. Essentially, the test procedure included applying a -1 MPa confining pressure, monitoring the axial strain rate, and initiating the constant axial strain rate when the monitored rate was less than the targeted test axial strain rate. The constant axial strain rate for each test was $-0.5 \cdot 10^{-7}$, $-1.0 \cdot 10^{-7}$, and $-2.0 \cdot 10^{-7}$ s^{-1}.

The crushed-salt constitutive model was incorporated into the thermomechanical finite element stress analysis program SPECTROM-32 (Callahan et al., 1989), Version 4.10. Thus, prediction of the constant strain-rate tests was performed using the finite element program with one element. For each test simulation, an element was initially subjected to the time-weighted average axial and lateral stress conditions realized in each test. The stress condition was designed to be a hydrostatic load of -1 MPa; however, the actual test conditions achieved were slightly different than the design condition. The stress conditions were maintained for the *hydrostatic* duration realized in the tests. Then, time-weighted average, time-wise kinematic boundary conditions were applied to the element that matched the constant strain-rate conditions achieved in the tests. Thus, the constant strain-rate test simulations used the average actual test conditions. Simulations were performed using both the shear database and combined database parameter sets given in Table 2.

Axial and lateral strain and axial stress simulation results obtained for the three constant strain-rate tests are shown in Figures 3 through 8 plotted with the laboratory data. Predicted results are shown for the two different parameter value sets given in Table 2. In the three strain comparison figures, the axial and lateral strain simulation results reproduce the laboratory data reasonably well; however, an offset is noticeable in the strains that appears to be generated by the hydrostatic portion of the tests. The laboratory data exhibits a curious lateral and axial strain response during the presumed hydrostatic portion of the tests. Normally, one would expect the axial and lateral strain to be the same during hydrostatic compression of a homogeneous material. However, the axial and lateral strains are noticeably different in all three of the constant strain-rate tests during the hydrostatic loading phase of the tests. Three possible explanations for this anomalous behavior are: (1) existence of a stress difference (i.e., the axial and lateral stresses were not identical in the tests), (2) other modes of deformation (cataclastic deformation and sliding along the particle interfaces upon loading), and (3) specimen inhomogeneity (dynamic compaction of the test specimens in three lifts). With the data available, the importance of these three factors' impact on the specimen deformation cannot be quantified. All three factors are probably contributing to the observed deformation behavior of the specimens.

The results show that the simulation results reproduce the laboratory strain data quite well. The axial strain rate is an imposed condition that should be reproduced exactly, and the fact that the simulation results parallel the axial laboratory data provides a check that the correct strain rate was used in the simulations. Figures 3, 5, and 7 show decreasing agreement between computed results and laboratory data as the strain rate increases. However, the prediction of the lateral strains is still quite good, even at the highest strain rate when one considers the fact that both the test data and load path for the constant strain-rate tests are outside of the database used to derive parameter values. The figures also show minor differences between the predicted results for the shear and combined database parameter value sets, even though some of the parameter values in the two sets appear to be quite different. Figures 4, 6, and 8 compare the computed axial stress values with the measured test data. In all cases, the computed axial stresses increase more rapidly than observed in the tests. As with the computed strains, the computed stress agreement decreases as the strain rate increases. Under constant strain-rate loading conditions, crushed salt appears to deform more easily than is predicted by the creep consolidation model.

SHAFT SEAL CALCULATIONS

The current seal designs for the WIPP specify the emplacement of various materials (asphalt, compacted clay, concrete, and crushed salt) in the shafts from the earth's surface to the repository horizon. Crushed salt will be emplaced within the shafts at

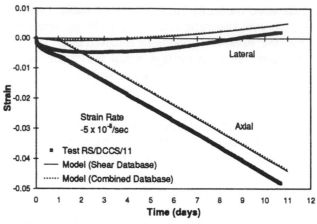

Figure 3. Laboratory and Model Strain Comparisons for Test RS/DCCS/11.

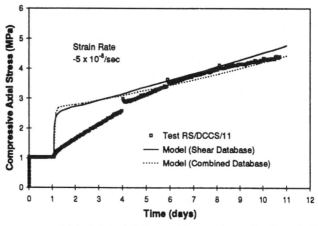

Figure 4. Laboratory and Model Axial Stress Comparisons for Test RS/DCCS/11.

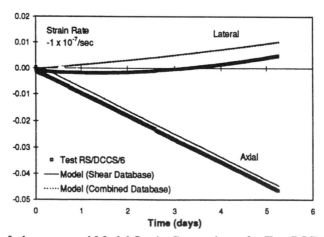

Figure 5. Laboratory and Model Strain Comparisons for Test RS/DCCS/6.

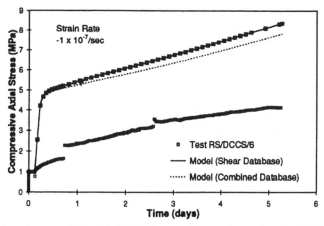

Figure 6. Laboratory and Model Axial Stress Comparisons for Test RS/DCCS/6.

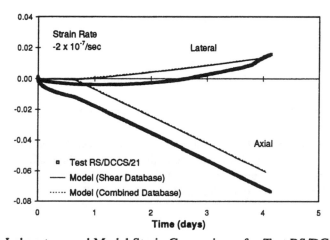

Figure 7. Laboratory and Model Strain Comparisons for Test RS/DCCS/21.

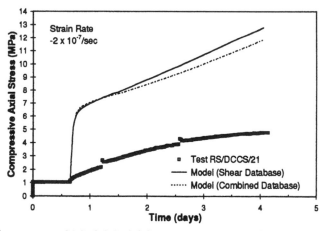

Figure 8. Laboratory and Model Axial Stress Comparisons for Test RS/DCCS/21.

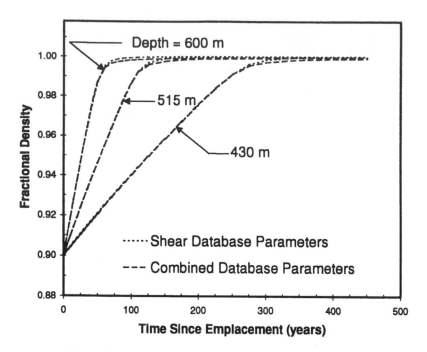

Figure 9. Crushed-Salt Fractional Density History in a Shaft.

depths ranging from 430 meters to 600 meters and dynamically compacted to a fractional density of about 0.90. The flow characteristics of the crushed-salt seal are of primary importance since the long-term function of the seal is to prevent the flow of water into the repository from overlying aquifers and to prevent the flow of gas and brine out of the repository. Closure of the shaft resulting from creep of the surrounding intact salt will consolidate the crushed salt over time. The consolidation process reduces the permeability of the crushed salt from its emplaced state to its intact state, creating an effective long-term seal.

A simple axisymmetric representation of the shaft and host rock was selected to illustrate the behavior of the crushed-salt seal. By changing the initial stress and temperature conditions, different depths may be examined. A condition of plane strain is assumed to exist with respect to the axial (vertical) direction. The initial stress state before shaft excavation is assumed to be lithostatic. The shaft is excavated into the lithostatic stress field at time zero. The shaft is assumed to be open for 50 years and then instantaneously filled with crushed salt dynamically compacted to a fractional density of 90 percent. The finite element program **SPECTROM-32** was used to predict the creep deformation of the host rock and the consolidation of the crushed-salt seal material. The outer boundary condition consisted of a constant traction boundary condition consistent with the initial stress level for each respective depth analyzed.

Figure 9 presents the results of six shaft seal analyses using the shear and combined database parameter value sets at three different depths. At depths of 600 m, 515 m, and 430 m, approximately 60 years, 110 years, and 240 years, respectively, are required to attain a crushed-salt fractional density of 99 percent. These results show that the responses for either parameter value set at each depth are similar up to a fractional density of about 98 percent. Thereafter, the model results diverge slightly. The results also indicate that the shear parameter value set predicts the fastest consolidation rate.

REFERENCES

Brodsky, N. S., 1994. *Hydrostatic and Shear Consolidation Tests With Permeability Measurements on Waste Isolation Pilot Plant Crushed Salt,* SAND93-7058, prepared by RE/SPEC Inc., Rapid City, SD, for Sandia National Laboratories, Albuquerque, NM.

Callahan, G. D., 1999. *Crushed Salt Constitutive Model,* SAND98-2680, prepared by RESPEC, Rapid City, SD, for Sandia National Laboratories, Albuquerque, NM.

Callahan, G. D., K. D. Mellegard, and F. D. Hansen, 1998. "Constitutive Behavior of Reconsolidating Crushed Salt," *Proceedings, 3rd North American Rock Mechanics Symposium,* Cancún, Mexico, June 3–5, published in *International Journal of Rock Mechanics and Mining Sciences,* Vol. 35, No. 4/5, Paper No. 029.

Callahan, G. D., M. C. Loken, L. D. Hurtado, and F. D. Hansen, 1996. "Evaluation of Constitutive Models for Crushed Salt," *Proceedings, Fourth Conference of The Mechanical Behavior of Salt,* École Polytechnique de Montréal, Mineral Engineering Department, Québec, Canada, June.17 and 18, M. Aubertin and H. R. Hardy Jr. (eds.), Penn State University, Trans Tech Publications, Clausthal, Germany, 1998, pp. 317–330.

Callahan, G. D., M. C. Loken, L. L. Van Sambeek, R. Chen, T. W. Pfeifle, J. D. Nieland, and F. D. Hansen, 1995. *Evaluation of Potential Crushed-Salt Models,* SAND95-2143, prepared by RE/SPEC Inc., Rapid City, SD, for Sandia National Laboratories, Albuquerque, NM.

Callahan, G. D., A. F. Fossum, and D. K. Svalstad, 1989. *Documentation of SPECTROM-32: A Finite Element Thermomechanical Stress Analysis Program,* DOE/CH/10378-2, prepared by RESPEC, Rapid City, SD, for Department of Energy, Chicago Operations Office, Argonne, IL, Vols. I-II.

Frane, J. W., L. Engelman, and J. Toporek, 1985. *BMDP Programmer's Guide and Subroutine Writeups: Documentation of the FORTRAN Source; Part I. Programmer's Guide; Part II. Subroutine and Common Block Writeups,* BMDP Tech. Report No. 55.

Hansen, F. D., Callahan, G. D., M. C. Loken, and K. D. Mellegard, 1998. *Crushed-Salt Constitutive Model Update,* SAND97-2601, prepared by RE/SPEC Inc., Rapid City, SD, for Sandia National Laboratories, Albuquerque, NM.

Holcomb, D. J. and D. W. Hannum, 1982. *Consolidation of Crushed Salt Backfill Under Conditions Appropriate to the WIPP Facility,* SAND82-0630, Sandia National Laboratories, Albuquerque, NM.

Holcomb, D. J. and M. E. Shields, 1987. *Hydrostatic Creep Consolidation of Crushed Salt With Added Water,* SAND87-1990, Sandia National Laboratories, Albuquerque, NM.

Mellegard, K. D., T. W. Pfeifle, and F. D. Hansen, 1998. *Laboratory Characterization of Mechanical and Permeability Properties of Dynamically Compacted Crushed Salt*, SAND98-2046, prepared by RESPEC, Rapid City, SD, for Sandia National Laboratories, Albuquerque, NM.

Munson, D. E., A. F. Fossum, and P. E. Senseny, 1989. *Advances in Resolution of Discrepancies Between Predicted and Measured In Situ WIPP Room Closures*, SAND88-2948, Sandia National Laboratories, Albuquerque, NM.

Pfeifle, T. W. and P. E. Senseny, 1985. *Permeability and Consolidation of Crushed Salt From the WIPP Site*, RSI-0278, prepared by RE/SPEC Inc., Rapid City, SD, for Sandia National Laboratories, Albuquerque, NM.

Spiers, C. J. and R. H. Brzesowsky, 1993. Densification Behaviour of Wet Granular Salt: Theory Versus Experiment, *Proceedings, Seventh Symposium on Salt*, Vol. I, pp.83–92, Elsevier Science Publishers B. V., Amsterdam.

Zeuch, D. H., D. J. Zimmerer, and M. E. Shields, 1991. *Interim Report on the Effects of Brine-Saturated and Shear Stress on Consolidation of Crushed, Natural Rock Salt From the Waste Isolation Pilot Plant (WIPP)*, SAND91-0105, Sandia National Laboratories, Albuquerque, NM.

Basic and Applied Salt Mechanics, Cristescu, Hardy, Jr & Simionescu (eds)
© 2002 Swets & Zeitlinger, Lisse, ISBN 90 5809 383 2

Mechanical and permeability properties of dynamically compacted crushed salt

F.D.Hansen
Sandia National Laboratories, Carlsbad, New Mexico, USA

K.D.Mellegard
RESPEC, Rapid City, South Dakota, USA

Thorough scientific and engineering characterization of crushed salt assures that access shafts will not become avenues for hazardous material migration from the Waste Isolation Pilot Plant (WIPP). Essential work to characterize engineering properties, construction methods, and constitutive modeling underpins the shaft seal design, construction, and analysis. A comprehensive shaft seal system design and analysis, independent technical review, and critical acceptance were milestones along the path to regulatory compliance. After approximately a decade of regulatory interactions, the WIPP began underground disposal operations in March 1999. This historic achievement is due in part to a rigorous, unassailable shaft seal system. The WIPP itself is a full-scale geologic repository located in the bedded salt of the Salado Formation at a depth of 655 meters. Four shafts connect the underground facility to the surface. Compacted crushed salt placed near the bottom of each shaft provides a primary seal element. The technical basis for crushed salt use as a seal component was established by extensive research undertaken by Sandia National Laboratories and RESPEC.

Dynamically compacted crushed salt provides chemical compatibility with the host salt rock and returns the excavated material to its natural environment. In addition, crushed salt is an attractive sealing material because its permeability decreases as the surrounding rock salt creeps inward and reduces the void space. Hansen and Ahrens (1998) demonstrated that crushed salt could be readily placed at 0.9 initial fractional density, where fractional density is the ratio of the current bulk density to intact salt density (i.e., 2.16 g/cm^3). During the shaft seal design and analysis exercises it became clear that limited engineering data were available for crushed salt possessing fractional densities above 0.9. Thus, a specific suite of investigations, including shear consolidation creep, permeability, and constant strain-rate triaxial compression, was executed. Laboratory results were expected to illuminate the phenomenology of crushed-salt deformation behavior and add test results to a database for estimating parameters in a crushed-salt constitutive model, which is presented in a companion paper to these proceedings (Callahan and Hansen, 1999).

Dynamically compacted crushed-salt specimens with a diameter of 100 mm and lengths up to 200 mm were derived from the full-scale compaction demonstration and from

---

[1] Work supported by U.S. Department of Energy (DOE) Contract DE-AC04-94AL85000.
[2] A multiprogram laboratory operated by Sandia Corporation, a Lockheed Martin Company, for the DOE.

a laboratory-scale dynamic-compaction study. Starting material was wetted to moisture contents of nominally 1.6 wt%. Test procedures included:

- Shear consolidation tests (effectively standard creep tests), which were initially loaded hydrostatically to various levels of constant confining pressure, were conducted. Axial stress difference was increased in each test by 4 MPa while the confining pressure was held constant. At test termination an unload/reload cycle was performed by reducing the stress difference to near zero and then reloading the specimen to a stress difference of 4 MPa before totally unloading the specimen. The ascending loading data from the reloading portion of the cycle were used for estimating elastic constants.

- Constant strain-rate tests were performed on the laboratory-scale dynamically compacted crushed salt. Tests were performed at 1 MPa confining pressure and axial strain rates of 0.5×

- 10^{-7}, 1.0×10^{-7}, and 2.0×10^{-7} s^{-1}.

- Permeability tests were conducted with nitrogen and brine. A hydrostatic stress of 1 MPa was applied to the external surfaces of the specimen by pressurizing the annulus between the specimen and the pressure vessel wall. The pressure at the inlet was limited to 0.34 MPa.

A constitutive model for crushed salt consolidation should predict a stress state at which radial strain rate would initially be positive (consolidation) and then reverse direction and become negative as the specimen density increases. This phenomenon was exhibited during experiments conducted at 4 and 5 MPa confining pressure, as shown in Figure 1. The radial strain rate in those two tests was initially positive (specimen diameter decreasing), and after some densification of the specimen the radial strain rate decreased and ultimately changed sign (specimen diameter increasing). This observation supports the original hypothesis that the direction of the radial strain response depends on both the state of stress and the fractional density of the specimen.

It has been postulated and confirmed that consolidation of granular rock salt occurs by two primary mechanisms: grain boundary pressure solution and dislocation creep. Further, a minor amount of added moisture greatly enhances densification (Spiers and Brzesowsky, 1993). Consideration of the fundamental deformational processes guided formulation of the constitutive model for reconsolidating crushed salt. As crushed salt is loaded, the principal densification mechanism of fluid-phase grain boundary solution/redeposition is rampant. As consolidation proceeds, the material attains sufficient density so that its response assumes the constitutive response of intact salt. Thus, the proper material law for reconsolidating crushed salt incorporates pressure solution and dislocation creep.

Test data suggest that permeability decreases as fractional density increases. Figure 2 plots volumetric strain as a function of time on the primary axis and brine flow as a function of time on the secondary axis. Buret measurement and volumetric strain are actual test measurements; the other two flow rates are calculated. Data indicate that the saturated specimen was consolidating, and a calculated brine quantity was expelled at the vented end. Inspection of Figure 2 reveals that permeability rapidly decreased to zero. A similar phenomenon was observed under constant conditions by Brodsky (1994), who attributed permeability decrease to localized precipitation.

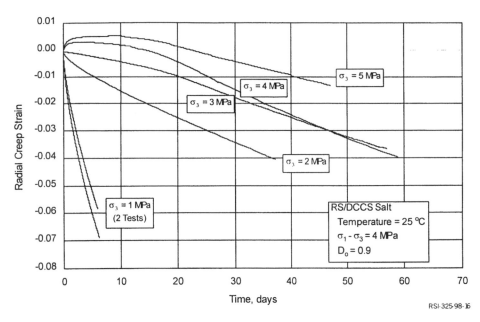

Figure 1. Radial creep strain as a function of time for laboratory-scale, dynamically compacted salt.

Permeability testing of the dynamically compacted crushed salt provided further evidence that the permeability decreases as the fractional density of the salt increases. This conclusion agrees with and augments previous results. Shear consolidation creep test results were added to a database of similar results for the purpose of estimating parameters in a constitutive model that represents the behavior of crushed salt (Callahan and Hansen, 1999). Current testing was performed at higher initial fractional densities (0.9) and stresses (1 to 5 MPa) than were used in previous programs to give better coverage of the range of conditions likely to be experienced by salt seal elements at the WIPP. The constitutive model predicted that stress states exist where the radial strain rate would initially be positive (consolidation) and then reverse direction and become negative as the specimen density increases. This phenomenon was clearly observed in multiple tests.

Constant axial strain-rate tests were performed but were not included in the database used for estimating the parameters in the constitutive model. The purpose of these tests was to provide results from a unique load path that could be used to evaluate the predictive capability of the constitutive model.

Observational microscopy supports the fundamental theory of densification behavior of wet granular salt. The constitutive model developed by Callahan and coworkers embodies the governing mechanisms as documented by microscopy. The experimental behavior is found to be consistent with theory, both in the mesoscopic scale of laboratory experiments and the microscopic scale where the processes occur. The complete understanding of reconsolidation processes—verified by laboratory experiments, parameterized in one database and predicted accurately in independent experiments, and comprehensively embodied in the mathematical formulation of the constitutive equations—provides the scientific basis and confidence for performance predictions of compacted-salt seal performance.

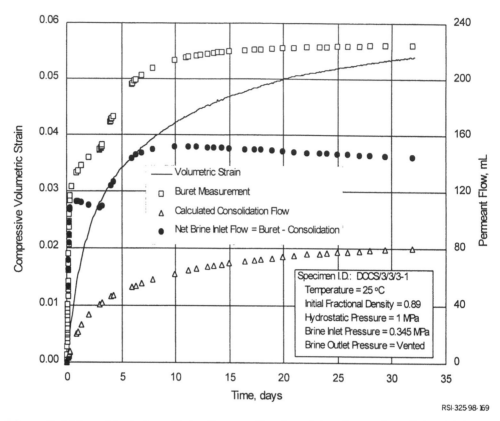

Figure 2. Volumetric strain and brine flow as a function of time for specimen DCCS3/3/3/3-1.

REFERENCES

Brodsky, N.S. 1994. *Hydrostatic and Shear Consolidation Tests With Permeability Measurements on Waste Isolation Pilot Plant Crushed Salt*. SAND93-7058. Prepared by RE/SPEC Inc., Rapid City, SD. Albuquerque, NM: Sandia National Laboratories.

Callahan, G.D. and F.D. Hansen. 1999. *Crushed-Salt Constitutive Model*. MECASALT5, University of Bucharest, Bucharest, Romania.

Hansen, F.D., and E.H. Ahrens. 1998. "Large-Scale Dynamic Compaction of Natural Salt," *The Mechanical Behavior of Salt, Proceedings of the Fourth Conference*, École Polytechnique de Montréal, Mineral Engineering Department, Québec, Canada, June 17-18, 1996. SAND96-0792C. Eds. M. Aubertin and H.R. Hardy, Jr. Clausthal-Zellerfield: Trans Tech Publications. 353-364.

Spiers, C.J., and R.H. Brzesowsky. 1993. "Densification Behavior of Wet Granular Salt: Theory Versus Experiment," *Seventh Symposium on Salt,* Kyoto, Japan, April 6-9, 1992. Eds. H. Kakihana, H.R. Hardy, Jr., T. Hoshi, and K. Toyokura. Amsterdam; New York, New York: Elsevier Science Publishers B.V. Vol. I, 83-92.

Basic and Applied Salt Mechanics, Cristescu, Hardy, Jr & Simionescu (eds)
© 2002 Swets & Zeitlinger, Lisse, ISBN 90 5809 383 2

Consolidation behavior of dry crushed salt: Triaxial tests, benchmark exercise, and in-situ validation

Ekkehard Korthaus
Institut fur Nukleare Entsorgungstechnik, Karlsruhe, Germany

ABSTRACT

The experimental data basis for the adjustment of the constitutive model for the reference backfill was extended regarding higher initial porosities (37– 38%) and permanent deviatoric conditions. The parameter set of the constitutive model was modified with respect to the effect of initial porosity. The experimental technique of the true triaxial testing device was verified within a benchmark exercise and with the aid of special tests concerning wall friction and anisotropy of the testing material. The results from the TSDE (formerly TSS) in-situ test in the Asse salt mine were evaluated for a first validation of the constitutive model developed for the reference backfill material. It was found that in the in-situ test the reference backfill behaves significantly more stiffly than predicted by the constitutive model adjusted to the laboratory test results. It is presumed that this discrepancy is mostly due to some demixing of the crushed salt during the slinger backfilling procedure.

INTRODUCTION

Present German concepts on nuclear waste repositories in rock salt formations envisage the use of crushed salt as backfill material for disposal and access drifts as well as for borehole plugs. To perform predictive calculations on the important safety-relevant effects of convergence and resulting backfill consolidation, a constitutive model for the crushed salt is needed. Whatever formulation is chosen, purely empirical or based on physical principles and mechanisms, the model is to describe the compaction and deformation behavior under hydrostatic and deviatoric stress loading and its dependency on temperature, porosity and initial porosity.

In general, such constitutive models contain some functions and free parameters which must be defined for a specified material. Detailed laboratory tests on the deformation behavior are needed for this, which should cover as largely as possible the conditions to be expected in a salt repository. As in-situ conditions can be established in laboratory tests in an approximate manner only, any adjusted constitutive model should be validated in large-scale in-situ tests.

This procedure was pursued to establish a constitutive model for dry crushed salt from the Asse salt mine, specified as reference backfill material for the investigations related to the German nuclear waste disposal projects. It had a broad grain size distribution and a maximum grain size of 31.5 mm.

Extensive laboratory tests were performed using a true triaxial testing apparatus (Korthaus and Schwarzkopf, 1993, Korthaus, 1996) specially developed for this purpose. The experimental technique was verified within a benchmark exercise on the crushed salt behavior. A constitutive model based on the work of Hein (1991) was fitted to the experimental data.

The results from the TSDE (Thermal Simulation of Drift Emplacement, formerly TSS) in-situ test in the Asse salt mine (Bechthold et. al., 1989) were evaluated for a first validation of the constitutive model developed for the reference backfill material.

LABORATORY TESTS

The true triaxial testing apparatus used for the tests has a cubical test cell of 250 mm side length and uses hydraulic pressure pads for triaxial stress loading. The pressure pads are made of thin stainless steel sheets and replaced in each new test. For tests at elevated temperatures the pressure pads are protected against corrosion by the use of envelopes made of titanium foils of 0.075 mm thickness.

The amount of hydraulic oil to be supplied to the pressure pads is used to determine the triaxial deformation of the sample (OMB method). As an additional technique of deformation measurement, inductive displacement transducers are used to measure the displacement of the central zones of the test sample surfaces (IWA method).

More details on the experimental device and procedures can be found in (Korthaus and Schwarzkopf, 1993, Korthaus, 1996).

Altogether, 16 triaxial tests have been performed on the reference backfill, including about 80 periods of constant hydrostatic or deviatoric stress adjusted by multi-step procedures. The duration of these periods typically was 4 – 6 days. Stress levels between 2 and 20 MPa and temperatures between 20 and 150°C were chosen, and low creep rates (1.E-5 – 5.E-3/d) were achieved during or near the end of the constant stress periods, i.e. representative conditions for repositories were approached. Initial porosities between 30 and 38% (most of them near 31%) were adjusted by different preconsolidations of the material during the filling procedure of the testing cell.

The results of 12 tests have been published earlier (Korthaus, 1996). In Table 1 the results needed for evaluation of the remaining 4 tests are given. They focused on the effect of higher initial porosity and the consideration of permanent deviatoric stress conditions. The first aspect is important for the modeling of in situ conditions where the initial porosity may vary substantially due do the technical backfilling procedure.

Participation in the crushed salt Benchmark CSCS (Comparative Studies on Crushed Salt) within the EC project BAMBUS (EC, 1996) made it possible to compare the true triaxial testing apparatus with two conventional (axisymmetric) triaxial testing devices (Korthaus, 1998). A hydrostatic multi-step test with six constant stress periods (Fig. 1) was performed at a temperature of 25°C. The testing material was produced from the reference backfill by removal of the fractions with grain sizes above 8 mm. The test was repeated using an additional technique in order to further reduce wall friction. Sheets of Viton of 0.5 mm thickness were placed on the front of the hydraulic pressure pads and provided with a thin film of grease in order to allow easy gliding of the Viton (Test 2).

Table 1   Stress Conditions and Results for the Creep Periods of the Experiments

Experiment / Init. Porosity	T (°C)	Stresses (MPa)			Porosity	Strain Rates (1/d)		
		σ_x	σ_y	σ_z		x	y	z
V160595	30	3.0	3.0	3.0	0.2655	4.0E-4	4.0E-4	4.0E-4
38%		2.4	2.4	2.4	0.2637	5.8E-5	5.8E-5	5.8E-5
		3.0	3.0	3.0	0.2611	1.5E-4	1.5E-4	1.5E-4
	100	3.0	3.0	3.0	0.2200	4.1E-4	4.1E-4	4.1E-4
		2.4	2.4	2.4	0.2169	5.3E-5	5.3E-5	5.3E-5
		5.0	5.0	5.0	0.1627	5.8E-5	5.8E-5	5.8E-5
		3.0	3.0	3.0	0.1617	7.5E-6	7.5E-6	7.5E-6
	80	3.0	3.0	7.0	0.1551	-4.0E-5	-4.0E-5	3.8E-4
V280396	150	8.6	6.4	7.5	0.1161	7.4E-4	2.95E-4	6.05E-4
31%		6.88	5.12	6.0	0.1140	1.2E-4	4.8E-5	1.04E-4
		15.0	11.0	13.0	0.0754	2.47E-3	5.5E-4	1.61E-3
V050996	50	5.0	5.0	5.0	.2266	7.4E-4	7.4E-4	1.21E-3
37.1%		4.0	4.0	4.0	.2247	1.8E-4	1.7E-4	2.8E-4
		4.0	4.0	4.0	.2229	1.5E-4	1.43E-4	2.2E-4
		8.7	8.7	8.7	.1798	5.8E-4	5.6E-4	8.3E-4
		7.0	7.0	7.0	.1776	1.4E-4	1.4E-4	2.0E-4
		7.0	7.0	7.0	.1756	1.2E-4	1.2E-4	1.7E-4
	90	8.7	8.7	8.7	.1554	1.12E-3	1.12E-3	1.55E-3
	50	8.7	8.7	8.7	.1454	7.6E-5	7.6E-5	1.0E-4
		6.6	6.6	6.6	.1437	3.0E-5	2.7E-5	3.4E-5
V240997	75	5.0	5.0	2.0	.1762	2.5E-4	2.5E-4	-1.1E-4
38%	50	5.0	5.0	2.0	.1748	6.7E-5	6.7E-5	-1.8E-5
	50	6.0	6.0	2.0	.1721	1.31E-4	1.31E-4	-9.3E-5
	50	4.0	4.0	1.6	.1717	2.8E-5	2.8E-5	-4.0E-6
	25	4.0	4.0	1.6	.1708	1.1E-5	1.1E-5	-2.0E-6

Comparing the results of both tests, it could be shown that wall friction only moderately affects the measurements with the true triaxial device (Fig. 2). The sample geometry during compaction, however, is found to behave practically ideally when the Viton sheets are used: There is no relevant difference between the two strain measurement methods (OMB and IWA), and also the post-test inspection of the sample showed that the surfaces were nearly perfectly plane.

The final porosity reached in the benchmark tests was additionally determined after removal of the samples from the test cell (13.5 and 12.4%). Very good agreement was found with the results of the in-line measurements (13.4 and 12.6%).

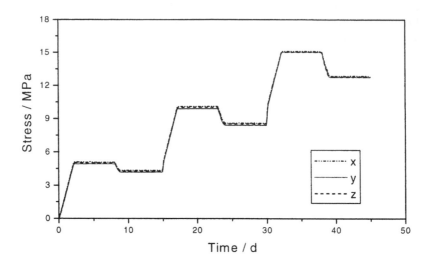

Figure 1 Stress History of Test V200198 (Benchmark Test 2)

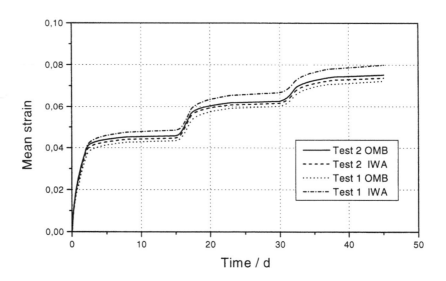

Figure 2   Strains of Tests V190297 and V200198 (Benchmark Tests 1 and 2)

An acceptable agreement exists with the final porosities determined after the tests by the other benchmark participants, when differences in the test parameters actually used (higher test temperature in one case) are taken into account (11.5%, 13.8%). The in-line measurements of the other benchmark participants, however, suffered from certain experimental problems concerning the determination of the absolute compaction or were degraded by some leakage effects. After suitable corrections for this, an acceptable agreement of the test results on the whole can be stated, too (Bechthold et al., 1999).

One item of consideration of the benchmark test was the stress exponent of the creep compaction to be derived from the material behavior before and after the stress stress drops. From the 2 tests within this work stress exponents between 5 and 7.5 were determined. The results of one of the other benchmark participants also allowed this kind of evaluation and gave values of about 5-6.5, i.e. a good agreement.

One special test was performed to clear up the effect that normally resulted in significant differences between horizontal and vertical deformation of the reference backfill under hydrostatic loading. It was argued that this must be attributed to the flat ㄱr oblong shape of the particles of this material and their non-isotropic orientation during the filling of the test cell. A special test was therefore performed on commercial salt with approximately cubical grains. Actually the differences mentioned above did not occur in this test. As a matter of fact, there was some smaller difference, but in the opposite direction (Fig. 3). From this result it can be concluded that the effect in question in fact is due to the special shape of the granular particles.

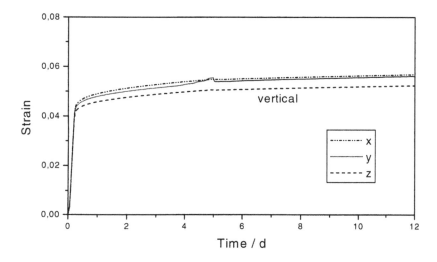

Figure 3   Strains from Hydrostatic Test V210898 on Fine-grained Commercial Salt

CONSTITUTIVE MODEL

The general deformation behavior of crushed salt is very complex, because it is composed of different mechanisms: Grain displacements and cracking (at high stresses) on the one hand, typical of normal granular materials, and, on the other hand, the visco-plastic contact zone deformation of the grains, which is characteristic for rock salt, and which results in creep strain deformations even at low constant stresses.

The constitutive model derived by Hein (1991) describes the general deformation behavior of crushed salt with two additive parts for the mechanisms of grain displacement and creep deformation of the grains. Under repository conditions with slow deformation rates and only moderate shear stresses or strains, the contribution of grain displacement is presumed to be small. This is why only the grain deformation part was adopted to model the deformation behavior of the reference backfill, being aware that this and the implicit assumption of a stationary creep law imply certain restrictions with respect to the interpretation and evaluation of short-term laboratory tests, in particular during or after stress raising.

The grain deformation part of Hein's model is defined as:

$$\dot{\varepsilon} = A \cdot e^{-Q/R/T} \cdot (h_1 \cdot p^2 + h_2 \cdot q^2)^{(n-1)/2} \cdot (h_1 \cdot p/3 \cdot \mathbf{1} + h_2 \cdot \mathbf{S})$$

with $\dot{\varepsilon}$: tensor of the inelastic strain rate Q :   activation energy
        p :   mean normal stress R :   universal gas constant
        $\mathbf{S}$: tensor of the deviatoric stress T :   absolute temperature
        q :   deviatoric stress invariant $\mathbf{1}$:   unit tensor
        n :   stress exponent

The functions h_1 and h_2 describe the dependency of porosity η. The formulation for h_1 proposed in (Hein, 1991) was replaced by the following one which has the advantage to vanish at zero porosity and also contains a suitable dependency of the initial porosity η_0, which was not considered in (Hein, 1991):

$$h_1 = a / (((\eta_0/\eta)^c - 1)/\eta_0^r)^m$$

The original formulation for h_2 was adopted with a modified normalization: $h_2 = b \cdot h_1 + 1$

This constitutive model was fitted to the experimental data now available for the reference backfill, using only the strain rates measured at the end of constant stress periods with compaction rates below 5.E-3/d, such that the effects of grain displacements and transient creep can be assumed to be small. Optimum fitting was found with values of Q and the stress exponent n not far from those corresponding to the stationary creep of Asse salt, i.e. 54.2 kJ/mol and n=5. These values were adopted exactly in order to be consistent in the boundary case of vanishing porosity, where the solid salt behavior is supposed to result.

The other fitting parameters were then found to be:

$A = 1.09 \cdot 10^{-6}/s/MPa^5$ $b = 0.9$
$m = 2.25$ $c = 0.1$
$a = 0.12$ $r = 0.85$

The value of A found here is not far from that one normally used for the stationary creep of Asse salt ($2.08 \cdot 10^{-6}/s/MPa^5$). These results correspond closely to those reported in

262

(Korthaus, 1996), but now with a better consideration of the effect of initial porosity, represented by a new value for the parameter r, which also requires a modification of the parameter a .

The parameter r had been set equal to c in the former fit. This choice was based on the assumption that at the limit of vanishing porosity the compaction rate is a function of porosity only and no longer depends on the initial porosity. In order to describe the results of the new tests with initial porosities of 37 or 38% satisfactorily, it was deemed necessary to drop this assumption. More experimental data are required with larger compaction and high initial porosity in order to verify the fit chosen here at porosities below 10%.

In Figure 4 it is shown how the experimentally determined compaction rates (or to be more precise, the mean strain rates) agree with the adjusted constitutive model. The scatter of the ratios, which is due to experimental errors and the limitations of the model in considering all details of the different load cases, has become somewhat larger after consideration of the results with high initial porosity. It can be seen that the tests with high initial porosity, V160595, V050996 and V240997 with η_0=37-38%, are now described with deviations similar to those having initial porosities of about 31%. It should be mentioned that deviations of about a factor of 100 can be found for these tests when the initial porosity is not adequately considered in the constitutive model.

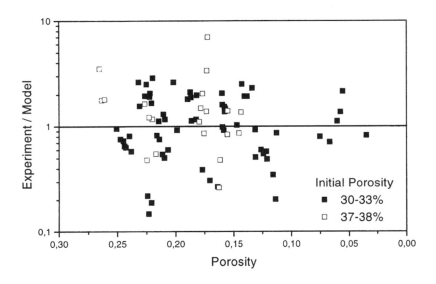

Figure 4   Ratio of Measured Consolidation Rates to the Predictions of
the Adjusted Constitutive Model

A similar result is found for the ratio of the equivalent strain rates from measurement and constitutive model. A better representation, however, can be obtained when the deviatoric behavior is described by an effective Poisson's ratio for creep deformations, defined in analogy to the Poisson's ratio for elastic deformations (Korthaus, 1996).

In Fig. 5 the ratio of the corresponding values derived from the experiment and the constitutive model is plotted versus the porosity. Some systematic discrepancies can be seen between those 'isotropic' experiments, in which a stress deviator was applied only temporarily within a predominantly hydrostatic consolidation process, and those 'anisotropic' ones, in which deviatoric loading was applied throughout the test. These discrepancies are explained by induced anisotropies in the tests of the latter type.

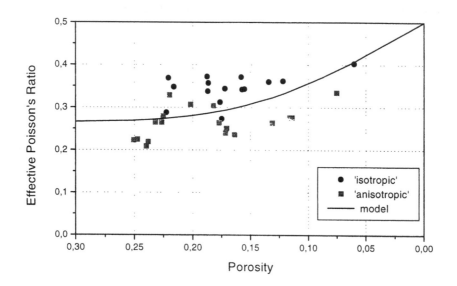

Figure 5   Effective Poisson's Ratio for Creep Deformation as Obtained from Measurements and Constitutive Model Prediction

The above representations demonstrate how the constitutive model can describe the dependency between strain rates and stresses, temperature and porosities independent of time. The direct comparison of measured and calculated strains as functions of time is also helpful to demonstrate the capabilities and limitations of the constitutive model.

In Fig. 6 this is shown for the hydrostatic benchmark tests (V190297, V200198) and in Figs. 7 – 8 for the deviatoric test V280396. In the first case, the underestimation of the strain rate during the stress raising periods can be observed, and the decrease with time of this discrepancy in the strains.

In the second case which represents a test with permanently deviatoric ("anisotropic") conditions, an overestimation can be noticed in the model calculation with regard to deviatoric response and mean strain (the relatively high measured strain in z-direction is due to the grain shape effect discussed above). A better agreement is obtained when the relatively low compaction rates (compared to the constitutive model) of this test are accounted for in the factor A and the parameter b equals 0.7, which is more appropriate for the description of "anisotropic" deviatoric tests ("corrected model").

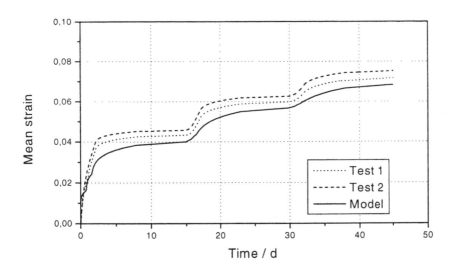

Figure 6   Benchmark Tests 1 and 2 (V190296,V200198): Measured Mean
Strains and Model Prediction

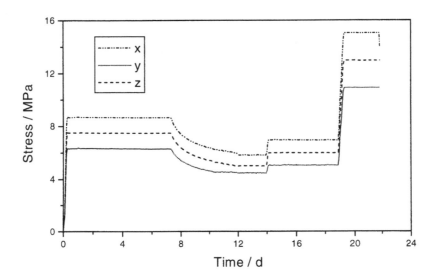

Figure 7   Stress History of Permanently Deviatoric Test V280396

Similar results are found when other tests with permanently deviatoric loading are
compared to the constitutive model prediction. In calculations for the behavior of
backfilled cavities in a repository, where deviatoric conditions will also be permanent,
the use of a parameter b~0.7 seems to be more suitable.

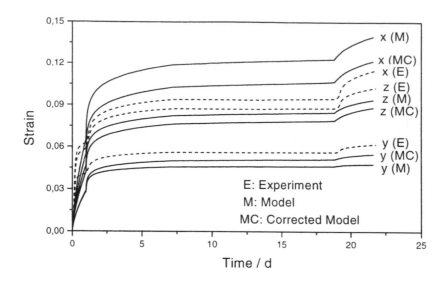

Figure 8   Strains in Test V280396: Measurement and Constitutive Model Predictions

IN-SITU VALIDATION

The in-situ test TSDE in the Asse salt mine (Bechthold et al., 1989) was designed to study thermomechanical and other effects relevant in drift disposal of spent fuel, in particular the behavior of crushed salt backfill. In two parallel drifts with 4.2 m width and 3.5 m height, six electrical heaters with the dimensions of Pollux casks (5.5 m length, 1.5 m diameter) were placed. The remaining volume was filled with reference backfill with an average initial porosity of about 35%.

The heating period lasted 8 years, with a thermal power of the heaters of about 6.4 kW each. Among others, temperatures, drift closure and stresses (at the drift walls) acting on the backfill were measured at several vertical cross-sections, i.e. quantities relevant for the backfill compaction in the drifts. Measurements of the porosity distribution over the drift height were performed, too, but the results are not very reliable. Therefore, the porosity reduction averaged over the drift cross-section was derived indirectly from the drift closure data.

Unfortunately, the backfill compaction achieved during the heating period was only modest (average porosity reduction from about 35 to 25% in the heated zone). Nevertheless, these data could be used for a first validation of the constitutive model for the reference backfill.

As no measurements of stresses and porosities inside the backfill were available and because of the non-uniform heating and compaction of the backfill in the drifts, a validation of the constitutive model was possible only in an integral way, i.e. with respect to the backfill behavior averaged over the drift cross-section. Special thermo-mechanical model calculations for the test were performed for this purpose.

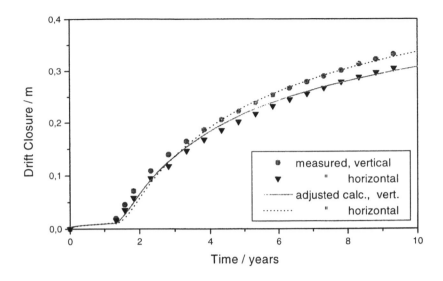

Figure 9   Drift Closure in TSDE Experiment: Measurement and Calculation Adjusted
for Validation of the Constitutive Model for the Reference Backfill

Hereby, the creep characteristics of the formation rock salt were adjusted such that
the drift closures, which had originally been overestimated in the calculations, agreed to
those observed in the experiment (Fig. 9). As measured and calculated temperatures
were in good agreement, too, the constitutive model validation could then be performed
by comparing measured and calculated stress buildup in the backfill at the drift walls.

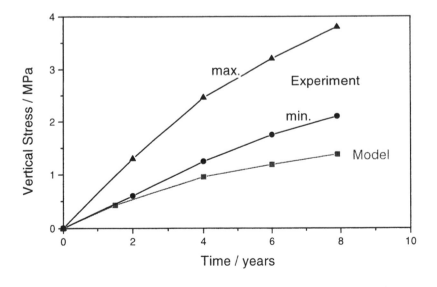

Figure 10   Stress at Drift Roof in TSDE: Measurement and Model Prediction

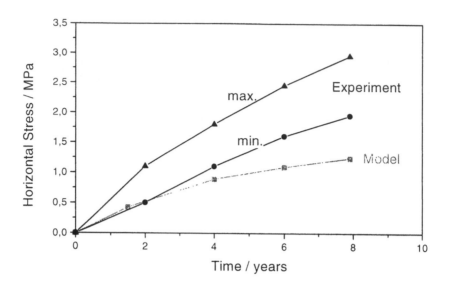

Figure 11   Stress at Drift Walls in TSDE: Measurement and Model Prediction

In Figs. 10 – 11 vertical and horizontal stresses at the drift walls, as obained from measurement and adjusted calculation, are shown for the central vertical cross-section. It can be seen that the backfill behaves significantly more stiffly in the in-situ experiment than in the calculation based on the constitutive model fitted to the laboratory tests. Taking mean values over the range of the experimental data, this discrepancy roughly amounts to a factor of 2 in the stresses. Due to a stress exponent of 5, this corresponds to a factor of 32 in the compaction rate. It is presumed that this discrepancy is due to some demixing of the crushed salt during the slinger backfilling procedure. Also some overestimation of the initial porosity may be partly responsible for this discrepancy.

The planned post-test investigations will furnish information on the grain size distributions actually present in the drifts and special laboratory tests will show the effect of such deviations from the reference backfill specifications. A check of the initial porosity is envisaged, too.

CONCLUSIONS

The recent tests performed with the true triaxial testing device confirmed the results reported earlier (Korthaus, 1996) with exception of the effect of initial porosity on the consolidation rate. Some improvement of the experimental method was shown to be possible by additional measures for the minimization of wall friction. When deviatoric experiments are performed on a material similar to the reference backfill, i.e. with flat or oblong shape of the particles, care should be taken of anisotropy effects. Significant discrepancies were found between the consolidation behavior in laboratory tests and in the TSDE in-situ test. They are not yet really understood, and they are hoped to be cleared by the planned post-test investigations for TSDE. Because of the relatively weak compaction reached in the TSDE test, other in-situ tests are required to validate the constitutive model at porosities below 20%.

Acknowledgment: This work was supported partly by the Commission of the European Communities within the framework of the R&D programme on the management and storage of radioactive waste.

REFERENCES

Korthaus, E., and Schwarzkopf, W. [1993], "Eine triaxiale Meßeinrichtung zur Untersuchung des Kompaktierungsverhaltens von Salzgrus", KfK 5211.

Korthaus, E., [1996], "Consolidation and Deviatoric Deformation Behavior of Dry Crushed Salt at Temperatures up to 150°C", 4th Conference on the Mechanical Behavior of Salt, Montreal, June 1996.

European Commission (EC), [1996]: "Nuclear Fission Safety (1994-1998)", Synopsis of the research projects (First deadline), EUR16980 EN, pp.77-78, Luxembourg.

Hein, H.J., [1991]
Ein Stoffgesetz zur Beschreibung des thermomechanischen Verhaltens von Salzgranulat, doctoral thesis RWTH Aachen.

Bechthold, W.; Heusermann, S.; Schrimpf, C.; Gommlich, G., [1989]
Large Scale Test on in-situ Backfill Properties and Behavior Under Reference Repository Conditions, NEA/CEC, Workshop on Sealing of Radioactive Waste Repositories, Braunschweig, Germany, May 22-25, 1989.

Korthaus, E., [1998], "Experiments on Crushed Salt Consolidation with True Triaxial Testing Device as a Contribution to an EC Benchmark Exercise", FZKA 6181.

Bechthold, W., et. al. [1999], "Backfilling and Sealing of Underground Repositories for Radioactive Waste in Salt, -The BAMBUS Project-", Final Report, to be published by the EC.

Part VI

Numerical modeling

Basic and Applied Salt Mechanics, Cristescu, Hardy, Jr & Simionescu (eds)
© 2002 Swets & Zeitlinger, Lisse, ISBN 90 5809 383 2

Stress distribution of rock around a horizontal tunnel

Iuliana Paraschiv-Munteanu
Department of Mechanics, University of Bucharest, Romania

ABSTRACT

Study of stress distribution during creep of the rock surrounding a circular horizontal tunnel is a very important problem, mainly for mining engineering. At big depths, an opening excavated in rock can close completely after time intervals which are of the order of several tens of years. For the design of underground cavities one must be able to predict quite accurately not only the stress and strain distribution around them, but also the apperance and possibly slow spreading in time of a microcraked domain produced just by the excavation. Since the microcraking is related to dilatancy, the irreversible volumetric changes, either dilatancy or compressibility have started to be studied too. If the stress is in the dilatancy domain damage by microcraking can develop steadily in time, ultimately leading to a major underground failure. That is why it is important to study the stress variation during creep when microcraks are also developing.

INTRODUCTION

In this paper we determine a complete numerical solution about behaviour of elasto-viscoplastic rock around circular horizontal tunnel. The obtained numerical approach is comparative with simplified solution (creep solution) and linear elastic solution (instantaneous response).

The problem was studied for the other author analysing different aspects. In some cases was used linear elasticity (Obert and Duvall [1967], Jumikis [1979]). Massier ([1994], [1997]) had used a linear viscoelastic model with plane state of deformations. A study of this problem using an elasto-viscoplastic model proposed by Cristescu [1994] was done by Cristescu and Hunsche [1998], Paraschiv and Cristescu [1993], supposed plane state of deformations and determined a creep solution. In this paper we determine a numerical solution without using hypothesis that stress state remains constant in time. For numerical solution we used the scheme proposed by Paraschiv-Munteanu in [1997], [1998] using finite element method for spatial integration and a complet implicit method for integration in time. In most cases we observed that in proximity of underground opening the stress becomes relaxed relatively to the moment of excavation. However, for short period of time the creep solution and the numerical solution are very close.

THE CONSTITUTIVE EQUATION

This is done in this paper using an elastic-viscoplastic nonassociated constitutive equation (see Cristescu [1989], Cristescu and Hunsche [1998]). The reference configuration, with respect to which the strains must be estimated, is the state *in situ* before excavation, there where the future excavation is envisaged. Using standard notation we have for the rate of deformation tensor:

$$\dot{\boldsymbol{\varepsilon}} = \frac{\dot{\boldsymbol{\sigma}}^{R}}{2G} + \left(\frac{1}{3K} - \frac{1}{2G}\right)\dot{\sigma}^{R}\mathbf{1} + k_{T}\left\langle 1 - \frac{W^{I}(t)}{H(\boldsymbol{\sigma})}\right\rangle\frac{\partial F}{\partial \boldsymbol{\sigma}}(\boldsymbol{\sigma}) + k_{S}\frac{\partial S}{\partial \boldsymbol{\sigma}}(\boldsymbol{\sigma}), \qquad (1)$$

where G and K are elastic moduli which may depend of the invariants of the stress and strain and possibly on the damage of rock. $H(\boldsymbol{\sigma}(t)) = W^{I}(t)$ is the equation of the stabilization boundary for the transient creep. $F(\boldsymbol{\sigma})$ and $S(\boldsymbol{\sigma})$ are the viscoplastic potentials for the transient and stationary creep respectively, k_{T} and k_{S} are the corresponding viscosity coefficients, $\mathbf{1}$ is the unit tensor and $\sigma = \frac{1}{3}(\sigma_{1} + \sigma_{2} + \sigma_{3})$ is the mean stress. We will use the notation τ for the octahedral shear stress:

$$\tau = \frac{1}{3}\left[(\sigma_{1} - \sigma_{2})^{2} + (\sigma_{2} - \sigma_{3})^{2} + (\sigma_{3} - \sigma_{1})^{2}\right].$$

In (1) the work-hardening parameter (the internal state variable) is the irreversible stress work per unit volume:

$$W^{I}(t) = W^{P} + \int_{0}^{t}\sigma(s)\dot{\varepsilon}_{v}^{I}(s)\,ds + \int_{0}^{t}\boldsymbol{\sigma}'(s)\cdot\left(\dot{\boldsymbol{\varepsilon}}^{I}\right)'(s)\,ds, \qquad (2)$$

where $t = 0$ is the moment of excavation and the W^{P} is the primary (initial) value of W^{I} (Cristescu [1989]). In (2) ε_{v}^{I} is the irreversible part of the volumetric deformation and $\left(\dot{\boldsymbol{\varepsilon}}^{I}\right)'$ is the deviator of the irreversible part of the rate of deformation tensor. Finally, the bracket from (1) has the meaning of the positive part the mentioned function, i.e.:

$$\langle A \rangle = A^{+} = \frac{1}{2}(A + |A|).$$

All constitutive functions and parameters involved in (1) are determined from experimental data. The constitutive equation (1) can describe the following mechanical properties exhibited by most rocks: transitory and stationary creep, work-hardening during transient creep, volumetric compressibility and/or dilatancy, as well as short-term failure. All these properties are incorporated into the constitutive equation via the procedure used to determine the constitutive functions (Cristescu [1989], [1994], Cristescu and Hunsche [1998]).

PROBLEM STATEMENT

The stress distribution just after excavation is obtained by exact elastic solution. Let a be the initial radius of the cavity and $m \in \mathbf{N}$, $m \geq m_0 \geq 5$, number of radius which defined the limits of the domain, $\Omega = [a, ma] \times [0, 2\pi)$. We assume that in all horizontal directions the primary stresses are the same, σ_h, and the depth is sufficient great to consider that σ_v, the vertical initial (primary) stress, is not variable in the domain Ω (σ_v corresponds for the axis of the tunnel). The conditions for $r \to \infty$ (in case of infinitely domain) has been considered on the external boundary of the domain Ω.

PROPOSITION 1. *If on the walls of the cavity, $\Gamma_1 = \{(a, \theta) \,|\, \theta \in [0, 2\pi)\}$, a pressure p is acting (due to various reasons and which may be constant or variable):*

$$\sigma_{rr}^S(a, \theta) = p, \quad \sigma_{r\theta}^S(a, \theta) = 0, \quad \forall\, \theta \in [0, 2\pi), \tag{3}$$

and on the external boundary of the domain Ω, $\Gamma_2 = \{(ma, \theta) \,|\, \theta \in [0, 2\pi)\}$, we have:

$$\begin{cases} \sigma_{rr}^S(ma, \theta) = \dfrac{1}{2}(\sigma_h + \sigma_v) + \dfrac{1}{2}(\sigma_h - \sigma_v)\cos 2\theta \\[2mm] \sigma_{\theta\theta}^S(ma, \theta) = \dfrac{1}{2}(\sigma_h + \sigma_v) - \dfrac{1}{2}(\sigma_h - \sigma_v)\cos 2\theta \quad, \\[2mm] \sigma_{r\theta}^S(ma, \theta) = -\dfrac{1}{2}(\sigma_h - \sigma_v)\sin 2\theta \end{cases} \tag{4}$$

then the stress state just after excavation is:

$$\tilde{\sigma}_{rr}^S(r, \theta) = 2A_1 + \frac{C_1}{r^2} + \left(-2A_2 - \frac{6C_2}{r^4} - \frac{4D_2}{r^2}\right)\cos 2\theta$$

$$\tilde{\sigma}_{\theta\theta}^S(r, \theta) = 2A_1 - \frac{C_1}{r^2} + \left(2A_2 + 12B_2 r^2 + \frac{6C_2}{r^4}\right)\cos 2\theta$$

$$\tilde{\sigma}_{r\theta}^S(r, \theta) = \left(2A_2 + 6B_2 r^2 - \frac{6C_2}{r^4} - \frac{2D_2}{r^2}\right)\sin 2\theta \tag{5}$$

$$\tilde{\sigma}_{zz}^S(r, \theta) = \nu\left[4A_1 + \left(12B_2 r^2 - \frac{4D_2}{r^2}\right)\cos 2\theta\right] + \sigma_h$$

where A_1, B_1, C_1, A_2, B_2, C_2, D_2 are constants :

$$A_1 = \frac{m^2}{4(m^2 - 1)}(\sigma_h + \sigma_v) - \frac{1}{2(m^2 - 1)}p,$$

$$C_1 = \frac{m^2 a^2}{m^2 - 1}\left[p - \frac{1}{2}(\sigma_h + \sigma_v)\right],$$

$$A_2 = -\frac{m^2\left(m^4 + m^2 + 4\right)}{4\left(m^6 - 1\right)}\left(\sigma_h - \sigma_v\right), \quad B_2 = \frac{m^2}{2a^2\left(m^6 - 1\right)}\left(\sigma_h - \sigma_v\right),$$

$$C_2 = -\frac{m^4 a^4\left(m^2 + 1\right)}{4\left(m^6 - 1\right)}\left(\sigma_h - \sigma_v\right), \quad D_2 = \frac{m^2 a^2}{2\left(m^2 - 1\right)}\left(\sigma_h - \sigma_v\right).$$

Proof. The components of stress are obtained from equilibrium equation using the Airy function and the constants result from conditions (3) and (4). □

From (5) it is easy to obtain:

PROPOSITION 2. *The components of deformation corresponding stres state (5) are:*

$$\tilde{\varepsilon}_{rr}(r,\theta) = \frac{1+\nu}{E}\left\{2(1-2\nu)A_1 + \frac{C_1}{r^2} + \left[-2A_2 - 12\nu B_2 r^2 - \frac{6C_2}{r^4} - \frac{4D_2}{r^2}(1-\nu)\right]\cos 2\theta\right\},$$

$$\tilde{\varepsilon}_{\theta\theta}(r,\theta) = \frac{1+\nu}{E}\left\{2(1-2\nu)A_1 - \frac{C_1}{r^2} + \left[2A_2 + 12(1-\nu)B_2 r^2 + \frac{6C_2}{r^4} + \frac{4D_2}{r^2}\nu\right]\cos 2\theta\right\}, \qquad (6)$$

$$\tilde{\varepsilon}_{r\theta}(r,\theta) = \frac{1+\nu}{E}\left(2A_2 + 6B_2 r^2 - \frac{6C_2}{r^4} - \frac{2D_2}{r^2}\right)\sin 2\theta,$$

and the components of the displacement are:

$$\tilde{u}_r(r,\theta) = \frac{1+\nu}{E}\left\{2(1-2\nu)A_1 r - \frac{C_1}{r} + \left[-2A_2 r - 4\nu B_2 r^3 + \frac{2C_2}{r^3} + \frac{4D_2}{r}(1-\nu)\right]\cos 2\theta\right\},$$

$$\tilde{u}_\theta(r,\theta) = \frac{1+\nu}{2E}\left[42A_2 r + 4(3-2\nu)B_2 r^3 + \frac{4C_2}{r^3} + \frac{4D_2}{r}(2\nu-1)\right]\sin 2\theta. \qquad (7)$$

Observation. It is easy to observe that in the case of infinitely domain, when $m \to \infty$, the stress, deformation and displacement components are the same like in papers of Cristescu [1989], Paraschiv and Cristescu [1993], because, when $m \to \infty$.

we have:

$$A_1 \to \frac{1}{4}\left(\sigma_h + \sigma_v\right), \quad C_1 \to a^2\left(p - \frac{1}{2}\left(\sigma_h + \sigma_v\right)\right)$$

$$A_2 \to -\frac{1}{4}\left(\sigma_h - \sigma_v\right), \quad B_2 \to 0, \quad C_2 \to -\frac{a^4}{4}\left(\sigma_h - \sigma_v\right), \quad D_2 \to \frac{a^2}{2}\left(\sigma_h - \sigma_v\right).$$

So, this result proves in one way that taking $m \geq m_0 \geq 5$ it is acceptable for moving the infinitely condition on the boundary $r = ma$.

Elastic solution is used as initial data for the integration in long time intervals using FEM. The general formulation of the problem of determination the stress distribution around a circular horizontal tunnel in elasto-viscoplastic rock, like a quasistatic problem, is:

find the displacement function $(u_r, u_\theta) : \mathbf{R}_+ \times \Omega \longrightarrow \mathbf{R}^2$, the stress function $\boldsymbol{\sigma} : \mathbf{R}_+ \times \Omega \longrightarrow \mathcal{S}_3$ and the irreversible stress work function $W^{\mathrm{I}} : \mathbf{R}_+ \times \Omega \longrightarrow \mathbf{R}$ such that:

$$\mathrm{Div}\ \boldsymbol{\sigma}^{\mathrm{R}}(t, r) = 0 \quad \text{in } \mathbf{R}_+ \times \Omega, \tag{8}$$

$$\dot{\boldsymbol{\sigma}}^{\mathrm{R}} = 2G\dot{e} + (3\mathrm{K} - 2\mathrm{G})\dot{\varepsilon}\mathbf{1} + \\ k_{\mathrm{T}} \left\langle 1 - \frac{W^{\mathrm{I}}(t)}{H(\boldsymbol{\sigma})} \right\rangle \left[\frac{(2G - 3\mathrm{K})}{3} \frac{\partial F}{\partial \boldsymbol{\sigma}} \mathbf{1} - 2\mathrm{G}\frac{\partial F}{\partial \boldsymbol{\sigma}} \right] + \\ k_{\mathrm{S}} \left[\frac{(2G - 3\mathrm{K})}{3} \frac{\partial S}{\partial \boldsymbol{\sigma}} \mathbf{1} - 2\mathrm{G}\frac{\partial S}{\partial \boldsymbol{\sigma}} \right] \quad \text{in } \mathbf{R}_+ \times \Omega, \tag{9}$$

$$\dot{W}^{\mathrm{I}} = k_{\mathrm{T}} \left\langle 1 - \frac{W^{\mathrm{I}}(t)}{H(\boldsymbol{\sigma})} \right\rangle \frac{\partial F}{\partial \boldsymbol{\sigma}} \cdot \boldsymbol{\sigma} \quad \text{in } \mathbf{R}_+ \times \Omega, \tag{10}$$

$$\left\{ \begin{array}{l} \left\{ \begin{array}{l} \sigma_{rr}^{\mathrm{R}}(t, a, \theta) = p - \sigma_{rr}^{\mathrm{P}}(\theta) \\ \sigma_{r\theta}^{\mathrm{R}}(t, a, \theta) = 0 \end{array} \right. , \quad (\forall)\, t > 0,\ \theta \in [0, 2\pi),\ (\text{on } \Gamma_1). \\ u(t, ma, \theta) = 0,\ (\forall)\, t > 0,\ \theta \in [0, 2\pi),\ (\text{on } \Gamma_2), \end{array} \right. \tag{11}$$

$$\left\{ \begin{array}{l} \boldsymbol{\sigma}^{\mathrm{S}}(0, r, \theta) = \boldsymbol{\sigma}^{\mathrm{P}}(\theta) + \check{\boldsymbol{\sigma}}(r, \theta) \\ (u_r, u_\theta)(0, r, \theta) = (\check{u}_r, \check{u}_\theta)(r, \theta) \quad , \quad (\forall)\, (r, \theta) \in \Omega, \\ W^{\mathrm{I}}(0, r, \theta) = H(\boldsymbol{\sigma}^{\mathrm{P}}(\theta)) \end{array} \right. \tag{12}$$

where $\check{\boldsymbol{\sigma}}$ and $(\check{u}_r, \check{u}_\theta)$ are the stress and, respectively, the displacement corresponding for the moment of excavation and $\sigma_{rr}^{\mathrm{P}}(\theta) = \frac{1}{2}(\sigma_h + \sigma_v) + \frac{1}{2}(\sigma_h + \sigma_v)\cos 2\theta$.

THE NUMERICAL APPROACH

For the problem (8)-(12) we determine a numerical solution based on some results presented by Ionescu and Sofonea [1993] using a complete implicit method for integration in time (see Paraschiv-Munteanu [1997]).

If $(u, \boldsymbol{\sigma}^{\mathrm{R}}, W^{\mathrm{I}})$, where $u = (u_r, u_\theta)$, is the solution of the problem (8)-(12) then we determine:

$$\bar{u} = u - \check{u}, \quad \bar{\boldsymbol{\sigma}} = \boldsymbol{\sigma}^{\mathrm{R}} - \check{\boldsymbol{\sigma}}^{\mathrm{R}}, \tag{13}$$

such that

277

$$\bar{u}(t,ma,\theta) = 0, \quad (\forall)\, t > 0,\ \theta \in [0,2\pi),$$

$$\mathrm{Div}\,\bar{\sigma}(t,r,\theta) = 0, \quad (\forall)\, t > 0 \ \text{and}\ (r,\theta) \in \Omega,\tag{14}$$

$$\bar{\sigma}(t,a,\theta)\mathbf{n} = 0, \quad (\forall)\, t > 0,\ \theta \in [0,2\pi).$$

So, we have to solve the problem:

find the displacement function $\bar{u} : \mathbf{R}_+ \times \Omega \longrightarrow \mathbf{R}^2$, the stress function $\bar{\sigma} : \mathbf{R}_+ \times \Omega \longrightarrow \mathcal{S}_3$ and the irreversible stress work function $W^{\mathrm{I}} : \mathbf{R}_+ \times \Omega \longrightarrow \mathbf{R}$ such that:

$$
\begin{aligned}
\dot{\bar{\sigma}} =\ & 2G\varepsilon(\dot{\bar{u}}) + (3\mathrm{K} - 2\mathrm{G})\,\varepsilon(\dot{\bar{u}})\mathbf{1} + k_{\mathrm{T}}\left\langle 1 - \frac{W^{\mathrm{I}}(t)}{H(\bar{\sigma} + \dot{\sigma} + \sigma^{\mathrm{P}})}\right\rangle \\
& \left[\frac{(2G - 3\mathrm{K})}{3}\frac{\partial F}{\partial \sigma}(\bar{\sigma} + \dot{\sigma} + \sigma^{\mathrm{P}})\mathbf{1} - 2G\frac{\partial F}{\partial \sigma}(\bar{\sigma} + \dot{\sigma} + \sigma^{\mathrm{P}})\right] + \\
& k_{\mathrm{S}}\left[\frac{(2G - 3\mathrm{K})}{3}\frac{\partial S}{\partial \sigma}(\bar{\sigma} + \dot{\sigma} + \sigma^{\mathrm{P}})\mathbf{1} - 2G\frac{\partial S}{\partial \sigma}(\bar{\sigma} + \dot{\sigma} + \sigma^{\mathrm{P}})\right] \\
& \text{in } \mathbf{R}_+ \times \Omega.
\end{aligned}
\tag{15}
$$

$$
\dot{W}^{\mathrm{I}} = k_{\mathrm{T}}\left\langle 1 - \frac{W^{\mathrm{I}}(t)}{H(\bar{\sigma} + \dot{\sigma} + \sigma^{\mathrm{P}})}\right\rangle \frac{\partial F}{\partial \sigma}(\bar{\sigma} + \dot{\sigma} + \sigma^{\mathrm{P}}) \cdot (\bar{\sigma} + \dot{\sigma} + \sigma^{\mathrm{P}})
\tag{16}
$$
$$\text{in } \mathbf{R}_+ \times \Omega,$$

$$
\left\{
\begin{aligned}
& \bar{\sigma}(t_0,r,\theta) = 0 \\
& \bar{u}(t_0,r,\theta) = 0 \\
& W^{\mathrm{I}}(t_0,r,\theta) = H(\sigma^{\mathrm{P}}(\theta))
\end{aligned}
\right.
\qquad (\forall)\,(r,\theta) \in \Omega.
\tag{17}
$$

In order to determine a numerical approach of the solution of the problem (15)-(17) we consider an interval $[0,T]$, $T > 0$ and $t_0 = 0$.

Let us note

$$\mathbf{V}_1 = \left\{ v = (v_1,v_2,0)\,|\,v_i \in \mathrm{L}^2(\Omega),\ v_i = v_i(r,\theta),\ v_i(ma,\theta) = 0\, i = 1,2 \right\},\tag{18}$$

$$\mathcal{V}_2 = \left\{ \sigma \in \left[\mathrm{L}^2(\Omega)\right]_s^{3\times3}\,|\,\sigma = \sigma(r),\ \mathrm{Div}\,\sigma = 0 \ \text{în } \Omega,\ \sigma(a,\theta)\mathbf{n} = 0 \right\}.\tag{19}$$

From (14) result that the solution $(\bar{u},\bar{\sigma},W^{\mathrm{I}})$ of the problem (15)-(17) has the properties:

$$\bar{u} \in \mathbf{V}_1, \quad \bar{\sigma} \in \mathcal{V}_2.\tag{20}$$

Let $M \in \mathbf{N}$, $M \geq 2$, $\Delta t = \dfrac{T}{M}$ be the step time and

$$t_0 = 0, \quad t_{n+1} = t_n + \Delta t, \quad n = \overline{0, M - 1}.\tag{21}$$

Let us consider $\mathbf{V}_h \subset \mathbf{V}_1$ a finite-dimensional subspace constructed using the finite element method. We determine $\left(\overline{u}_h^n, \overline{\sigma}_h^{n+1}, \left(W^I\right)_h^{n+1}\right)$ approach of the solution $\left(\overline{u}, \overline{\sigma}, W^I\right)$ on the moment t_n.

Let $\mathcal{B} = \{\varphi_1, \ldots, \varphi_I\} \subset \mathbf{V}_h$ be a base of $\mathbf{V}_h$, dim $\mathbf{V}_h = I$. Taking $\overline{u}_h^0 = 0$ we determine $\overline{u}_h^{n+1} \in \mathbf{V}_h$, $n \geq 0$, such that:

$$\overline{u}_h^{n+1} = \sum_{j=1}^{I} \alpha_j^{n+1} \varphi_j, \tag{22}$$

where the constants α_j^{n+1}, $j = \overline{1, I}$ are the solution of a linear system. For the stress approach and irreversible stress work approach we consider $\overline{\sigma}_h^0 = 0$ and $\left(W^I\right)_h^0 = H(\sigma^P)$ and we determine $\overline{\sigma}_h^{n+1}$ and $\left(W^I\right)_h^{n+1}$, $n \geq 0$, using the following implicite scheme:

$$
\begin{aligned}
\overline{\sigma}_h^{n+1} &= \overline{\sigma}_h^n + 2G\left[\boldsymbol{\varepsilon}(\overline{u}_h^{n+1}) - \boldsymbol{\varepsilon}(\overline{u}_h^n)\right] - (3K - 2G)\left[\varepsilon(\overline{u}_h^{n+1}) - \varepsilon(\overline{u}_h^n)\right]\mathbf{1} + \\
&\quad \Delta t\, k_{\mathrm{T}} \left\langle 1 - \frac{\left(W^I\right)_h^{n+1}}{H(\overline{\sigma}_h^{n+1} + \tilde{\sigma} + \sigma^P)} \right\rangle \\
&\quad \left[\frac{(2G - 3K)}{3}\frac{\partial F}{\partial \sigma}(\overline{\sigma}_h^{n+1} + \tilde{\sigma} + \sigma^P)\mathbf{1} - 2G\frac{\partial F}{\partial \boldsymbol{\sigma}}(\overline{\sigma}_h^{n+1} + \tilde{\sigma} + \sigma^P)\right] + \\
&\quad \Delta t\, k_S \left[\frac{(2G - 3K)}{3}\frac{\partial S}{\partial \sigma}(\overline{\sigma}_h^{n+1} + \tilde{\sigma} + \sigma^P)\mathbf{1} - 2G\frac{\partial S}{\partial \boldsymbol{\sigma}}(\overline{\sigma}_h^{n+1} + \tilde{\sigma} + \sigma^P)\right],
\end{aligned}
\tag{23}
$$

and, respectively,

$$
\begin{aligned}
\left(W^I\right)_h^{n+1} &= \left(W^I\right)_h^n + \Delta t\, k_{\mathrm{T}} \left\langle 1 - \frac{\left(W^I\right)_h^{n+1}}{H(\overline{\sigma}_h^{n+1} + \tilde{\sigma} + \sigma^P)} \right\rangle \\
&\quad \frac{\partial F}{\partial \sigma}(\overline{\sigma}_h^{n+1} + \tilde{\sigma} + \sigma^P)(\overline{\sigma}_h^{n+1} + \tilde{\sigma} + \sigma^P).
\end{aligned}
\tag{24}
$$

For numerical solution we use the scheme proposed by Paraschiv-Munteanu [1997] using FEM for spatial integration and a complet implicit method for integration in time. For short period of time the creep solution and the numerical solution are very close. Thus deformation by creep and stress variation can simultaneously be described. The similar results for deep boreholes are obtained by Paraschiv-Munteanu [1997], [1998].

REFERENCES

Cristescu, N., [1989], "Rock Reology", Kluwer Academic Publ., Dordrecht.

Cristescu, N., [1994], "Viscoplasticity of Geomaterials. Time-Dependent Behaviour of Geomaterials", Edited by Cristescu N. and Gioda G., Springer, Wien.

Cristescu, N., Hunsche, U., [1998], "Time Effects in Rock Mechanics", John Wiley & Sons, Chichester-New York-Weinheim-Brisbane-Singapore-Toronto.

Ionescu, I.R., Sofonea, M., [1993], "Functional and Numerical Methods in Viscoplasticity". Oxford University Press.

Jumikis, A.R., [1979], "Rock mechanics", Trans. Tech. Publ., Clausthal-Zellerfeld.

Massier, D., [1994], "Fluajul unei roci linear vâscoelastice în jurul unei cavităţi cilindrice circulare executate în masiv cu tensiuni primare oarecari", Studii şi cercetări de mecanică aplicată, Bucharest, Romania, vol. 53, 5, 427-442.

Massier, D., [1997], "Mecanica Rocilor", University of Bucharest, Romania.

Obert, L., Duvall, W., [1967], "Rock mechanics and the design of structures in rock", John Wiley & Sons, Inc., New York.

Paraschiv, I., Cristescu, N., [1993], "Deformability response of rock salt around circular mining excavations", Rev. Roum. Sc. Tech. - Mécanique Appliquée, vol. 38, 3, 257-276.

Paraschiv-Munteanu, I., [1997], "Metode Numerice în Geomecanică", Ph.D. Dissertation, University of Bucharest, Romania.

Paraschiv-Munteanu, I., [1998], "A numerical analysis of behaviour of the elasto-viscoplastic rock around vertical boreholes", Analele Univ. Bucureşti. Mat., vol. XLVII, 83-92.

Basic and Applied Salt Mechanics, Cristescu, Hardy, Jr & Simionescu (eds)
© 2002 Swets & Zeitlinger, Lisse, ISBN 90 5809 383 2

Comparative study of the convergence of a tunnel excavated in rock salt: Mathematical modelling, in-situ measurement, and numerical solution

Mariana Nicolae
General Division for Forecasts, Ministry of Finance, Bucharest, Romania

ABSTRACT

One of the most important investigation methods of the geomechanical phenomena around the underground mining excavation is the mathematical modelling method. It allows the correlation of theoretical models with both the laboratory data and the field measurements.

This paper presents the constitutive equations of Cristescu type [1]-[2], determined from a large number of distinct laboratory tests. The aim is to describe the time behaviour of rock salt around underground mining excavations.

Generally, in the mathematical models used today one is making several simplifying assumptions, which could yield to some discrepancies when the results are compared with the measurements "in situ". Thus, mathematical models, which would describe more accurately the mechanical properties exhibited by rock salt, are needed.

Constitutive Equations for Rock Salt

In order to describe the slow deforming in time rock salt, taking into account the volumetric dilatancy and/or compressibility, it's mechanical properties will be modelled using an elaso-viscoplastic constitutive equation [1]:

Based on the experimental data, let us formulate some constitutive assumptions.
- *The rock salt is homogeneous and isotropic*;
- *The strain rates are additive*:

$$\dot{\varepsilon} = \dot{\varepsilon}^{E} + \dot{\varepsilon}^{I}$$

where ε is the deformation tensor, ε^{E} is the elastic component of the deformation tensor, ε^{I} is the irreversible component of the deformation tensor, σ is the stress tensor, and **1** is the Kronecker tensor of the second order, K is the bulk modulus, G is the shearing modulus, and upper " · " denotes time derivatives.

- The irreversible part $\dot{\varepsilon}^{I}$ of the strain rate tensor, is obtained from:

$$\dot{\varepsilon}^{I} = k(\sigma, \bar{\sigma}, \Delta)\left\langle 1 - \frac{W^{I}(t)}{H(\sigma, \bar{\sigma}, \Delta)}\right\rangle \frac{\partial F(\sigma, \bar{\sigma}, \Delta)}{\partial \sigma}$$

where function F is *a viscoplastic potential*, $H(\sigma, \bar{\sigma}, \Delta)$ is a yield function, and k>0 is a *viscosity coefficient*;

- The equation of *the stabilisation boundary* is written in the form: $H(\sigma, \bar{\sigma}, \Delta) = W^I(t)$, where $W^I(t)$ is the work-hardening parameter;

- From these assumptions, the *constitutive equation* of quasi-linear type Cristescu, [1] is obtained as:

$$\dot{\varepsilon} = \frac{\dot{\sigma}}{2G} + \left(\frac{1}{3K} - \frac{1}{2G}\right)\dot{\sigma}\mathbf{1} + k\left\langle 1 - \frac{W^I(t)}{H(\sigma)}\right\rangle \frac{\partial F}{\partial \sigma}$$

The irreversible part of the volumetric strain rate, is obtained from:

$$\dot{\varepsilon}_v^I = k\left\langle 1 - \frac{W^I(t)}{H(\sigma)}\right\rangle \frac{\partial F}{\partial \sigma} \cdot \mathbf{1}$$

Determination of constitutive parameters

The *yield function* $H(\sigma, \tau)$ was determined from the experimental data obtained in triaxial creep tests. The yield function was decomposed into the sum of two terms: $H(\sigma) = H^H(\sigma) + H^D(\sigma, \bar{\sigma})$, where $H^H(\sigma)$ is determined from the irreversible stress work of the volume change W^{IH}, and $H^D(\sigma, \bar{\sigma})$ is determined by the data of the irreversible stress work corresponding to the change in shape W^{ID} (we assume that at stabilisation $H(\sigma(t)) = W^I(t)$).

For rock salt we obtain [5]:

$$H^H(\sigma) = \frac{a\left(\dfrac{\sigma}{\sigma_*}\right)}{\left(\dfrac{\sigma}{\sigma_*}\right)^2 + b^2},$$

with a=0,0249 MPa ; b=2,1794495; $\sigma*$=1 MPa

For the function $H^D(\sigma, \tau)$ we obtain:

$$H^D(\sigma_2, \tau) = A(\sigma_2)\left(\frac{\tau}{\sigma_*}\right)^7 + B(\sigma_2)\left(\frac{\tau}{\sigma_*}\right)^2 + C(\sigma_2)\left(\frac{\tau}{\sigma_*}\right),$$

the coefficients are given by the expressions:

$$A(\sigma_2) = a_0 \exp\left[a_1\left(\frac{\sigma_2}{\sigma_*}\right)^2 + a_2\left(\frac{\sigma_2}{\sigma_*}\right) + a_3\right]$$

$$B(\sigma_2) = b_1 \exp\left[b_2\left(\frac{\sigma_2}{\sigma_*}\right)\right] \qquad ; \qquad C(\sigma_2) = \frac{c_1}{\left(\dfrac{\sigma_2}{\sigma_*}\right) + c_3} + c_2$$

282

and the constants have the values:

$a_0 = 1$ MPa; $\quad\quad a_1 = 0.1000432$; $\quad\quad a_2 = -2.2535811$; $\quad\quad a_3 = -11.380055$

$b_1 = 0.027$ MPa; $\quad\quad b_2 = -1.23$; $\quad\quad c_1 = -0.0016766$ MPa;

$c_2 = 0.00962$ MPa; $\quad\quad c_3 = -0.0080848$

Finally, the viscoplastic potential, is obtained as $F(\sigma,\tau) = F_1(\sigma,\tau) + g(\tau)$, where the function $F_1(\sigma,\tau)$ has the expression:

$$F_1(\sigma,\tau) = \frac{d_1\left[A(\tau) - \frac{\tau}{\sigma_*}D(\tau)\right]}{2} ln\left[\left(\frac{\sigma}{\sigma_*}\right)^2 + d_2\right] + \frac{d_2}{\sqrt{d_2}}\left[B(\tau) - \frac{\tau}{\sigma_*}E(\tau)\right]arctg\frac{1}{\sqrt{d_2}}\left(\frac{\sigma}{\sigma_*}\right) -$$

$$-\left(\frac{\tau}{\sigma_*}\right)\frac{d_1}{m}P(\tau)ln\left[m\left(\frac{\sigma}{\sigma_*}\right) + n\right] + d_1\left[C(\tau) - \frac{\tau}{\sigma_*}G(\tau)\right]ln\left[\left(\frac{\sigma}{\sigma_*}\right) + Z(\tau)\right]$$

with $\quad A(\tau) = \dfrac{d_2}{Z^2(\tau) + d_2}$ $\quad;\quad$ $B(\tau) = -\dfrac{Z(\tau)d_2}{Z^2(\tau) + d_2}$ $\quad;\quad$ $C(\tau) = 1 - \dfrac{d_2}{Z^2(\tau) + d_2}$

$$D(\tau) = \frac{pd_2[n + mZ(\tau)] + (q - d_2)[nZ(\tau) - md_2]}{d_2[n + mZ(\tau)]^2 + [nZ(\tau) - md_2]^2}$$

$$E(\tau) = \frac{d_2\{(q - d_2)[n + mZ(\tau)] + p[nZ(\tau) - md_2]\}}{d_2[n + mZ(\tau)]^2 + [nZ(\tau) - md_2]^2}$$

$$P(\tau) = \frac{n\{[mqZ(\tau) + d_2n - md_2p][n + mZ(\tau)] + p[Z(\tau)(n - mp) - mq][nZ(\tau) - md_2]\}}{[n - mZ(\tau)]\{d_2[n + mZ(\tau)]^2 + [nZ(\tau) - md_2]^2\}}$$

$$G(\tau) = \frac{Z(\tau)\{[-md_2Z(\tau) - qn + md_2p][n + mZ(\tau)] + [-nZ(\tau) + mq + np][nZ(\tau) - md_2]\}}{[n - mZ(\tau)]\{d_2[n + mZ(\tau)]^2 + [nZ(\tau) - md_2]^2\}}$$

$$Z(\tau) = -r_1\tau - r_2\tau^6 - \tau_0$$

Comparison of the convergence of a tunnel excavated in rock salt as predicted by the elasto-viscoplastic mathematical model, 'in situ' measurements, and the numerical solution

In order to validate the results of convergence of the excavation predicted by the elasto-viscoplastic mathematical model, a measurement of the convergence "in situ" was also done [4].

The initial stress state "in situ" is obviously non-zero. It is due to the overburden and to the tectonic stresses.

This stress state, varies with depth and orientation in horizontal directions. In most of the cases one considers the stress state "in situ" to be described by two components: a vertical one σ_v and a mean horizontal one σ_h. Since these stress components have a smooth variation, we assume them to be constant in the neighbourhood of the opening and equal to their value at the tunnel axes.

The location where the convergence measurements were done is in a non disturbed zone, outside the influence of any other mining excavations. Surface markers located at two opposite sides of the gallery cross-section were connected between themselves. Also, the convergence of the wall was measured by connecting marks placed at the two ends of a "diameter" (the cross-section of the tunnels considered are either non-circular, or circular). As the mathematical model shows, the strongest convergence takes place at the corners of the cross sections. This fact is confirmed by the in field observations of underground galleries. Even if the tunnels are designed with either a square transversal cross section or a rectangular or a circular one, generally after excavation, quite often additional work is needed in order to detach some pieces of rock from the tunnel's roof.

From the "in situ" measurements, we get the conclusion that the convergence seems to be apparently linear in time.

A comparison between the "in situ" measurements and the model prediction shows some small discrepancies, witch may be due to either a non-perfect determination of the model, or to the disregarding of the steady-state creep term, or to an inaccuracy of the experimental data, or to a combination of thereof.

The results obtained by model prediction (in which the stresses were assumed to be constant during creep) and by *in situ* measurements were compared with the ones obtained through the finite element method without any simplifying assumption. Around the underground gallery we have traced a network of triangles. This allows us to define correctly the conditions on the gallery boundary as well as for a finite integration domain. We have determined the strain and the stress in the rock salt in each of the network knots. The comparison of thus obtained results with the theoretical prediction and with the ones obtained by "in situ" measurements allowed us to get to the conclusion: that the strain and the stress data obtained through the numerical solution are pretty close to those obtained "in situ" measurements.

Conclusions

The comparison between the experimental results, the theoretical predictions, obtained with the mathematical model and the numerical solution, lead us to the following conclusions:

- the elasto-viscoplastic mathematical model gives a good description of the rock salt mechanical properties from both *qualitative* and *quantitative* point of view;

- for the mining excavations considered here, in the zones neighbouring the opening no failure by compression stresses were encountered;

- at the roof of the excavation a tensile strength failure can occur for certain primary stresses ratio,

- the model prediction compared with the "in situ" measurements for both the circular cavity and the square with rounded corners are matching quite well. The differences can be attributed to not long enough time of "in situ" measurements, or to the disregarding of the steady-state term in the constitutive equation;

- the convergence in time of the underground gallery as described by the model using the finite element method shows a good matching of these results with the ones obtained by measurements "in situ". These suggest us to use all the three methods to investigate the convergence of the underground mining excavations, in order to obtain a good description of the real phenomena.

References

1   Cristescu, N.D., *Rock Rheology.* Kluwer Academic, Dordrecht, 1989.
2   Cristescu, N.D., and Hunsche, U., *Time Effects in Rock Mechanics,* J. Wiley &
     Sons, Chichester – New York – Weinheim – Brisbane – Toronto, 1998.
3   Nicolae, M, D. Massier, *Rev. Rum. Sci. Tech. Mec. Appl.,* 1993, **38,** 521.
4   Nicolae, M. *Research work concerning the characterisation of the rheological
     Behavior of salt in the slanic-Prahova,* 3rd Conference Mechanical Behavior of Salt,
     Paris, Trans Tech Publ., 1996, 559.
5.   Nicolae, M. *Mathematics Modelling of the Geomechanics Phenomena round the
     opening feasible in salt massif.* Ph.D. Thesis, Bucharest University, 1997.

Basic and Applied Salt Mechanics, Cristescu, Hardy, Jr & Simionescu (eds)
© 2002 Swets & Zeitlinger, Lisse, ISBN 90 5809 383 2

Numerical modelling in viscoplastic rock salt

Simona Roatesi
Department of Mathematics, University of Bucharest, Bucharest, Romania

ABSTRACT

This paper deals with the analysis of the numerical results for underground openings in viscoplastic rock salt. Despite their complexity, the elasto-viscoplastic criteria are often integrated using analytical means for simple geometry e.g. a circular tunnel and under constant loading. But for general conditions there is a great interest in the formulation of these models for use with numerical methods such as the finite element method. The results in this paper are obtained with a finite element code called CESAR made in LCPC-Paris [1992]. The viscoplastic module is codded and implemented in finite element code CESAR by the author and presented in the paper as well. Since it was performed, the numerical code was tested through experimental data and analytical solution available for different boundary problems by the author in previous papers.

1. INTRODUCTION

Significant efforts have been devoted in the last decades to the viscoplastic approach in finite element method (FEM). Among the great number of valuable published papers in this field, let us mention here only a few. The numerical stability of schemes used in viscoplasticity has been investigated by Zienkiewicz and Cormeau [1974], Cormeau [1975], Ionescu and Sofonea [1993], while practical applications using FEM for viscoplastic materials has been performed by many authors in different fields of solid mechanics but not so common by in the soil and rock mechanics field: Gioda [1993], Bernaud [1991], Cristescu et al [1994], etc.

Most commonly used viscoplastic equations are simple ones e.g. Von Mises with a power law (Perzyna [1966]). Associated flow rule is also often used. However for realistic simulation of complex behaviour, a more comprehensive yield criteria and flow rule has to be used. This is usually non-associated. A class of this type of viscoplastic models has been introduced by Cristescu [1989]. The models introduced included one for salt.

Despite their complexity, these elasto-viscoplastic criteria are often integrated using analytical means for simple geometry e.g. a circular tunnel and under constant loading. However, these conditions are not realised so there is a recent interest (Roatesi [1997], [1998], Roatesi and Chan [1998]) in the formulation of these models for use with numerical methods such as FEM.

Backward Euler with direct substitution and backward Euler with Quasi-Newton method.have been introduced for the numerical integration of the constitutive relations. The Quasi-Newton method used was the Powell hybrid method using a forward difference approximation to estimate the jacobian.

For the numerical examples studied, the backward Euler scheme with either the direct substitution or the Powell method seems to be unconditionally stable. However as the flow criteria and the yield function are highly non-linear, a mathematical proof is not readily available.

With the backward Euler method, the gradient quantities are defined at the resulting stress state at the end of the increment. Simple integration scheme for the incremental visco-plastic strain and plastic work done was originally used. However as the size of time steps reached the order of 100,000 seconds, the accuracy of the numerical integration of the visco-plastic strain and plastic work becomes a significant factor in the accuracy.

This lack of accuracy also leads to the instability of the scheme. The lack of accuracy has been tackled using the analytical scheme to calculate the visco-plastic strain and plastic work done within a time step. This has highly enhanced the accuracy of the overall scheme.

On the other hand, because of the predictive nature of the backward Euler scheme, the stress state may reach locations in the stress space which is out-of-bound in the original model. This out-of-bound region has the same meaning as the zone outside the failure surface for classical plasticity theory. However, for the numerical implementation of a continuum elasto-viscoplastic model, such zone should not exist as the stress state should be allowed to reach any location of the stress space and failure is signified by a high plastic flow which would never stabilise. In order to allow for this numerical difficulty, the constitutive model was reformulated so that no zone in the stress space is forbidden. Nevertheless, for the zone outside the failure surface i.e. the "unstable" zone, even though the stress state is allowed to enter it, the strain rate will be so high that the stress state will never stabilise within that region.

2. THE CONSTITUTIVE MODEL

A constitutive law of the following form will be considered, see Cristescu [1989]:

$$\dot{\varepsilon} = \left(\frac{1}{3K} - \frac{1}{2G}\right)\dot{\sigma}1 + \frac{1}{2G}\dot{\sigma} + k(\sigma, d)\left\langle 1 - \frac{W^{I}(t)}{H(\overline{\sigma}, \sigma)}\right\rangle \frac{\partial F}{\partial \sigma} \tag{1}$$

$$W^{I}(t) = \int_{0}^{T} \sigma(t) \cdot \dot{\varepsilon}^{I}(t)dt \tag{2}$$

is the irreversible stress work that is the hardening parameter or internal state variable and

$$H(\sigma, \overline{\sigma}) = W^{I}(t) \tag{3}$$

the equation of the stabilization boundary.

In the previous equation σ is the mean stress and

$$\overline{\sigma}^{2} = \sigma_{1}^{2} + \sigma_{2}^{2} + \sigma_{3}^{2} - \sigma_{1}\sigma_{2} - \sigma_{2}\sigma_{3} - \sigma_{3}\sigma_{1} \tag{4}$$

the equivalent stress (or the octahedral shear stress $\tau = \sqrt{2}\big/3\,\overline{\sigma} = \sqrt{2}\big/\sqrt{3}\,II_{\sigma'}$ with $II_{\sigma'}$ being the second invariant of the stress deviator), the first two invariants of the stress tensor σ

$H(\sigma)$ and $F(\sigma)$ represent the loading function (generally a function of stress tensor σ) and the plastic potential, respectively.

The bracket $< >$ represents the positive part of respective function: $\langle A \rangle = (A + |A|)/2 = A^+$

For the model describing the salt behavior Cristescu [1994], the constitutive functions and material constants used are:

$H(\sigma, \tau) = H_H(\sigma) + H_D(\sigma, \tau)$ (the yield surface), with

$$H_H(\sigma) = \left\{ \begin{array}{ll} h_0 \sin(\omega \dfrac{\sigma}{\sigma_*} + \varphi) + h_1 & \text{if } \sigma \leq \sigma_0 \\[2mm] h_0 + h_1 & \text{if } \sigma \geq \sigma_0 \end{array} \right. \tag{5}$$

$$H_D(\sigma, \tau) = A(\sigma)\left(\frac{\tau}{\sigma_*}\right)^{14} + B(\sigma)\left(\frac{\tau}{\sigma_*}\right)^{3} + C(\sigma)\left(\frac{\tau}{\sigma_*}\right)$$

with the coefficients A,B,C thoght to be dependent on the confining pressure, Using the data, these coefficients are obtained of the following form:

$$A(\sigma) := a_1 + \frac{a_2}{\left(\dfrac{\sigma}{\sigma_*}\right)^6}, \quad B(\sigma) := b_1 \frac{\sigma}{\sigma_*} + b_2, \quad C(\sigma) := c_2 + \frac{c_1}{\left(\dfrac{\sigma}{\sigma_*}\right)^3 + c_3},$$

The expression for the viscoplastic potential $F(\sigma, \tau)$ is much more involved:

$$k_1 F(\sigma, \tau) := \sigma_* \left\{ \frac{f_1 p_1}{4}[Y(\sigma, \tau)]^4 + \left[-\frac{4}{3}f_1 p_1 Z(\tau) + \frac{f_2 p_1 + f_1 p_2}{3} \right] [Y(\sigma, \tau)]^3 + \right.$$

$$+ \left[3f_1 p_1 [Z(\tau)]^2 - \frac{3}{2}(f_2 p_1 + f_1 p_2)Z(\tau) + \frac{1}{2}\left(f_2 p_2 + f_1 p_3 - \frac{\tau}{\sigma_*}p_1\right) \right][Y(\sigma, \tau)]^2 + \tag{6}$$

$$+ \left[-4f_1 p_1 [Z(\tau)]^3 + 3(f_2 p_1 + f_1 p_2)[Z(\tau)]^2 - 2\left(f_2 p_2 + f_1 p_3 - \frac{\tau}{\sigma_*}p_1\right)Z(\tau) \right] Y(\sigma, \tau) +$$

$$+ \left[f_1 p_1 [Z(\tau)]^4 - (f_2 p_1 + f_1 p_2)[Z(\tau)]^3 + \left(f_2 p_2 + f_1 p_3 - \frac{\tau}{\sigma_*}p_1\right)[Z(\tau)]^2 - \right.$$

$$\left. -\left(f_2 p_3 - \frac{\tau}{\sigma_*}p_2\right)Z(\tau) - \frac{\tau}{\sigma_*}p_3 \right] \ln Y(\sigma, \tau) + \left(f_2 p_3 - \frac{\tau}{\sigma_*}p_2\right)\frac{\sigma}{\sigma_*} \right\}[G(\tau) + 1] + g(\tau),$$

with $Y(\sigma, \tau), Z(\tau), G(\tau), g(\tau)$, defined by:

$$Y(\sigma, \tau) := -r\frac{\tau}{\sigma_*} - s\left(\frac{\tau}{\sigma_*}\right)^6 + \tau_0 + \frac{\sigma}{\sigma_*},$$

$$Z(\tau) := -r\frac{\tau}{\sigma_*} - s\left(\frac{\tau}{\sigma_*}\right)^6 + \tau_0 = Y(\sigma, \tau) - \frac{\sigma}{\sigma_*},$$

$$G(\tau) := u_1 \frac{\tau}{\sigma_*} + u_2\left(\frac{\tau}{\sigma_*}\right)^2 + u_3\left(\frac{\tau}{\sigma_*}\right)^3 + u_4\left(\frac{\tau}{\sigma_*}\right)^8,$$

$$g(\tau) := g_0 \frac{\tau}{\sigma_*} + \frac{g_1}{2}\left(\frac{\tau}{\sigma_*}\right)^2 + \frac{g_2}{4}\left(\frac{\tau}{\sigma_*}\right)^4,$$

h_0 =0.116 MPa, h_1=0.103 MPa, ω=2.91 , φ =-64 , σ_0 =53 Mpa, a_1=7e-21 Mpa, a_2=6.73e-12 Mpa, b_1=1.5732e-6 Mpa, b_2=1.766e-5 Mpa, c_1=26.123 Mpa, c_2=-0.00159 Mpa, c_3=3134, r=0.91, s=1.025e-8, τ_1=1.82, p_1=-9.83e-7 1/s, p_2=-5.226e-5 1/s, p_3=9.84e-5 1/s, u_1=0.0365, u_2=-0.00265, u_3=5.256e-5, u_4=1.576e-12, g_0 = 0.0108 MPa/s, g_1 = -6.582e-5 MPa/s, g_2 = 5.954e-6 MPa/s, f_1=-0.01697, f_2=0.8996, q=1.5e-8 1/s, k=2.5e-6 1/s (the viscosity coefficient) and the elastic constants are: E = 30 GPa, v = 0.27.

Note that the yield criteria are both first order Euler function in stresses.

Within the constitutive formulation the damage parameter for dilatant rocks, d_f is introduced, being able to describe the damage history.

The primary stress of the massive is dependent of the depth at which the opening is excavated. It is considered that the lining of the cavity is a distinct material with specific mechanical properties. It is studied the influence of the lining mounting.

The applicable domain for the constitutive equation is considered for compressive stresses (positive) and bounded by the failure surface which can may be incorporated in the constitutive equation.

The damage parameter for dilatant rocks can be introduced within the constitutive law, being able to describe the damage history. For that purpose let's consider the irreversible stress work split in two parts corresponding to the volume deformation and to the shape change, denoted W^I_v and W^I_d respectivelly.

During an uniaxial or triaxial tests, for most rocks W^I_v increase firstly, in the compressibility period, then deceases in the dilatancy period. The total decreasing of W^I_v from its maxim value W^I_v (max) will be used as a damage measure due to the loading, by dilatancy. So, let's define: d(t) = W^I_v (max) - W^I_v (t) the damage parameter at time t. It can be mentioned that t is a moment during dilatancy period, so t > t_{max} , with t_{max} the moment when W^I_v is reaching its maximum.

The parameter that corresponds to the failure threshold will be considered as the total release energy due to microcracking during the whole dilatancy process, and it is: d_f = W^I_v (max) - W^I_v (failure) and it can be estimated by laboratory tests.

For many types of rocks W^I_v (max) is small when they are in an advanced range of dilatancy, so the damage parameter can be: d_f (t) = - W^I_v (t) and the parameter corresponding to the failure threshold will be: d_f = - W^I_v (failure).

3. THE NUMERICAL FORMULATION

The numerical formulation of the boundary problem is then presented. to overcome the stability-related time increment restrictions inherent in the explicit methods, an implicit time-integration scheme requiring constitutive iteration has been proposed. The backward Euler method is used for the numerical integration of the constitutive equation.

A few procedures for integrating the evolution equations of stress and viscoplastic work done over a generic time increment of length Δt in order to be implemented in finite element analysis are presented here.

Using backward Euler scheme, the constitutive equation (1) was also written as:

$$\dot{\sigma} = f(\sigma, W^I) = \lambda \, tr \dot{\varepsilon} \, 1 + 2\mu\dot{\varepsilon} - k\left\langle 1 - \frac{W^I(t)}{H(\sigma,\tau)} \right\rangle \left[\lambda \, tr \frac{\partial F}{\partial \sigma} 1 + 2\mu \frac{\partial F}{\partial \sigma} \right].$$

(7)

Let $[0,T] \subset \mathbf{R}$ be the time interval of interest. At time $t_n \in [0,T]$ it is assumed that the total strain increment, the viscoplastic strain, the stress state, the internal variable which is the viscoplastic work done and the time increment Δt are known, that is: $\{\Delta\varepsilon, \varepsilon_n^{vp}, \sigma_n, W_n^I, \Delta t\}$ given data at t_n.

By applying an implicit backward Euler scheme to the evolution equation (7) and making use of the initial condition, one is led to the following discrete non-linear coupled system for every iteration i:

$$\begin{cases} \sigma_{n+1}^{i+1} = \sigma_n + f\left(\sigma_{n+1}^i, W_{n+1}^{Ii}\right)\Delta t \\ W_{n+1}^{Ii+1} = W_n^I + \sigma_{n+1}^{i+1}\, \Delta\varepsilon_{vpn+1}^{i+1} \end{cases}$$

where $\Delta\varepsilon_{vp}^{i+1} = \Delta\varepsilon - D_e^{-1}(\sigma_{n+1}^{i+1} - \sigma_n)$ has been used to calculate the evolution of stress state and viscoplastic work done.

The extension of the numerical model

As the constitutive domain is bounded by the failure surface it is worthwile to analyse the cases when the stress state is outside the constitutive domain during the calculation of a boundary problem.

The stress state may reach locations in the stress space which are out-of-bound in the constitutive domain because either of the primary (initial) conditions of the problem, either caused by the loading or during the iteration. In the first two cases the reason is of phisical nature, but the third case is purely numerically. In that case, a numerical procedure has to be adopted in order to avoid the outside locations.

In our constitutive models, as the failure surface is defined as $Y(\sigma,\tau)=0$ and due to restrictions to the constitutive functions we obtained as a condition of being located inside the constitutive domain as $Y(\sigma,\tau)>0$. That restriction is tested during the calculations and if it is not fulfilled, a numerical procedure is used in order to bring the stress state back into the domain. For the salt model, taking into account that the failure surface is defined as:

$$Y(\sigma,\tau):= -r\frac{\tau}{\sigma_*} - s\left(\frac{\tau}{\sigma_*}\right)^6 + \tau_0 + \frac{\sigma}{\sigma_*} = 0,$$

and the values of the constitutive constants we can obtain a condition for the stress state to be inside the onstitutive domain.

4. NUMERICAL APPLICATION

The proposed boundary problem is as follows:

It is assumed that the rock mass is an infinite body in which circular opening is made. Therefore only plane strain condition is considered.

Assuming further that the underground opening is at a certain depth where the horizontal and vertical components of the primary (initial) stress σ_h and σ_v are known and, generally, distinct. Cylindrical coordinate system is chosen for convenience with axis Oz being the symmetry axis of the opening.

It will be considered that the primary stress depends on the depth at which the opening is excavated. Let us consider a Cartesian coordinate system with the Oz-axis directed verticaly downward. One gets: $\sigma_{xx}(h) = \sigma_h(h)$, $\sigma_{yy}(h) = \sigma_h(h)$, $\sigma_{xy} = \sigma_{yz} = \sigma_{xz} = 0$, $\sigma_{zz} = \int \rho g\, dz = \sigma_v(h)$, with h=constant the depth of the excavation.

Sometimes a mean value for the density is used, so: $\sigma_v = K\rho h = \gamma h$ with $K = 9.86 \times 10^{-6}$, the density ρ expressed in Kg/m^3, the depth h in metres, the stress σ_v in MPa and the volumetric weight γ in MPa/m. The evaluation $\sigma_h = n\sigma_v$, $n \in [0.3,3]$ will be considered. Due to the fact that the primary stress varies with the depth, a half of the domain has to be considered due to the Ox-axis simmetries (not a quarter of the domain as in the case of the assumption of constant primary stress in the entire domain).

The influence of the Earth surface can be neglected or not depending on the depth under consideration. An internal pressure acting on the tunnel walls could be considered or not.

In the case of numerical approach, the domain to be discretised and the boundary conditions are (see figure 1):

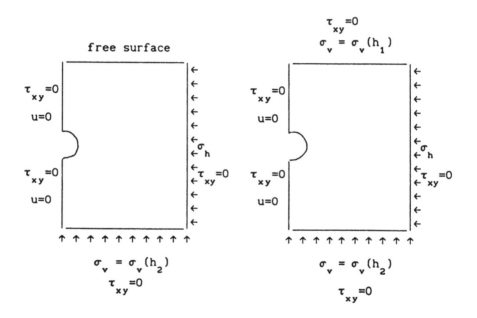

Figure1–The domain and the boundary conditions for the numerical formulation

It has been denoted: u, v the horizontal and vertical component of displacement, respectively, τ_{xy} the tangential component of stress, h_1 and h_2 the depth of superior and inferior horizontal edges of domain, respectively.

In the calculations performed displacement boundary conditions were used.

The numerical modelling uses a mesh of 6-nodded triangles containing 524 nods and 221 elements. The mesh used is refined around the opening.

The opening radius is r = 1 m, the depth h = 258.5m and the ratio n = σ_h / σ_v = 0.3.

Further some interesting sets of applications are performed in order to emphesize the importance of different factors of the analysis. The stress, the displacement, the viscoplastic and the damage zones around the opening are studied. Unfortunately, graphical output for these various studies did not copy satisfactorily in black & white and only a few examples are presented in this paper.

The influence of the ratio n = σ_h / σ_v

The influence of the ratio n is one of the most important. The localization of the zones of fracture, damage, high compression etc around the opening is given by that factor. The calculations were repeated, for instance, for $n=1.8$. In this case the zones of damage are located in the floor and ceiling (in the floor being bigger due to the fact that the primary stress is depending on the depth) (see figure 2), while for the case $n=0.3$ these zones are located in the wall of the opening (see figure 3). In both figures Δt is 150 days.

The zones of high compression stress are located in the same manner. The compression stress is higher when the depth is higher. Eventually fracture due to tensile stress could appear in the wall in the case $n=1.8$ and in the ceiling in the case $n=0.3$.

Regarding the vertical displacement the unsymmetry can be seen by the different values in the ceiling and in the floor. For $n=0.3$ the zones of isovalues are lenghtened, while for the case of $n=1.8$ they are horizontal.

The influence of the depth h

Concerning the influence of the depth another set of calculations was performed in which a depth of 1000m was considered. Examinating for instance the isovalues for the vertical displacement much bigger values were observed for the case of greater depth.

The same conclusion is drawn in the case of damage zones indicating dilatancy.

Concerning the hoop stress we found that the compression stress occurs at great depth, while tensile stress appears at small depth.

The influence of the pressure p

The calculations were first performed with no pressure acting on the opening wall, and then repeated with a nonzero pressure p acting on the wall, namely; $p=2$Mpa. Considering the isovalues for the octahedrical stress, it was seen that the presence of a pressure on the opening wall results in a smaller compression stress and reduces the risk of a tensile stress that could lead to a wall fracture. Basically, the influence of the pressure is on the magnitude of the entities under consideration.

FINAL REMARKS

The data presented points out the features of the numerical solution and its concordance with the available analytical solutions and in situ measurements. It is important to mention that there is no more symmetry as in the case of constant primary stress in the whole domain, due to the fact that the p[rimary stress varies with depth.

This loss of symmetry is in accordance with the practical observations in which for instance, for that state of primary stress, the ceiling of the tunnel goes down less than the floor goes up. Time increases this effect.

It is well known that the choice of a time step Δt is quite important in viscoplasticity problems. This choice must be correlated with the viscosity parameter k as well, so that the value of kΔt used insures a good accuracy of the numerical scheme.

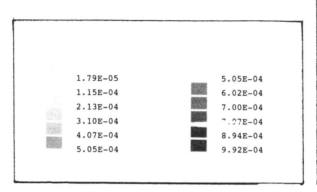

Fig.2–Contours of damage parameter
evolution of a tunnel in rock salt for:
p=0Mpa, n=1.8, h=258.8m, Δt=150days

1.79E-05	5.05E-04
1.15E-04	6.02E-04
2.13E-04	7.00E-04
3.10E-04	7.97E-04
4.07E-04	8.94E-04
5.05E-04	9.92E-04

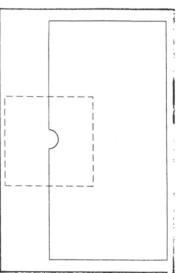

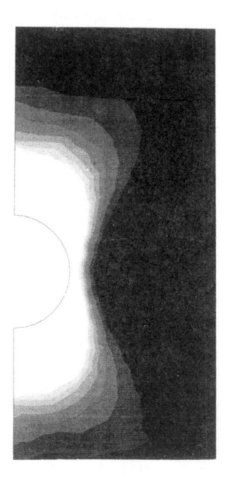

Fig.3-Contours of damage parameter
evolution of a tunnel in rock salt in the case:
p=0Mpa, n=0.3, h=258.8m, Δt=150days

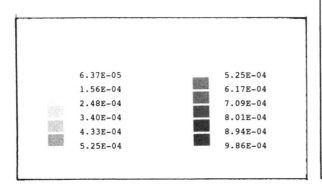

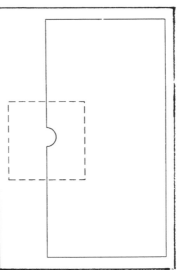

Basic and Applied Salt Mechanics, Cristescu, Hardy, Jr & Simionescu (eds)
© 2002 Swets & Zeitlinger, Lisse, ISBN 90 5809 383 2

Numerical methods in rate-type viscoplasticity

Ioan Rosca
Faculty of Mathematics, University of Bucharest, Bucharest, Romania

ABSTRACT

This paper deals with initial and boundary value problem describing the quasistatic evolution of rate–type viscoplastic material with internal state variables sumitted to contact boundary condition. Two variational formulation are proposed and some existence and uniqueness results are obtained. The numerical approach of the solutions and a concrete algorithm based on an iterative method is proposed. Some numerical results in the one–dimensional case are also presented.

INTRODUCTION

In order to model (to describe) the behavior of real materials like rubbers, metals, pastes, rocks and so on, for which the plastic rate of deformation depends both in stress and strain tensors, constitutive laws of the following form are considered

$$\dot{\sigma} = \mathcal{A}(\sigma, \varepsilon)\dot{\varepsilon} + \mathcal{G}(\sigma, \varepsilon) \tag{1}$$

in which σ denotes the stress tensor, ε represents the small strain tensor, $\mathcal{A}$ and $\mathcal{G}$ are constitutive given functions.

The first constitutive law (1) with $\mathcal{A}(\sigma, \varepsilon) = \mathcal{E}$, $\mathcal{G}(\sigma, \varepsilon) = -\sigma$ has been proposed by Maxwell in 1867, and this explains why certain authors are calling Maxwellian, materials that can be described by a constitutive equation (1).

A semilinear rate–type constitutive equation has proposed by Sokolovskii in 1948 with $\mathcal{A}(\sigma, \varepsilon) = \mathcal{E}$, and

$$\mathcal{G}(\sigma, \varepsilon) = \begin{cases} 0 & \text{if } |\sigma| < \sigma_Y \\ (-sgn\,\sigma)F(\sigma| - \sigma_Y) & \text{if } |\sigma| \geq \sigma_Y, \end{cases}$$

where $\sigma_Y > 0$ is the yield limit and $F : \mathbf{R}_+ \to \mathbf{R}$ is a smooth function with $F(0) = 0$, $F'(r) > 0$, for all $r > 0$.

Malvern in 1951 has generalized Sokolovskii idea by assuming the function $\mathcal{G}$ depends on both state variables ε and σ. The function $\mathcal{A}$ is the same as above, $\mathcal{A} = \mathcal{E}$, while $\mathcal{G}$ is given by

$$\mathcal{G}(\sigma, \varepsilon) = \begin{cases} 0 & \text{if } 0 \leq \sigma \leq f(\varepsilon) \\ -kF(\sigma - f(\varepsilon)) & \text{if } \sigma > f(\varepsilon) \end{cases} , \varepsilon \geq 0$$

where F has the same properties as above,

The general form (1), of the quasistatic rate–type constitutive equations, has been proposed by Cristescu [1], [2] and used afterwards by many authors. Consideration of a non-constant function $\mathcal{A}(\sigma, \varepsilon)$ is a necessity that is experimentally justified.

For geomaterials the following constitutive law

$$\dot{\varepsilon} = \frac{\dot{\sigma}}{2G} + \left(\frac{1}{3K} - \frac{1}{2G}\right)\dot{\sigma}\mathbf{1} + k_T\left\langle 1 - \frac{W(t)}{H(\sigma, \tau)}\right\rangle\frac{\partial F}{\partial \sigma} + k_S\frac{\partial S}{\partial \sigma} \qquad (2)$$

has been considered by Cristescu and Hunsche [3], [4], [5]. Here G and K are shear and bulk moduli respectively (generally not constant), $H(\sigma, \tau)$ is the yield function, $F(\sigma, \tau)$ is the viscoplastic potential, $S(\sigma, \tau)$ is potential for the steady–state creep k_T and k_S are two corresponding viscosity coefficients, and $W(t)$ is the irreversible stress work per unit volume.

The purpose of this paper is to investigate some abstract systems arising from quasistatic rate–type viscoplastic model involving different boundary conditions.

MECHANICAL PROBLEM

Let us consider an elastic–viscoplastic body whose material particles fulfil a bounded domain $\Omega \subset \mathbf{R}^d$ and whose boundary Γ, assumed to be sufficiently smooth, is partitionned into three disjoint measurable parts Γ_1, Γ_2 and Γ_3. Let $meas(\Gamma_1) > 0$ and let $T > 0$ be a time interval. We assume that displacement field vanishes on $\Gamma_1 \times (0, T)$, that surface tractions g act on $\Gamma_2 \times (0, T)$ and that body forces b act on $\Omega \times (0, T)$. We also assume contact conditions on $\Gamma_3 \times (0, T)$ modeled by the subdifferential $\partial\varphi$ of function φ, we consider the case of quasistatic processes. With these assumptions we have:

$$\dot{\sigma} = E(\lambda)\dot{\varepsilon}(u) + G(\lambda, \sigma, \varepsilon) \quad \text{in} \quad \Omega \times (0, T) \qquad (3)$$

$$Div\sigma + b = 0 \quad \text{in} \quad \Omega \times (0, T) \qquad (4)$$

$$u = 0 \quad \text{on} \quad \Gamma_1 \times (0, T) \qquad (5)$$

$$\sigma\nu = g \quad \text{on} \quad \Gamma_2 \times (0, T) \qquad (6)$$

$$-\sigma\nu \in \partial\varphi(u) \quad \text{on} \quad \Gamma_3 \times (0, T) \qquad (7)$$

$$u(0) = u_0, \quad \sigma(0) = \sigma_0 \quad \text{on} \quad \Omega, \qquad (8)$$

in which $u = (u_i) : \Omega \times (0, T) \to \mathbf{R}^d$ is the displacement function, $\sigma = (\sigma_{ij}) : \Omega \times (0, T) \to M^d$ is the stress, and $\lambda : \Omega \times (0, T) \to \mathbf{R}^m$ is an internal parameter function. Here M^d denotes the set of second order symmetric tensor on $\mathbf{R}^d$, $\nu = (\nu_i)$ represents the unit outward normal on Ω, u_0 and σ_0 initial data, and finally E and G are constitutive given functions.

We now present two examples of boundary conditions included in the abstract formulation (7) (see also Rochdi et al. [8]). For this, let us denote in the sequel

$$u_\nu = u_i \nu_i, \quad u_{\tau i} = u_i - u_\nu \nu_i, \quad u_\tau = (u_{\tau i}), \quad \sigma_\nu = \sigma_{ij}\nu_j\nu_i, \quad \sigma_{\tau i} = \sigma_{ij}\nu_j - \sigma_\nu \nu_i, \quad \sigma_\tau = (\sigma_{\tau i}).$$

EXAMPLE 1. (Signorini's contact conditions). We consider the boundary conditions

$$u_\nu \le s, \quad \sigma_\nu \le 0, \quad \sigma_\tau = 0, \quad \sigma_\nu(u_\nu - s) = 0 \quad \text{on} \quad \Gamma_3 \times (0, T) \tag{9}$$

where $s \ge 0$ on Γ_3 (see Figure 1).

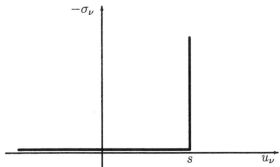

Figure 1: The graph of the contact boundary condition of example 1.

These conditions model the frictionless contact between a deformable body and a rigid foundation S; here for every point $x \in \Gamma_3$, $s(x)$ stands for the distance between x and S in the direction of the normal ν, before the forces are applied to the body. The inequality $u_\nu \le s$ describe the impenetrability condition between the body and the foundation S. It is easily seen that (9) are a special case of (7) with

$$\varphi(v) = \begin{cases} 0 & \text{if } v_\nu \le s \\ +\infty & \text{if } v_\nu > s. \end{cases}$$

EXAMPLE 2. The graph in Figure 1 is not the graph of a mapping since neither σ_ν is a function on u_ν, nor u_ν is a function on σ_ν. A penalized form of this graph is presented in Figure 2. It corresponds to the following boundary condition:

$$\sigma_\nu = \begin{cases} 0 & \text{if } u_\nu \le s \\ -\dfrac{1}{\mu}(u_\nu - s)^p & \text{if } u_\nu > s \end{cases}, \sigma_\tau = 0 \text{ on } \Gamma_3 \times (0, T) \tag{10}$$

where $0 < p \le 1$ and $\mu > 0$. This conditions model a contact problem involving a normal displacement condition with friction; here $1/\mu$ may be interpreted as a *normal friction coefficient*; the case $\mu = 0$ corresponds to an infinite normal friction coefficient i.e. to the impenetrability condition (formally (10) becomes (9) when $\mu \to 0$). It is easily seen that (10) are a special case of (7) with

$$\varphi(v) = \begin{cases} 0 & \text{if } v_\nu \le s \\ \dfrac{1}{\mu(p+1)}(v_\nu - s)^{p+1} & \text{if } v_\nu > s. \end{cases}$$

In some applications the parameter function λ in (3) is needed to be considered as an unknown function whose evolution is given by

$$\dot{\lambda}(t) = \psi(\lambda(t), \sigma(t), \varepsilon(u(t)), \quad \text{for all } t \in [0, T] \tag{11}$$

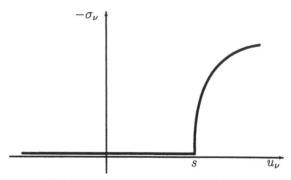

Figure 2: The graph of the contact boundary conditions of exemple 2

where ψ is a constitutive function. In this case, the mechanical problem may be formulated as follows.

Find the displacement field $u : \Omega \times [0, T] \to \mathbf{R}^d$, the stress field $\sigma : \Omega \times [0, T] \to M^d$, and the internal state variable $\lambda : \Omega \times [0, T] \longrightarrow \mathbf{R}^m$ such that

$$\dot{\sigma} = E(\lambda)\dot{\varepsilon}(u) + G(\lambda, \sigma, \varepsilon(u)) \quad \text{in} \quad \Omega \times (0, T) \tag{12}$$

$$\dot{\lambda} = \psi(\lambda, \sigma, \varepsilon(u)) \quad \text{in} \quad \Omega \times (0, T) \tag{13}$$

$$Div\sigma + b = 0 \quad \text{in} \quad \Omega \times (0, T) \tag{14}$$

$$u = 0 \quad \text{on} \quad \Gamma_1 \times (0, T) \tag{15}$$

$$\sigma\nu = g \quad \text{on} \quad \Gamma_2 \times (0, T) \tag{16}$$

$$-\sigma\nu \in \partial\varphi(u) \quad \text{on} \quad \Gamma_3 \times (0, T) \tag{17}$$

$$u(0) = u_0, \quad \sigma(0) = \sigma_0, \quad \lambda(0) = \lambda_0 \quad \text{on} \quad \Omega \tag{18}$$

VARIATIONAL PROBLEM

In order to obtain the variational formulation for the problem (3)-(8), let us consider the following notations:

$$H = L^2(\Omega)^d, \quad H_1 = H^1(\Omega)^d, \quad \mathcal{H} = [L^2(\Omega)]_s^{d \times d}, \quad \mathcal{H}_1 = \{\tau \in \mathcal{H}; Div\,\tau \in H\}.$$

Let V be a closed subspace of H_1 defined by all the functions which vanish on Γ_1 in the trace sense. We denote in the sequel by ”$\cdot$” the inner product on the space $\mathbf{R}^d$ and M^d, by $|\cdot|$ the Euclidean norms on these spaces, and let $\langle \cdot, \cdot \rangle_{\mathcal{H}}$ represent the inner product on the space $\mathcal{H}$. Let us introduce the following notations:

$$L(t, v) = \int_\Omega b(t) \cdot v d\Omega + \int_{\Gamma_2} g(t) \cdot v d\Gamma$$

$$j(v) = \begin{cases} \int_{\Gamma_3} \varphi(v) d\Gamma & \text{if} \quad \varphi(v) \in L^1(\Gamma_3) \\ +\infty & \text{otherwise .} \end{cases} \tag{19}$$

Using Green's formula, we obtain that if the couple (u, σ) is a regular solution for the mechanical problem (3)-(8), then

$$\dot{\sigma}(t) = E(\lambda)\dot{\varepsilon}(u(t)) + G(\lambda, \sigma(t), \varepsilon(u(t))), \quad \text{a.e. } t \in (0, T) \tag{20}$$

$$\left.\begin{array}{l} u(t) \in V, \\ \langle \sigma(t), \varepsilon(v) - \varepsilon(u(t)) \rangle_{\mathcal{H}} + j(v) - j(u(t)) \geq \\ L(t, v - u(t)), \ \forall \ v \in V, \end{array}\right\} \ \forall \ t \in [0, T] \tag{21}$$

$$u(0) = u_0, \quad \sigma(0) = \sigma_0 \tag{22}$$

Let us remark that the converse of the previous afirmation also holds; indeed, if (u, σ) is a regular solution of (20)-(22), after a formal computation it can be proved that (u, σ) is a solution of (3)-(8). For this reason we may consider (20)-(22) as a variational formulation of the mechanical problem (3)-(8).

Also, to the problem (12)–(18) we can associate a variational problem, namely: find $u : [0, T] \longrightarrow H_1$, $\sigma : [0, T] \longrightarrow \mathcal{H}$, $\lambda : [0, T] \longrightarrow L^2(\Omega)^m$ such that

$$\dot{\sigma}(t) = E(\lambda)\dot{\varepsilon}(u(t)) + G(\lambda, \sigma(t), \varepsilon(u(t))), \quad \text{a.e. } t \in (0, T) \tag{23}$$

$$\dot{\lambda}(t) = \psi(\lambda(t), \sigma(t), \varepsilon(u(t))), \quad \text{a.e. } t \in (0, T) \tag{24}$$

$$\left.\begin{array}{l} u(t) \in V, \\ \langle \sigma(t), \varepsilon(v) - \varepsilon(u(t)) \rangle_{\mathcal{H}} + j(v) - j(u(t)) \geq \\ L(t, v - u(t)), \ \forall \ v \in V \end{array}\right\} \ \forall \ t \in [0, T] \tag{25}$$

$$u(0) = u_0, \quad \sigma(0) = \sigma_0, \quad \lambda(0) = \lambda_0. \tag{26}$$

In order to obtain an existence and uniqueness result, let us consider the following assumptions:

The function $E : \mathbf{R}^m \times \Omega \longrightarrow M^d$ is a symmetric and positively definite tensor, i.e.

(E_1) $E(\lambda, x)\sigma \cdot \tau = \sigma \cdot E(\lambda, x)\tau, \ \forall \ \lambda \in \mathbf{R}^m, \sigma, \tau \in M^d$ a.e. in Ω;

(E_2) there exists $\alpha > 0$ such that $E(\lambda, x)\sigma \cdot \sigma \geq \alpha|\sigma|^2$, for all $\lambda \in \mathbf{R}^m, \sigma \in M^d$, a.e. in Ω;

(E_3) there exist $L > 0$ such that $|E_{ijkh}(\lambda_1, x) - E_{ijkh}(\lambda_2, x)| \leq L|\lambda_1 - \lambda_2|$ for all $\lambda_1, \lambda_2 \in \mathbf{R}^m, i, j, k, h \in \{1, \ldots, d\}$ a.e. in Ω;

(E_4) $x \to E_{ijkh}(\lambda, x)$ is a measurable function with respect to the Lebesgue measure on Ω, for all $\lambda \in \mathbf{R}^m$ and $i, j, k, h \in \{1, \ldots, d\}$;

(E_5) there exists $\beta > 0$ such that $|E_{ijkh}(\lambda, x)| \leq \beta$ for all $\lambda \in \mathbf{R}^m$ and i, j, k, h in $\{1, \ldots, d\}$, a.e. in Ω.

The function $G : \Omega \times \mathbf{R}^m \times M^d \times M^d \to M^d$ is Lipschitz continuous in the last three variables, i.e.

(G_1) there exists $\tilde{L} > 0$ such that

$$|G(x, \lambda_1, \sigma_1, \varepsilon_1) - G(x, \lambda_2, \sigma_2, \varepsilon_2)| \leq \tilde{L}(|\lambda_1 - \lambda_2| + |\sigma_1 - \sigma_2| + |\varepsilon_1 - \varepsilon_2|)$$

for all λ_1, λ_2 in $\mathbf{R}^m$ and $\sigma_1, \sigma_2, \varepsilon_1, \varepsilon_2$ in M^d, a.e. in Ω;

(G_2) $x \to G(x, \lambda, \sigma, \varepsilon)$ is a measurable function with respect to the Lebesgue measure on Ω, for all λ in $\mathbf{R}^m$ and σ, ε in M^d;

(G_3) the function $x \to G(x, 0, 0, 0)$ belongs to $L^2(\Omega)^d$.

The data $b, g, \varepsilon, \varphi\sigma_0, u_0$ satisfied some compatibility conditions:

$$b \in C^1(0, T; H), \quad g \in C^1(0, T; L^2(\Gamma_2)^d) \tag{27}$$

$$u_0 \in H_1, \quad \sigma_0 \in \mathcal{H}_1 \tag{28}$$

$$\langle \sigma_0, \varepsilon(v) - \varepsilon(u_0) \rangle_{\mathcal{H}} + j(v) - j(u_0) \geq L(0, v - u_0) \ \forall \ v \in V \tag{29}$$

The constitutive function $\psi : \Omega \times \mathbf{R}^m \times M^d \times M^d \longrightarrow \mathbf{R}^m$ has following properties

(ψ_1) there exists $L' > 0$ such that

$$|\psi(x, \lambda_1, \sigma_1, \varepsilon_1) - \psi(x, \lambda_2, \sigma_2, \varepsilon_2)| \leq L'(|\lambda_1 - \lambda_2| + |\sigma_1 - \sigma_2| + |\varepsilon_1 - \varepsilon_2|),$$

for all $\lambda_1, \lambda_2 \in \mathbf{R}^m$ and $\sigma_1, \sigma_2, \varepsilon_1, \varepsilon_2 \in M^d$, a.e. in Ω

(ψ_2) $x \to \psi(x, \lambda, \sigma, \varepsilon)$ is a measurable function with respect to the Lebesgue measure on Ω, for all σ, ε in M^d and λ in $\mathbf{R}^m$.

(ψ_3) the function $x \to \psi(x, 0, 0, 0)$ belongs to $Y = L^2(\Omega)^m$.

The assumptions (E1)-(E5), (G1)-(G3) and $(\psi_1) - (\psi_3)$ allow us to consider three operators denoted again by E, G and ψ such that
$E : Y \times \mathcal{H} \to \mathcal{H}, \quad G : Y \times \mathcal{H} \times \mathcal{H} \to \mathcal{H}, \quad \psi : Y \times \mathcal{H} \times \mathcal{H} \to Y$ and

$$E(\lambda, \sigma)(\cdot) = E_{ijkh}(\cdot, \lambda(\cdot))\sigma_{kh}(\cdot), \quad \forall \ \lambda \in Y, \sigma \in M^d,$$

$$G(\lambda, \sigma, \varepsilon)(\cdot) = G(\cdot, \lambda(\cdot), \sigma(\cdot), \varepsilon(\cdot)) \ \forall \ \sigma, \varepsilon \in \mathcal{H}, \lambda \in Y \text{ a.e. } \Omega,$$

$$\psi(\lambda, \sigma, \varepsilon) = \psi(\cdot, \lambda(\cdot), \sigma(\cdot), \varepsilon(\cdot))), \quad \forall \ \sigma, \varepsilon \in \mathcal{H}, \lambda \in Y, \text{ a.e. in } \Omega.$$

The first result of this section is given by:

THEOREM 1. *Suppose that hypetheses (E1)-(E5), (G1)–(G3), (27)-(29) are fulfilled. Then there exists $(u, \sigma) \in C^1(0, T; H_1) \times C^1(0, T; \mathcal{H}_1)$ a unique solution of the variational problem (20)-(22).*

Proof. As in Djabi and Sofonea [6] we may obtain the proof in two steps, namely:
i) Using standard arguments of elliptic variational inequalities theory, it results that for all $\eta \in C^0(0, T; \mathcal{H})$ there exists (u_η, σ_η) in $C^1(0, T; H^1) \times C^1(0, T, \mathcal{H}_1)$ a unique solution for the variational problem

$$\sigma_\eta(t) = E(\lambda)\varepsilon(u_\eta(t)) + z_\eta(t) \tag{30}$$

$$u_\eta(t) \in V, \langle \sigma_\eta(t), \varepsilon(v) - \varepsilon(u_\eta(t)) \rangle_{\mathcal{H}} + j(v) - j(u_\eta(t)) \geq L(t, v - u_\eta(t)), \ \forall \ v \in V \tag{31}$$

for all $t \in [0, T]$, where

$$z_\eta(t) = \int_0^t \eta(s)ds + \sigma_0 - E(\lambda)\varepsilon(u_0). \tag{32}$$

ii) The operator $\mathcal{R} : C^0(0, T, \mathcal{H}) \to C^0(0, T; \mathcal{H})$ defined by

$$\mathcal{R}\eta = G(\lambda, \sigma_\eta, \varepsilon(u_\eta)), \quad \forall \eta \in C^0(0, T, \mathcal{H}) \tag{33}$$

has a unique fixed point $\eta^* \in C^0(0, T, \mathcal{H})$. Moreover, the couple $(u, \sigma) = (u_{\eta^*}, \sigma_{\eta^*})$ is the unique solution of the problem (20)–(22). $\square$

A second result of this section is given by:

THEOREM 2. *Suppose that assumptions (E1)–(E5), (G1)–(G3), $(\psi_1) - (\psi_3)$ and (27)–(29) are fulfilled. Then there exists (u, σ, λ) (with u in $C^1(0, T; H_1)$, σ in $C^1(0, T; \mathcal{H}_1)$, and λ in $C^1(0, T; Y)$) a unique solution of the variational problem (23)–(26).*

Proof. The proof is made also in two steps.
(i) For each $\eta = (\eta^1, \eta^2) \in C^0(0, T, \mathcal{H} \times Y)$ we introduce the function $z_\eta = (z_\eta^1, z_\eta^2)$ in $C^1(0, T, \mathcal{H} \times Y)$ defined by

$$z_\eta = \int_0^y \eta(s)ds + z_0, \ \forall \ t \in [0, T] \tag{34}$$

where $z_0 = (\sigma_0 - E(\lambda_0)\varepsilon(u_0), \lambda_0)$. As in proof of Theorem 1 we can obtain that for all $\eta \in C^0(0, T; \mathcal{H} \times Y)$ there exist $u_\eta : [0, T] \to H_1$ and $\sigma_\eta : [0, T] \to \mathcal{H}_1$ such that

$$\sigma_\eta(t) = E(z_\eta^2(t))\varepsilon(u_\eta) + z_\eta^1(t)$$

$$\langle \sigma_\eta(t), \varepsilon(v) - \varepsilon(u_\eta(t)) \rangle_{\mathcal{H}} + j(v) - j(u_\eta(t)) \geq L(t, v - u_\eta(t)), \ \forall \ v \in V.$$

We denote by the function defined by $\lambda_\eta = z_\eta^2$.
(ii) The operator $\mathcal{R} : C^0(0, T; \mathcal{H} \times Y) \to C^0(0, T, \mathcal{H} \times Y)$ defined by

$$\mathcal{R}\eta = (G(\lambda_\eta, \sigma_\eta, \varepsilon(u_\eta)), \psi(\lambda_\eta, \sigma_\eta, \varepsilon(u_\eta))), \quad \forall \ \eta \in C^0(0, T, \mathcal{H} \times Y)$$

has a unique fixed point. Moreover $(u_{\eta^*}, \sigma_{\eta^*}, \lambda_{\eta^*})$ is the unique solution of variational problem (23)–(26). $\square$

A NUMERICAL APPROACH

In this section a numerical algorithm, which can be directly run on a computer is considered, in order to approximate the solution (u, σ) of the variational problem (20)–(22). This algorithm is based on the approximation of the unknowns in space and time. As it follows from above, the existence and uniqueness of the solution

303

for the problem (20)–(22) may be obtained in two steps: i) the study of the elliptic problem (35)–(36) defined for every element η of $C(0,T;\mathcal{H})$; ii) the fixed point property of the operator $\mathcal{R}$ defined by (33). For this reason, in order to approximate (20) - (22) we consider the discret version of (30)–(32) and the discret version of the above fixed point property. These considerations lead us to consider the following algorithm:

find $u_h^n : [0,T] \to V_h$ and $\sigma_h^n : [0,T] \to \mathcal{H}$ such that

$$\sigma_h^n = E(\lambda_h)\varepsilon(u_h^n) + z_h^n \tag{35}$$

$$\left. \begin{array}{c} u_h^n(t) \in V_h, \\ \langle \sigma_h^n(t), \varepsilon(v) - \varepsilon(u_h^n(t)) \rangle_{\mathcal{H}} + j(v) - j(u_h^n(t)) \geq L(t, v - u_h^n(t)), \ \forall \ v \in V_h \end{array} \right\} \tag{36}$$

for all $t \in [0,T]$, where: V_h is a finite dimensional subspace of V (constructed for instance using the finite element method):

$z_h^n \in C^1(0,T,\mathcal{H})$ is defined by

$$z_h^n(t) = \int_0^t \eta_h^n(s)ds + \sigma_0 - E(\lambda_h)\varepsilon(u_0), \quad \forall \ t \in [0,T], \tag{37}$$

η_h^n is recursively given by the equality

$$\eta_h^n = G(\lambda_h, \sigma_h^{n-1}, \varepsilon(u_h^{n-1})), \quad \forall n \in \mathbf{N} \tag{38}$$

$\eta_h^0 = \eta_0$ is an arbitrary element of $C(0,T;\mathcal{H})$.

Let us remark that, for a given $t \in [0,T]$ the algorithm considered here requires the resolution of the elliptic variational inequality (36) in order to compute the unknowns u_h^n (for numerical approximation of elliptic variational inequality see for example Glowinski et al. [7]). In the particular case when $\Gamma_3 = \emptyset$, $j(v)$ vanishes and the previous algorithm leads to the resolution of a sequence of linear algebraic systems in the finite dimensional space V_h.

Let us now consider $\Delta_M = \{t_k\}_{k=0}^M$ a partition of the time interval $[0,T]$, with $t_0 < t_1 < \cdots < t_M = T$. Using (36) and (37) we are able to compute the elements $u_h^n(t_k)$, $\sigma_h^n(t_k)$ for every $n \in \mathbf{N}$ and $k \in \{0,\ldots,M\}$. Indeed, let us denote by $P_h(n,k)$ the set defined by

$$P_h(n,k) = \{\eta_h^n(t_k), z_h^n(t_k), u_h^n(t_k), \sigma_h^n(t_k)\} \tag{39}$$

For all $n \in \mathbf{N}$ and $k \in \{0,\ldots,M\}$, we split the computing of $P_h(n,k)$ in the following steps:

a) **Computing the set $P_h(n,0)$** . For every $\eta_h^0 = \eta_0 \in C(0,T;\mathcal{H})$ by (37) we get $z_h^n(0) = z_0$ for every $n \in \mathbf{N}$, hence by (35), (36) and $\eta_h^n(0)$ for all $n \in \mathbf{N}$.

b) **Computing the set $P_h(0,m)$**. Since η_h^0 is given, the values $\eta_h^0(t_k)$ are known for all $m \in \{0,M\}$. The elements $z_h^0(t_k)$ can be obtained using the trapezoidal rule in order to approximate (37):

$$z_h^0(0) = z_0, \ z_h^0(t_k) = z_h^0(t_{k-1}) + \frac{t_k - t_{k-1}}{2}[\eta_n^0(t_k) + \eta_h^0(t_{k-1})] \tag{40}$$

for all $m \in \{1, M\}$ and finally $u_h^0(t_k)$, $\sigma_h^0(t_k)$ are determined from (35), (36) and (40) for all $t \in \{0, M\}$.

c) **Computing the set** $P_h(n, k)$. Let us suppose that the sets $P_h(n + 1, k - 1)$, $P_h(n, k)$ are known for a given $n \in \mathbf{N}$ and $m \in \{1, \ldots, M\}$. Using (38) we get

$$\eta_h^{n+1}(t_k) = G(\lambda_h(t_k), \sigma_h^n(t_k), \varepsilon(u_h^n)(t_k)))$$

and using again the trapezoidal rule, from (37) we obtain

$$z_h^{n+1}(t_k) = z_h^{n+1}(t_{k-1}) + \frac{t_k - t_{k-1}}{2}[\eta_h^{n+1}(t_k) + \eta_h^{n+1}(t_{k-1})].$$

Finally, $u_h^{n+1}(t_k)$, $\sigma_h^{n+1}(t_k)$ can be obtain by (35), (36). Using now the steps a), b), and c) we compute the set $P_h(n, k)$ for all $n \in \mathbf{N}$ and $k \in \{0, \ldots, M\}$ in this way the aproximative solution $u_h^n(t)$, $\sigma_h^n(t)$ is computed for all $t = t_k$, $k \in \{0, \ldots, M\}$.

As in Roşca and Sofonea [9] we can prove that (u_h^n, σ_h^n) converge to (u, σ), the unique solution of problem (20)–(22), when $\mathrm{dist}(V_h, V) \to 0$ and $n \to \infty$.

NUMERICAL RESULTS

In order to illustrate the previous algorithm two one–dimensional examples are considered, in this section.

EXAMPLE 3. Let us consider a problem of the form (3)-(8) in the following context: $\Omega = (0, 1)$, $\Gamma_1 = \{0\}$, $\Gamma_3 = \emptyset$, $b(x, t) = 0$, $g(t) = 15$, $u_0(x) = 3.0375x - 0.0375x^2$, $\sigma_0(x) = 15$, $T = 3$, $E = 20$, $G(\sigma, \varepsilon) = -10(\sigma - f(\varepsilon))$ with

$$f(\varepsilon) = \begin{cases} 10\varepsilon & \text{if } \varepsilon \leq \varepsilon \leq 2 \\ -5\varepsilon + 30 & \text{if } 2 < \varepsilon < 4 \\ 10\varepsilon - 30 & \text{if } \varepsilon \geq 4. \end{cases}$$

In this case, problem (3)-(8) represents a classical displacement traction problem and, as it was already pointed up, the algorithm proposed in previous section leads to the resolution of a sequence of linear algebraic systems. In order to apply this algorithm, let $V_h \subset V$ be the finite element space constructed with polynomial function of degree 1, Ω being divided into finite elements; the initial value considered was $\eta_0 = 0$ and the number of iterations was $n = 10$ (the numerical experiments show that for $n \geq 10$ the numerical solution stabilized); the time step of the uniform partition Δ_M for the interval $[0, T]$ is 0.05. The computed solution $\varepsilon(u_h^n)$ obtained using the algorithm (35)-(38) is ploted in the Figure 3.

The numerical results obtained agree with the theoretical ones: indeed in this particular case the exact solution u, σ can be computed and it follows

$$\lim_{t \to \infty} \varepsilon(x, t) = \begin{cases} 4,5 & \text{if } 0 \leq x \leq 0.5 \\ 3 & \text{if } x = 0.5 \\ 1.5 & \text{if } 0.5 < x \leq 1. \end{cases}$$

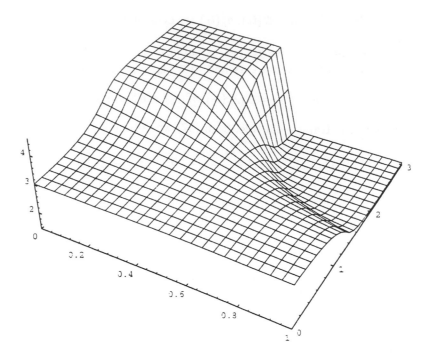

Figure 3: Computed deformation field $\varepsilon(u_h^n)$ for example 1.

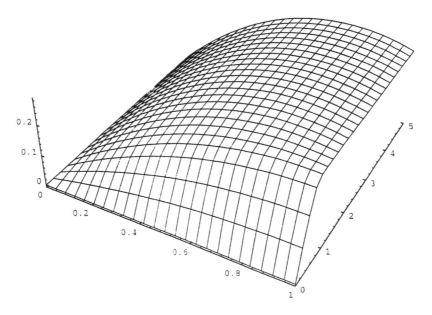

Figure 4: Computing displacement field u_h^n.

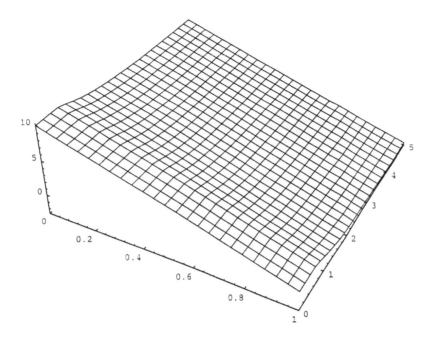

Figure 5: Computing stress field σ_h^n.

EXAMPLE 4. Let us consider now a Signorini's contact problem of the form (3)-(8) in the following context: $d = 1, \Omega = (0,1), \Gamma_1 = \{0\}, \Gamma_2 = \emptyset, \Gamma_3 = \{1\}, b(x,t) = 10,$ $s = 0.25, u_0(x) = 0, \sigma_0(x) = 10 - 10x, T = 5, E = 10, G(\sigma, \varepsilon) = -(\sigma - 10\varepsilon).$

Using (9), the boundary conditions for $x = 1$ are given for all $t \in [0, T]$ by :

$$u(1,t) \le s, \quad \sigma(1,t) \le 0, \quad \sigma(1,t)(u(1,t) - s) = 0 \qquad (41)$$

We applied the algorithm (35)-(38) in the following context: V_h is the finite element space constructed with polynomial function of degree $1, \Omega$ being divided into 100 finite elements, $\eta = 0, n = 10$. In order to approximate (37) we used trapezoidal rule with the time steps 0.05. The displacement field obtained is ploted in Figure 4 and the stress field results using numerical algorithm is ploted in Figure 5.

The above numerical result obtained agree with the theoretical ones; indeed, they point out a contact moment $t_c < 1$ such that $u(1,t) = s, \sigma(1,t) < 0$, for all $t > t_c$.

REFERENCES

[1] N. D. Cristescu, On the propagation of elastic–plastic waves in metallic rods, *Bull. Acad. Pol. Sci.*, **11**(1963), 129–133.

[2] N. D. Cristescu, Some problems on the mechanics of extensible strings, in *Stress Waves in Anelastic Solids*, eds. H. Kolsky & W. Prager, Springer Verlag, 1964, 118–132.

[3] N. D. Cristescu & U. Hunsche, A constitutive equation for salt. *Proc.* 7th *Int. Congr. on Rock Mechanics*, Aachen, 16–20 Sept. 1991, Balkema, Rotterdam, vol. 3, 1993, 1821–1830.

[4] N. D. Cristescu & U. Hunsche, *Time Effects in Rock Mechanics*, John Wiley & Sons 1997

[5] N. D. Cristescu & U. Hunsche, Determination of a nonassociated constitutive equation for rock salt from experiments, *Proc. IUTAM Symp-1991*, Hanover, 1992.

[6] S. Djabi & M. Sofonea, A fixed point method in quasistatic rate–type viscoplasticity, *Appl. Math. and Comp. Sci.*, **3**(1993), 269–279.

[7] R. Glowinski, J.–L. Lions, R. Trémolières, *Analyse Numérique des Inéquations Variationnelles*, Dunod, Paris, 1976.

[8] M. Rochdi, I. Roşca & M. Sofonea, On the existence, behaviour, and numerical approach of the solution for contact problems in rate–type viscoplasticity, *Proceedings of the "3rd Biennial Joint Conference On Engineering Systems, Design and Analysis", (ESDA'96)*, PD–73, **ASME**(1996), 87–92.

[9] I. Roşca & M. Sofonea, Error estimates of an iterative method for a quasistatic elastic–visco–plastic problem, *Appl. Maths.* **39**(1994), 401–414.

Part VII

Storage and disposal projects

Basic and Applied Salt Mechanics, Cristescu, Hardy, Jr & Simionescu (eds)
© 2002 Swets & Zeitlinger, Lisse, ISBN 90 5809 383 2

Geological and geomechanical conditions of depth location and effective capacity of natural gas storage caverns in polish salt domes

Stanislaw Branka and Jaroslaw Slizowski
CHEMKOP, Krakow, Poland

SUMMARY

Geologic, geophysical and geomechanical investigations have been carried out relative to the design and constriction of the first natural gas storage caverns in Poland located in a salt deposit. Only a small part of the Zechstein saliferous formation, which spreads out under the most territory of Poland, is suitable for mining activity. Unique geologic structure of salt domes enables one to distinguish the "regional form" of the Middle-Poland Zechstein. The very complicated structure of diapir salt formation in Poland forces the storage cavern location on different depths even within one site. Thus each cavern has different parameters.

Laboratory tests of the mechanical properties have been performed for salt core samples from each cavern. The results obtained have been adapted for use in rock salt constitutive equations. For some cavern boreholes numerically calculated convergence has been found to fit the values of field measurements over a several month period. Applicability of laboratory tests and the correctness of an accepted numerical model has been considered in relation to potential mining practice.

Basic and Applied Salt Mechanics, Cristescu, Hardy, Jr & Simionescu (eds)
© 2002 Swets & Zeitlinger, Lisse, ISBN 90 5809 383 2

Salt formations of Ukraine: Storage and disposal projects

D.Khrushchov & S.Shehunova
Institute of Geological Sciences NASU, Kiev, Ukraine

In Ukraine there are seven salt formations with age from Devonian to Neogene. On accessible depth salt formations are located: in Transcaroathian depression as a salt domes (diapirs), in Neardobroudgea foredeep as a bedded deposits, in Dnieper-Doniets depression and north-western Donbass as a bedded and salt dome forms, in Nearcarpathian foredeep – as intensively dislocated fragmented bed-lens like bodies. Devonian (Upper Franian and Famenian) formations, Nikitovska suites of Lower Permian in Dnieper-Doniets depression and north-western Donbass, Neogene formation in Transcarpathian depression, Badenian formation in Nearcarpathian foredeep and Jurassic formation in Neardobroudgea foredeep are characterised with almost monomineral (halite) salt composition. Kramatorsk suite in Dnieper-Doniets depression and north-western Donbass as well as Prebadenian deposits of Neogene in Nearcarpathian foredeep (i.e. Vorotyshcha series and Balich suite) contain thick deposits of potassium-magnesia salts. The spread of these salts is not favourable for underground constructions.

The major part of salt formations are in exploitation for rock salt (and one for potassium and magnesia salts) excavation.

Besides it these formations are regarded as underground storages construction media. The storages for oil and oil products there exist in salt domes in Dnieper-Doniets and Transcarpathian depressions. Due to economic reasons most of them do not function nowadays. The selection of salt domes and sites for storages location was completed basing upon the special methodology. The selection of salt domes-candidates is carrying out on the basis of two groups of criteria: economic and mining-geological. Mining-geological group comprises the following criteria: structural, neotectonical, lythological, hydrogeological, geomechanical. The selection of intervals favourable for storages location in salt domes is completed on the basis of structural, lythological and geomechanical models set having been developed as a result of boreholes cores comprehensive atudies and logging as well as surface and interboreholes geophysical investigation data interpretation.

General principle of selection on intervals favourable for cavities construction considers the preference of the most thick salt beds with minimum thickness of nonsaline rocks and rocks with high content of insoluble admixtures and potassium-magnesia salts. In the case of bedded salt formation this task is relatively simple, however insufficient thickness of salt beds forces to vary the form of cavities or to use garland system of cavities. In salt domes the general regularities of diapirs internal construction have to be taken into consideration. The salt domes of small area are unfavorable because of presence of salt layers being characterized with low integrity and high permeability as well as danger of salt plug walls dissolution. The selection of favorable intervals is the most complicated task in the case of folded formation construction due to spread of low integrity and high fracturing zones in folds crests.

The methodology of sites exploration depends on geological construction of salt formation, cavities creation methods and storage target. The depth over 900m is favorable due to creep manifestation. The methodology of rock salt beds strength properties evaluation and low integrity zones identification was developed. The methodology is based upon the rock salt massif model considering interdependencies of lithological peculiarities and physic-mechanical properties of rock salt. The vertical zonality of salt massif strength properties also has to be taken into consideration. The main groups of damage situations are revealed, namely: decrease of storage useful volume; collapse of cameras walls and ceiling; damage of exploitation tubes; loss of the product; change of product quality. The methodology of sites selection application as well as mentioned above models set elaboration has to secure the absence of geological factors capable to induce any damages of the storage. On the moment of the methodology mentioned above the estimation of the cavity long term stability has been rated basing on common initial data but only for restricted time intervals according to exploitation period being projected. The future fate of these cavities has not been considered. The peculiarities of sites and favorable intervals selection for different target oriented storage (i.e., oil, gas, chemical products etc.) and different technologies (i.e. dissolution, explosion, nuclear explosion) differ.

For instance, principles of sites and favorable intervals selection and technical criteria for construction of compressed and liquid gases storage first of all have to secure cameras tightness. One has to avoid layers and zones of high permeability ("sandstone like" rock salt, zones of fractured and fragmented salt, slide planes etc.). The depth of storage has to secure lythostatic pressure exceeding the latter of gas being stored (usually 600 – 700m). The condition of tightness is especially rigid storage of high penetrating capacity gases (helium).

Technical criteria for geological conditions of construction of oil products have to secure stability, preservation of camera useful volume, configuration and tightness in the degree preventing essential leakage or hiding of a product, preservation of exploitation installation integrity and absence minerals capable to change the quality of the product being stored.

For construction of air accumulated cameras for gas turbine power plants the selection of site and salt formation interval has to secure the possibility for construction of large cameras on the depth providing sufficient contrapressure according to projected technical parameters. Taking into consideration the exploitation of essential pressure fluctuations sufficient rock salt mechanical properties also have to be secured.

The project of radioactive waste disposal in salt formations has been considered as a possible (alternative) variant in the Programme of radioactive waste management and

Programme of radioactive waste deep geological disposal in Ukraine. In the course of the preparatory stage of the latter Programme realization eight salt domes have been pre-selected as favorable for wastes disposal in Dnieper-Doniets depression. In Donbass region two zones of bedded salts spread have been selected as potentially favorable. The selection of sites-candidates has been carried out on the basis of the methodology specially elaborated. This methodology comprises two tools for sites selection and characterization, namely: criteria of selection and ranking and set of ecological-geological models. The criteria of selection are ranking. The criteria of exclusion are also used. The set of ecological-geological models aimed at sites characterization comprises the following models: structural, geomorphologic, neotectonic, hydrogeological (hydrodynamical and hydrochemical), geochemical, geomechanical and thermophysical. The methods of geological and geophysical studies being used for models development have certain peculiarities for bedded salt formations and salt domes. The target oriented studies of salt mines reworked cameras are also planned. Special methodology comprising a special set of selection-ranking criteria and ecological-geological models set for sites characterization has been elaborated.

According to methodology mentioned the next step of R&D is specialized large scale and detailed characterization of site-candidates (i.e. reference sites in the vicinities of Chernobyl exclusion zones or alternative sites in salt formations, including attractive from the economical point of view variant of abandoned salt mines workings). This stage conventionally corresponds to scales 1:50000 – 1:25000 and 1:10000 and larger. It comprises phases: prospecting, preliminary exploration and detailed exploration. The characterization of old mines comprises specific studies corresponding these phases. In the case of characterization positive results the terminal stage, i.e. experimental studies in underground research laboratory, has to be started.

Basic and Applied Salt Mechanics, Cristescu, Hardy, Jr & Simionescu (eds)
© 2002 Swets & Zeitlinger, Lisse, ISBN 90 5809 383 2

Feasibility study of large volume gas storage in rock salt caverns by numerical method

Ming Lu
SINTEF, Civil & Environmental Engineering. Trondheim, Norway

Einar Broch
Norwegian University of Science and Technology, Trondheim, Norway

Friedhelm Heinrich & Axel Hausdorf
Freiberg University of Mining and Technology, Freiberg, Germany

Wolfgang Schreiner & Armin Lindert
Institute fur Gebirgsmechanic GmbH, Leipzig, Germany

ABSTRACT

Finite Element analyses are performed to study the feasibility of large volume gas storage caverns in rock salt created by leaching. A number of operation phases have been studied including leaching, gas storage and an accidental gas blowout. The computing results have confirmed the general stability of the cavern and predicated 7.5% volumetric convergence in 30 years service. Numerical problems are encountered and corresponding solutions are also addressed.

INTRODUCTION

A comprehensive research program to study the feasibility of storing natural gas in large volume caverns in rock salt, created by leaching, has been carried out since 1994. As a part of the program, a research project, including laboratory tests and numerical analyses, has been performed by a team of researchers from a number of universities and research organizations in both Norway and Germany. The goal of the project is to establish a proper numerical model that handles the mechanical responses of the cavern under various loading conditions. Special attention is paid to cavern convergence resulting from the creep nature of the rock salt.

Laboratory triaxial tests have been performed to estimate the creep and strength parameters of the rock salt. Special creep tests have also been carried out to simulate the cavern real loading conditions. Tests studying the rock salt behavior under high temperature are underway. In order to investigate the adaptability of the creep laws a numerical pre-study was conducted on both a test specimen-like model and on axisymmetric models with the Finite Element code ANSYS (Hausdorf 1996 and 1997). Finally, 3-D analyses were undertaken. In addition to the major work with the Finite Element code ABAQUS, computations on some special cases were also performed by using the Finite Difference code FLAD^{3D} (Lindert 1996 and 1997). This paper focuses on the analysis with ABAQUS. The geological data are taken from a salt bed site in central Germany.

As a mater of fact, the analysis was performed for the first time in 1997 (Lu 1997, Lu & Broch 1997) and then reanalyzed with a new model in 1998 due to problems detected. The motivations of the re-analysis include, among others, to solve the problem of adopting the correct creep law and to improve the accuracy. This paper presents the results of the reanalysis.

NUMERICAL MODELS

Geometric Model

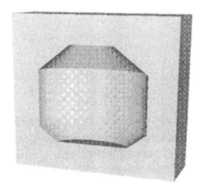  

Figure 1. Cavern shape.   Figure 2. Cavern dimensions.

The cavern geometry is simplified as a 170 m high cylinder with elliptic horizontal cross sections as shown in Figures 1 and 2. The major/minor axes at the cavern center is 200/100 m, which reduce to 80/40 and 130/65 m at roof and floor, respectively, making a total volume about 2.3 million m^3. As shown in Figure 3, the FE model includes only a quarter of the cavern, considering the geometric symmetry. The model consists of 3960 20-node elements and 18773 nodes.

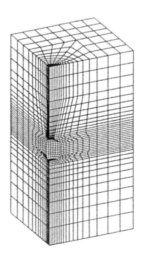

Figure 3. 3-D Finite Element mesh.

Material Model and In-Situ Rock Stress

The rock mass in the cavern site consists of a 510 m thick rock salt deposit. The cavern is situated entirely within the rock salt leaving only 40 m margin to the adjacent sandstone underneath. The thickness of the rock salt layer in overburden is 300 m. Measurements of the in-situ stress are performed and a lithostatic stress state is suggested for the rock salt formation. The elastic constants and in-situ stresses of all rock masses are given in Table 1.

Table 1. Elastic constants and in-situ stresses

Stratum	Depth [m]	E [GPa]	v	ρ [kg/m^3]	σ_h/σ_v
Sandstone	0-190	35	0.30	2600	0.5
Anhydride	90-280	35	0.25	2700	0.7
Rock salt	280-790	25	0.30	2140	1.0
Basal Anhydride	690-1430	25	0.30	2600	0.7

A hyperbolic sine law with strain hardening and a power law with time hardening are proposed for the primary and secondary creep, respectively:

$$\dot{\varepsilon}_{cr} = A \sinh \frac{\sigma}{\sigma_0} \exp(\frac{-\varepsilon_{cr}}{\varepsilon_0}) \; [1/h] \qquad (1)$$

$$\dot{\varepsilon}_{cr} = M\sigma^N \quad [1/h] \qquad (2)$$

where ε_{cr} and $\dot{\varepsilon}_{cr}$ are creep strain and creep strain rate, constants A=9.0E-7, σ_0 =2.5MPa, ε_0=3.0E-3, M=5.8E-12 and N=4.6. A triaxial strength criterion is also proposed by IfG based on laboratory triaxial tests:

$$\sigma_1 = \sigma_D (1 + \frac{k\sigma_3}{\sigma_z})^{\frac{1}{k}} \qquad (3)$$

where σ_D=27MPa is the uniaxial compressive strength, constants k=3.1 and σ_z=2.3MPa. Considering the scale effect, the Hoek-Brown criterion is also used in estimating the shear yielding.

$$\sigma_1 = \sigma_3 + \sqrt{m\sigma_c\sigma_3 + s\sigma_c^2} \qquad (4)$$

The parameters for the H-B criterion are estimated as follows: m=1.801, s=0.0155 and σ_c=27 MPa. Figure 4 shows the two yielding criteria and the definition of the factor of satefy against shear yielding.

319

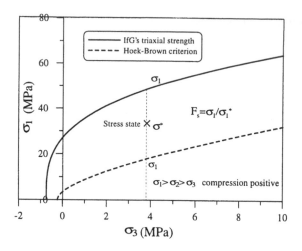

Figure 4. Empirical shear strength criteria and definition of factor of safety against shear yielding.

Computational Aspect

Computations initiate with the explicit integration in time domain, then shift to implicit integration when the code feels necessary, i.e. the time increments are too small in comparison to the step time period, restricted by the stability requirements. The implicit integration is unconditionally stable, so large time increments can be used. The computation accuracy is controlled by the maximum difference in the creep strain increment in two consecutive time increments. In most cases that have been computed the automatic time incrementing provided by ABAQUS is adopted. The geometric non-linearity is incorporated for all cases studied.

Based on the problems detected in the first analysis, a numerical procedure has been developed for identifying and automatically applying the creep law currently applicable for the material point under consideration. This procedure is used in the analysis and a comparison to the traditional way of selecting creep law is given later in the paper.

CONTENTS OF THE ANALYSIS

The main concerns from rock mechanics point of view are cavern stability and convergence resulting from the creep of the rock salt. The stability problem may occur when the internal gas pressure is low due to lacking of support. On the other hand, however, if the gas pressure is too high, fracturing or jacking may happen. This was presumed to be critical for the first a few years of storage and maybe even critical when an accidental gas-blowout takes place. On the contrary, the convergence is a long term problem. Based on this understanding the following simulations are carried out:

1. Initial stress field: The essential task is to generate a shear stress free initial stress field for the rock salt formation.

2. Leaching: Leaching is simulated in five stages, in which the elements to be leached out are removed and hydrostatic pressure of brine is applied on the surface created. Figure 5 sketches the leaching process.

3. First three years of gas storage. In these computations the brine pressure is replaced by the gas pressure, which follows the real annual variation cycle recorded at the chosen site, as shown in Figure 6. The maximum and minimum gas pressure is 9.0 and 2.5 MPa, respectively.

4. Next 27 years of gas storage. Considering the long term effect, the constant annual mean gas pressure of 6.0 MPa is adopted instead of the varying annual pressure cycles.

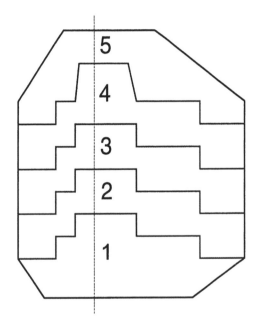

Figure 5. Five-stage leaching process.

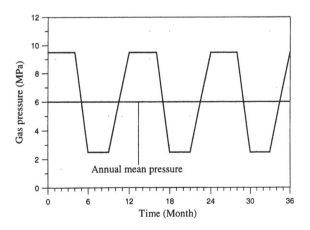

Figure 6. Annual variation of gas pressure.

321

5. An accidental gas blowout, during which the gas pressure drops linearly to zero within 12 days and the pressure-free state remains for three months for repairing work.

COMPUTATION RESULTS

Displacements, creep strain, Mises stress, and factor of safety against shear yielding at representative locations including, cavern floor center, cavern roof center, cavern mid-wall at major and minor axes, are shown in Figures 7-10. Detailed description of the computation results can be found in Lu (1999).

Deformation and Convergence

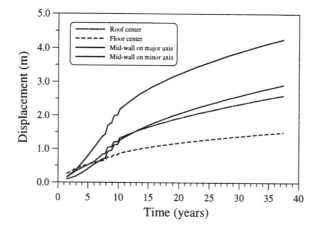

Figure 7. Time history of inward displacements.

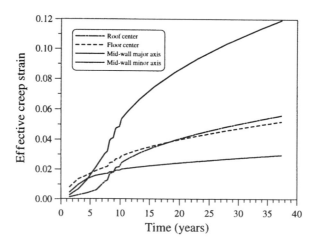

Figure 8. Time history of effective creep strains.

322

The displacements on the minor axis are always larger than those on the major axis. In other words, the aspect ratio of the elliptic cross section becomes larger and larger. The maximum predicted displacement after 30 years is 4.26 m, at the middle of the cavern wall. The relative shortening of the cavern diameters is 8.5 and 2.5 % in the minor and major axis, respectively. The displacements occurred during leaching constitute up to 20 - 40 % of the total displacements. The accidental gas blowout creates little additional deformation. The relative cavern volume convergence is 7.5 % up to 30 years of storage operation, counting from the beginning of storage operation. In other words, 0.17 Mm^3 of storage volume will be lost within 30 years operation. The total ground surface subsidence after 30 years storage is predicted as 14.5 cm at the location above the cavern center.

Contrary to the displacement, the creep strain on the major axis is larger than that on the minor axis. The maximum effective creep strain reaches at 16.2 % in 30 years storage.

Stress and Factor of Safety

As shown in Figure 9 the stresses are disturbed by the gas pressure fluctuations. In other words, stresses follow the pattern of the internal gas pressure variation. Tensile stress appears in the limited region of the cavern close vicinity when the gas pressure is low. This situation worsens when the gas blowout occurs. The maximum tensile stress during normal operation and gas blowout is 2.1 and 5.2 MPa, respectively. The predicted maximum compressive stress is over 30 MPa when the gas pressure is low. In general, the Mises stress decreases with the distance from the cavern surface and also decays slowly with time. After 30 years the Mises stress at the cavern roof is still 8.0 MPa.

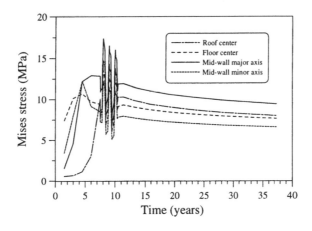

Figure 9. Time history of Mises stresses.

Figure 10 shows distributions of the factor of safety against shear yielding along the central vertical line above the cavern roof evaluated from both IfG and Hoek-Brown criteria. Distinct difference has been revealed from these curves. If the IfG shear strength criterion is used, no yielding will take place. If the H-B criterion is adopted, however, yielding occurs during normal operation when gas pressure is low. The yielding region

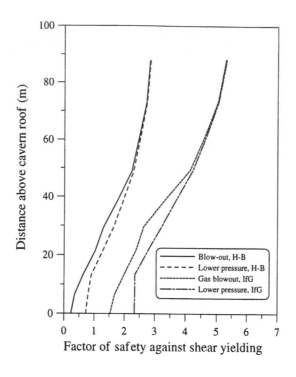

Figure 10. Factor of safety against shear yielding.

above the cavern roof is about 17 m deep. More severe yielding will take place during blowout, when the yielding zone will extend about 20 m into the rock mass. This crucial state indicates the period of an extremely low gas pressure must be kept limited.

RESULT ANALYSIS AND DISCUSSIONS

Cavern Stability

In the computations, the rock mass is actually considered to be visco-elastic material. Plastic behavior is not really simulated. The factor of safety against shear yielding is calculated based on the elastic stresses and the shear strength of the rock salt, as illustrated in Figure 4. The strength proposed by IfG is derived from the laboratory tests performed on small specimens. Although it is accepted that the scale effect for rock salt may not be as significant as the jointed hard rock masses, it for sure exists. The parameters, m and s, for the Hoek-Brown criterion are derived mainly from the RMR index of the rock mass classification system. It is not certain that the parameters are over- or under-estimated. But, the difference created from the IfG and H-B criteria is obvious.

Despite the difference between the two criteria, yielding appears only when the gas pressure is at the lowest level during the normal operation or blowout happens, and the region of yielding is limited in the cavern vicinity, although it will expand when more reliable elasto-plastic analysis is performed.

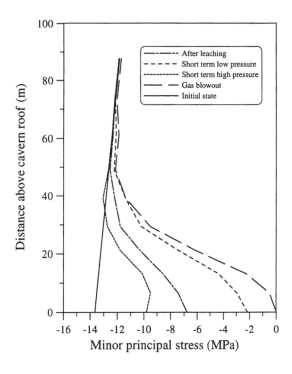

Figure 11. Distributions of the minor principal stress above the cavern roof.

Tension has also appeared. The maximum tensile stress during normal operation and gas blowout is 2.1 and 5.2 MPa, respectively. But the tensile stress occurs only within an extremely limited region in the cavern roof and floor. Considering the actual tensile strength of the rock salt may be well beyond the 2.1 MPa, it is not anticipated that the tension failure will become a major problem during the normal operation. If gas blowout takes place, however, the potential of rockfalls in the cavern roof will increase. One of the major problems associated with the pressurized storage in unlined rock caverns is the potential for fracturing and/or jacking. Jacking may cause significant disaster in terms of both stability and leakage of the stored medium. Figure 11 clearly demonstrates that for the cavern under investigation the potential of jacking virtually does not exist, since the minor principal stress is always compressive. A Norwegian rule of thumb for preventing hydraulic jacking for hydropower tunnels is that the internal water pressure should not be greater than the in-situ minor principal rock stress (Broch 1982). Förster (1981) has drawn the same conclusions from the pneumatic fracturing tests in deep borehole in rock salt. In our case, the in-situ rock stress in the cavern roof is 13.7 MPa, which is much higher than the maximum gas pressure of 9.5 MPa. The horizon where the in-situ rock stress is equivalent to the maximum gas pressure locates over 100 m above the cavern roof, where the influence of the gas pressure has disappeared.

Gas Pressure

As mentioned above the gas pressure follows the annual cycle for the first three years of operation and the mean pressure is used for the rest 27 years. The computing

325

results have indicated that the pressure fluctuations have much less influence on the deformation than on the stresses. Since the computing burden is dramatically enhanced if the real gas pressure variation is actually simulated, it is acceptable to use the mean pressure, if only the cavern convergence is requested. However, if the stresses and stability are also questioned, the real pressure variation has to be used.

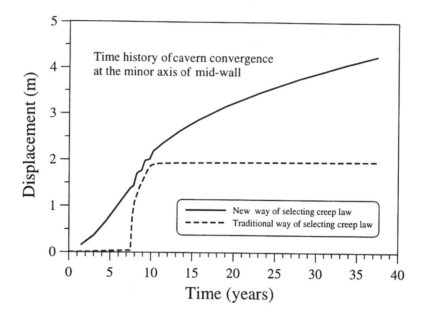

Figure 12. Comparison of two ways of selecting creep laws.

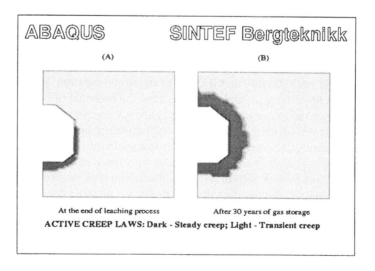

Figure 13. Distributions of applied creep laws: (a) At the leaching end. (b) After 30 years gas storage.

<u>Active Creep Laws</u>

The numerical technique developed in SINTEF Rock and Mineral Engineering has been used in the analysis. The technique detects and automatically selects the applicable or active creep law for each integration point of each element at any computing time. A computational procedure has also been developed to handle the numerical difficulties that may appear. See Lu & Broch (1997) for a detailed description. As mentioned above the constant creep law based on loading condition was used in the previous analysis, namely, the secondary creep for the leaching and long term analysis; the transient creep for the first three years of storage operation and gas blowout. Figure 12 illustrates a comparison of the displacement results from the two distinct ways of selecting the creep law. The irrational sharp turnings on the dashed line are resulting from the way of specifying constant creep law. On the other hand, the solid line has improved the result greatly.

Figures 13 shows the distributions of the active creep law at the end of leaching and after 30 years gas storage. The features of the distribution are:

- The transient creep is always dominating for the majority region;

- The steady creep law is active only in the cavern vicinity;

- The region where the steady creep law applies expands with time and by 30 years of storage the steady creep only applies to a region about 50 - 80 m around the cavern.

CONCLUSIONS

1. The maximum tensile and compressive stresses under the most unfavorable loading conditions have been predicted as 2.1 and 30.3 MPa, respectively. The shear yielding will take place in the vicinity of the cavern.
2. The maximum inward displacement is 426 cm occurring at the cavern mid-wall on the minor axis; the corresponding displacement on the major axis is 291 cm. This implies that the cavern cross section will become thinner and thinner with time.
3. It is predicted that the maximum volumetric convergence in 30 years of storage will be 7.5% of the original storage volume, or 0.17 Mm^3.
4. The maximum predicted ground surface subsidence is 14.5 cm above the center of the cavern.
5. Based on the Norwegian experience on the unlined pressure tunnels for hydropower developments, under the designed gas pressures the potential for significant gas leakage is very limited.
6. From the rock mechanics point of view, it is possible to create such a large volume gas storage cavern by means of leaching. The cavern is generally stable under the designed gas pressures, in spite that shear yielding and tension may appear in a limited region around the cavern.
7. The cavern stability is generally demonstrated even during the accidental gas blowout, when the gas pressure drops to zero. However, the 20 m deep shear yielding zone in the cavern roof is indeed a potential for failure such as rockfalls. More reliable visco-plastic analysis is needed to analyze the stability of the cavern roof during gas blowout.

Finally, the analysis shows no evidence that such a large volume gas storage cavern in rock salt is infeasible.

ACKNOWLEDGEMENTS

The authors of this paper would like to express their sincere thanks to the sponsors of the project, Statoil in Norway and VNG in Germany, for their generous financial support and permission for this publication.

REFERENCES

Broch, E., [1982], "The Development of Unlined Pressure Shafts and Tunnels in Norway," Proc. Int. Symp. Rock mech. Related to Caverns and Pressure Shafts, Aachen, vol. 2, W. Wittke (ed.), 545-554. Rotterdam: A.A.Balkema.

Förster, S., et. al. [1981], "Pneumatic and Hydraulic Fracturing Tests in Boreholes for Proof of Tightness and for the Stress Sounding in the Salt Rock Mass," (in German), Neue Bergbautechnik, Volume 11, No. 5.

Hausdorf, A., [1996], "Simulation The Conditions of Rock Mechanical Tests Numerically and Comprising the Results With the Test Results to Assess the Quality of the Used Creep Laws," Report, Freiberg University.

Hausdorf, A., [1997], "Model Generation and Testing as well as Studying Influence of Material Laws and Creep Laws on the Stress and Deformation State," Report, Freiberg University.

Lindert, A., [1996], "Studying Influence of Loading and Strength Parameters on Stability," Report, IfG.

Lindert, A., [1997], "Continuation of Stability Studies at Large Volume Caverns," Report, IfG.

Lu, M., [1997], "Three Dimensional Finite Element Analysis of Gas Storage in Rock Salt: Simulations for Single Cavern," SINTEF report, STF22 F97050

Lu, M., [1999], "3-D Analysis of Gas Storage in Rock Salt Cavern by Finite Element Method," SINTEF report, STF22 F98117.

Lu, M., and Broch, E., [1997], "Study of Large Volume Gas storage in Rock Salt Caverns by FEM," Proc. of the 9th International Conference on Computer Methods and Advances in Geomechanics, Wuhan, vol. 2. Yuan (ed.), 1489-1494. Rotterdam: A.A.Balkema.

Basic and Applied Salt Mechanics, Cristescu, Hardy, Jr & Simionescu (eds)
© 2002 Swets & Zeitlinger, Lisse, ISBN 90 5809 383 2

Development of a statistical technique for prediction of ground conditions at the WIPP site

Hamid Maleki
Maleki Technologies, Inc., Spokane, WA

Lokesh Chaturvedi,
Environmental Evaluation Group, Albuquerque.NM

ABSTRACT

The Waste Isolation Pilot Plant (WIPP) is an underground repository for the permanent disposal of transuranic radioactive waste generated by defense activities. Sandia National Laboratory and Westinghouse Electric Corp. completed detailed test mining and geotechnical monitoring in several experimental areas and are in the process of developing constitutive models for rock salt. In view of the limitations for unambiguous modeling of the complete strata deformation process, this paper incorporates the extensive deformation data into a statistical analysis to help predict changes in ground conditions at the site.

The statistical analysis approach consisted of (1) characterization of twelve mining, bolting, and geologic variables at 121 instrumented locations, (2) development of a mathematical model (3) identification of important variables influencing convergence and (4) application of the model for prediction of convergence and stability.

INTRODUCTION

The WIPP site is located about 40 km east of Carlsbad, New Mexico in southwestern USA. Its mission is to permanently dispose of both contact and remotely handled transuranic waste in underground workings (panels) located 655 m below the surface within a nearly 610-m thick Salado Formation that consists primarily of halite (DOE 1995).

Project architects developed structural designs for underground workings using experience from seven neighboring potash mines and applying both closed-form solutions and numerical models (DOE 1984). To account for differences in the depth of cover between the WIPP site and the potash mines (320 m), a lower extraction ratio was used for the waste panels and the workings were developed gradually while completing geotechnical evaluations. These designs consisted of 30- by 90-m pillars in a seven-entry waste panel system using 10-m wide rooms. Considering the lower extraction ratio and the large dimensions of the pillars, it was assumed that the salt pillars would experience creep deformation and enclose the waste packages gradually without significant fracturing.
Sandia National Laboratory and Westinghouse Electric Corp. completed detailed test mining and geotechnical monitoring in several experimental areas at two different stratigraphic positions and with different entry configurations. The intent was to characterize the site and develop predictive techniques to assess repository performance. Sandia's work focused on

development of site-specific constitutive models for rock salt. These models successfully predicted room closures during early stages of steady state creep deformation but departed from measured values as fractures formed in the mine roof. The lack of agreement between the calculated and the measured deformation is due to finite element continuum formulation (Munson 1997). In view of these limitations for unambiguous modeling of complete strata deformation process (Maleki and Chaturvedi 1997), a statistical technique was used in the present work as a tool for prediction of ground conditions. Extensive deformation data collected by Westinghouse Electric Corp. (Westinghouse 1998) has been used for this purpose.

Development of underground workings has taken place in phases. Proceeding each phase, preliminary designs were prepared by the project architect, followed by test mining and careful geotechnical evaluations for characterizing the site and verifying the preliminary designs. The E140 drift, the one main artery of the facility for air supply and access, was mined during 1983, followed by mining and design verification in the "site and preliminary design validation" (SPDV) area. This area was designed to have geometric and stratigraphic conditions similar to those in the waste panels. Panel 1 was mined between 1986 and 1988 (Figure 1).

GEOLOGIC SETTING AND ROOF SUPPORT

The WIPP underground facility is located in the Salado Formation, a horizontally bedded formation consisting primarily of halite, with layers of anhydrite, polyhalite, and clay. The strata near the storage horizon are composed of many layers of anhydrite, clay, and halite (Figure 2). The immediate roof is a 2-m thick layer of relatively pure halite overlain by clay G and anhydrite layer b, which is 10 cm thick. There is a prominent 1-m thick, fractured anhydrite nonsalt bed below the storage level, referred to as MB139 (Stormont 1990).

The mechanical and hydraulic properties of the halites and anhydrites were measured by Sandia National lab, but data on the properties of the clay layers are limited (Beauhiem and others 1993). There is a large difference between the strength and stiffness of the halites and anhydrites. Anhydrites are much stronger and stiffer than halites and thus concentrate stresses. The uniaxial compressive strength of rock salt varies from 14 to 31 Mpa. Clay seams are thought to provide minimum shear resistance, if any. Observations of boreholes have shown that some clays are moist, discharging water to boreholes near the excavations.

Before 1986, very few areas within the facility were supported with roof bolts. Between 1986 and 1991, most areas had been systematically bolted with 2- to 3-m long, mechanically anchored bolts. In 1991, a secondary support system consisting of wire mesh, expanded metal channel steel, and point-anchored threaded rebar was installed in room 1, panel 1, to help extend the life of this room. Between 1992 and 1996, other secondary support systems, including mechanical bolts, resin point-anchored threaded rebar (with and without slip nuts), and cable mesh, were installed in portions of panel 1 and E140. The operator designed a tertiary bolting system and installed it in S1950 and room 7 during 1997 and 1998.

To monitor ground conditions and evaluate support performance, Westinghouse carried out an intensive geotechnical monitoring program. This program consisted of monitoring strata deformation (figure 1), bolt loads, locations where bolts failed, strata fracturing, and lateral offsets at clay G and other horizons. These measurements have been very helpful in improving understanding of the strata behavior.

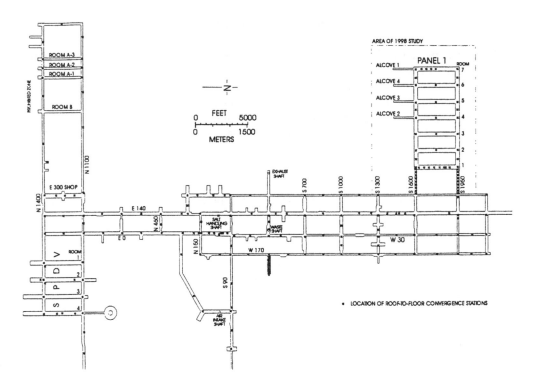

Figure 1. Location of convergence measurements over WIPP facilities.

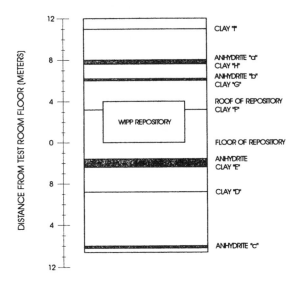

Figure 2. Anhydrite and clay interbeds near the repository horizon.

ANALYSIS TECHNIQUE

There are two methods used by engineers and researchers as tools to help predict future conditions: statistical and computational. Starfield and Cundall (1988) identify rock mechanics problems as "data-limited," that is, one seldom knows enough about a rock mass to use computational models unambiguously. Statistical methods, on the other hand, are uniquely capable of being applied where there are good data but a limited understanding of certain natural phenomena, such as the creep of rock salt. This is the case at the WIPP site, where there are sufficient data on convergence in excavations of different ages and geometries to allow a mathematical model to be built that can be used to identify significant variables and estimate future deformation. Examples of the use of both statistical and computational techniques to evaluate stability and performance are given by Maleki and others 1986, Maleki (1992), Parson and Dahl (1971), Munson and others (1993), Maleki and White (1997) and Maleki and Chaturvedi (1997). The Multiple regression analysis technique is used in this study.

The multiple regression procedure consisted of entering the independent variables one at a time into the equation using a forward selection methodology (SPSS, 1995). In this method, a variable is entered into the equation using the largest correlation with the dependant variable. If a variable fails to meet entry requirements, it is not included in the equation. If it does meets the criteria, the second variable with the highest partial correlation will be selected and tested for entering into the equation. This procedure is very desirable when there are hidden relationships among the variables.

The roof-floor convergence rate was used as the response variable after consideration of several other factors including horizontal deformation rate at clay G (Maleki 1998), measured fracture density in the mine roof, relative roof deformation, and total roof-floor convergence. The convergence rate was selected as the response variable because (1) it is a good measure of stability (Maleki and Chaturvedi 1997) and (2) it is available over the whole facility at 121 measurement locations.

DATA ANALYSIS

The first step in developing a statistical model was to identify variables that express variations in geologic, and mining conditions. The second step was to examine any correlation among independent variables by creating a bivariate correlation matrix. The presence of correlatable variables influenced the selection of forward stepwise regression procedures. The third step was to develop a multivariate regression model and identify significant factors that contribute to entry convergence and stability.

The distribution of roof-floor convergence rate (Figure 3) has been used to identify independent variables that are included in the regression analyses. For example, since roof-floor convergence rates are higher in wider rooms and in areas of higher overall extraction (such as panel 1), entry span and excavation ratio are identified as pertinent variables for including in the regression analyses.

The distribution of roof-floor convergence rates near panel 1 is used to elaborate on load transfer and interactions between E140 (access drift) and room 1, of panel 1. As illustrated in Figure 3, the convergence rate in panel 1 is highest in room 1. This was interpreted to be caused by the sequence of excavation of the access drifts with respect to panel 1 and load transfer from the access drifts toward panel 1. The access drifts (including the 7.5-m wide E140 drift) were mined 3 years before panel 1 was mined. The existence of large load transfer distances in evaporates is supported by both stress analyses (Maleki and Chaturvedi 1996; Maleki and Hynes 1991) and measurements (Able and others 1984; Maleki and Hynes 1991; Maleki and Hollberg 1995).

Figure 3 also shows that the convergence rate generally increased along a portion of the E140 drift located next to panel 1 (for a minimum distance equal to the width of panel 1). As panel 1 was mined, loads were transferred from panel 1 toward the access drifts. This load transfer increased floor heave, caused lateral movements within the roof beam, and contributed to fracturing of the roof beam in the wide entries (E140). The convergence rates are minimal in the narrow S1600 access drift as a result of the favorable design of this entry.

MULTIPLE REGRESSION RESULTS

Initial studies during 1997 (Maleki and Chaturvedi 1997) utilized available data from all accessible areas between 2 to 12 years in age. Linear regression analyses produced the best fit while identifying important factors that influenced stability. In the present study, additional data from unstable experimental rooms (SPDV) and new data for a portion of Panel 1 were included to extend the analyzed age of the openings by 2 years. New closure rate data from the SPDV area included accelerated rates prior to monitored roof falls in barricaded SPDV rooms 1 and 2 where minimal support was used. The following variables were included in these analyses.

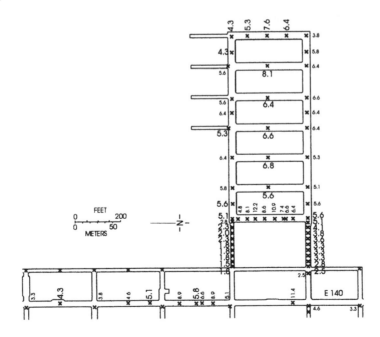

Figure 3. Roof-floor convergence rate (cm/yr) near panel 1 for 1996 review.

- *Roof span:* Measured convergence depends on roof span, which varies between 4 and 10 m.
- *Roof shape:* Entries are either rectangular or arched. This variable takes values of 0 to 1.
- *Roof beam thickness:* This variable measures the distance between the roof and clay G or H.
- *Entry height:* Height generally varies between 3 and 5 m.
- *Age:* Time from excavation year to present.
- *Excavation ratio:* This variable relates to higher vertical stresses in areas of higher extraction.
- *Bolt length:* Roof bolts vary in length from 0.3 to 4 m.
- *Bolt pattern: This* variable relates to the density of roof bolts.
- *Excavation sequence: This* variable identifies locations influenced by load transfer from previous excavations (such as room 1, panel 1). It varies between 0 to 1.

Nonlinear analyses produced a slightly better fit than linear regression analysis because nonlinear rates from the SPDV area are included in the present study. Using the square root of the convergence rate as a dependent variable, a multiple correlation coefficient (R) for the last step is 0.834. R is a measure of goodness-of-fit. Based on an examination of standardized regression coefficients (Maleki and Chaturvedi 1997), the following variables best explain the variations in the convergence rate.

- *Excavation ratio:* The convergence rate is higher as the excavation ratio (and the associated vertical stresses) increases.
- *Entry shape: Arched* entries converge less.
- *Excavation sequence:* Sequence of excavation influences convergence rate.
- *Age:* Convergence rate increases as entries age.
- *Span:* Increasing the span results in an increase in convergence rate.
- *Bolting pattern: Bolting* controls strata movement and prevents excessive convergence rates.

Table 1 presents a comparison of the 1997 analysis results and the present studies. It is very interesting that *beam thickness* and *entry height* do not add significantly to the goodness-of-fit and thus are not included in the final equation in the present study. Past and present bivariate correlations (Maleki and Chaturvedi 1997) indicate that beam thickness is

Table 1.-Important variables and statistical significance for 1997 and present studies.

1997 study		Present study	
Variable	T-significance	Variable	T-significance
Span	0.0006	Excavation ratio	0.0002
Excavation ratio	0.0000	Entry shape	0.0031
Beam thickness	0.0007	Excavation sequence	0.0751
Entry height	0.0288	Age	0.0005
Age	0.0294	Span	0.0299
		Bolting pattern	0.0000

correlated with convergence rate, but because this variable is correlated with other variables that are now in the equation, beam thickness is excluded from the present model. Similar argument is valid for the *entry height* as well. *Bolting pattern* is now identified as an important variable because minimally bolted areas in the SPDV rooms 1 and 2 collapsed while the closure rate for similar areas in Panel 1 are controlled through installation of supplementary support. Thus by including the data from SPDV areas in the present study, the importance of roof bolting and maintenance activities on reducing the rate of convergence is taken into consideration. Results are influenced by the structure of the existing data and may alter as additional data become available. This emphasizes the need for updating these analyses to improve the accuracy of these models in predicting ground conditions for the anticipated geologic conditions and functional requirements in future waste panels.

CONCLUSIONS

Based on analyses of extensive deformation measurements and observations of ground conditions, geotechnical factors that influence the stability of large underground openings in the deep workings of the WIPP repository have been identified. These factors include (1) entry geometry, (2) extraction ratio within panels, (3) sequence of excavation and load transfer between adjacent panels, (4) support density, and (5) age of excavations. Regression equations are used to help predict entry closure and stability depending on geologic and geometric conditions.

REFERENCES

Able J.F., and F. Jahangeri, 1984. Application of Performance Data from Evaporite Mines to Salt Nuclear Waste Repository Design. Inter. J. Min. Eng., Vol 2, pp.323-340.

Beauheim R., R.M. Roberts, T.F. Dale, 1993. Hydraulic Testing of Salado Formation Evaporites at the WIPP. Sand92-0533, p 228, Sandia National Laboratories, Albuquerque, NM.

DOE, 1984. Design Criteria, Waste Isolation Pilot Plant, Revised Mission Concept. DOE/WIPP 71, Feb 1984.

Maleki H., and L. Chaturvedi, 1997. Prediction of Room Closure and Stability of Panel 1 in the Waste Isolation Pilot Plant. Int. J. Rock Mech. & Min. Sci. 34:3-4, paper 186.

Maleki H., and L. Chaturvedi, 1999. Geotechnical Factors Influencing Stability in a Nuclear Repository in Salt. Proc. 37th US Rock Mech. Symp., Vail, Co.,University of Colorado.

Maleki H. and B. White, 1997. Geotechnical Factors Influencing Violent Failure in U.S. mines. Proc. Int. Sympo. Rock Support, Oslo, Norway, 1997.

Maleki, H., and H. Brest van Kampen, 1986. Impact of Mechanical Bolt Installation Parameters on Roof Stability. *In* 27th U.S. Symp. on Rock Mechanics.

Maleki, H., 1998. Stability Evaluation of Panel 1 During Waste emplacement. EEG-71. Environmental Evaluation Group, Albuquerque and Carlsbad, NM.

Maleki, H., and P. Hynes, 1991. Application of Microcomputers to Geotechnical Design and Stability Monitoring at the Tg Soda Ash Trona Mine. Proc. Longwall USA, Pittsburgh, Pen., pp 92-110.

Maleki, H., and K. Hollberg , 1993. A Case Study of Monitoring Changes in Roof Stability. Intern. J. of Rock Mech., vol. 30, no. 7.

Munson, D. E., J. Weatherby and K. DeVries, 1993. Two- and Three-Dimensional Calculation of Scaled In Situ Test Using the M-D Model of Salt Creep. *In* Proc., 34th U.S. Rock Mechanics Symp., Madison, WI.

Munson, D.E., 1997. Constitutive Model of Creep in Rock Salt Applied to Underground Room Closure. Int. J. Rock Mech. & Min. Sci. Vol 34, No. 2, pp 233: 247. Elseviere Science Ltd.

Parson and Dahl, 1971. A Study of Causes for Roof Instability in the Pittsburgh Coal Seam. *In* Proc. 7th Canadian Rock Mechanics Symp.

Starfield and Cundall, 1988. Towards a Methodology for Rock Mechanics Modeling. Inter. J. Rock Mech. Min. Sci. & Geomech. Abst., vol. 25, no.3, pp. 99-106.

Stormont, 1990. Discontinuous Behavior Near Excavations in a Bedded Salt Formation. Sandia 89-2403.uc-721.

SPSS, 1995. Statistical Packages for the Social Sciences, Chicago, IL.

Westinghouse, 1998. Annual Ground Control Operating Plan for the Waste Isolation Pilot Plan. Westinghouse Geotechnical Engineering, Feb. 1998.

Basic and Applied Salt Mechanics, Cristescu, Hardy, Jr & Simionescu (eds)
© 2002 Swets & Zeitlinger, Lisse, ISBN 90 5809 383 2

Computation of the extend of the disturbed rock zone surrounding long-term excavations in rock salt

J.D.Miller & L.D.Hurtado
WIPP Regulatory Compliance and Repository Isolation Systems Departments, Sandia National Laboratories, Albuquerque, New Mexico, USA

The United States Department of Energy (DOE) is responsible for implementing an effective, environmentally-sound solution to the problem of radioactive waste disposal. A large volume of radioactive waste from US defense programs exists in the form of transuranic (TRU) waste. The solution envisioned by the DOE for disposal of the defense program TRU waste is to emplace it into a deep underground facility expressly engineered for this purposed in bedded rock salt. The DOE project that is now accomplishing this goal is the Waste Isolation Pilot Plan (WIPP) located near Carlsbad, New Mexico.

The WIPP facility is the world's first modern full-scale, mined geological repository designed for the safe management, storage, and disposal of TRU radioactive wastes. After being certified by the U.S. Environmental Protection Agency (EPA), the first shipments of radioactive waste have been received and emplaced at the WIPP. The TRU waste being emplaced at the WIPP facility is limited, by law, to waste generated by US government defense programs. Certification by the EPA was achieved by demonstrating and documenting compliance with all applicable regulations, through a several-decade-long research program that investigated the site, its hydrological, chemical, and mechanical properties, and numerous other aspects of the conditions under which TRU waste could be safely interred. Performance assessments were conducted to evaluate the long-term (10,000 year) post-emplacement site characteristics, including all of the scenarios judged to be credible that could lead to any transport of radionuclides to the surrounding environment.

The DOE continues to conduct research to reduce conservatism in the predictions of performance, in preparation for the five year post-opening recertification of the WIPP. The program to be pursued after WIPP opens is called the Disposal Phase Experimental Program (DPEP). Part of the ongoing research and development program involves international collaborations and partnerships in a variety of radioactive waste management and disposal disciplines. International cooperation is being pursued in order to provide mutually beneficial progress and research cost savings.

1— This work performed for the United States Department of Energy under Contract DE-AC04-94AL85000

Collaboration between DOE and the German waste disposal program is particularly valuable because both programs are actively pursuing radioactive waste disposal in rock salt. Asse and Morsleben are two former salt mines located in Germany at which research is currently being performed to investigate the safe disposal of radioactive waste. Since 1965, the former Asse salt mine has been used for extensive research and development studies. In 1970, the Bartensleben salt mine near Morsleben was selected for the disposal of low- to medium-level radioactive waste. Due to the common interest in storing radioactive waste in rock salt, geomechanical scoping analyses based on finite element computations are being performed at Sandia National Laboratories to further the collaborative process and to improve the technical basis for performance assessment studies. Initial analyses of the German sites concentrate on the vertical shafts from the surface down to the mining horizon, with particular emphasis on the extent of disturbance of the intact salt surrounding each shaft as functions of depth and time. The objective is to investigate the extend of the distributed rock zone (DRZ) surrounding the mine opening. This is important because the DRZ is more permeable than intact rock salt, which could be an important factor for designing effective sealing systems.

The finite element is being used to calculate the extend of disturbed rock zone surrounding the mine openings. JAS3D is the finite element code that is being used for the analysis. JAS3D is a general-purpose analysis code developed at Sandia National Laboratories to model the mechanistic behavior of three-dimensional solids, and incorporates material models and solution algorithms from several older solid mechanics finite element analysis codes including SANTOS and JAC. Both SANTOS and JAC were developed at Sandia and have been used to compute salt creep deformation effects in WIPP and other geologic systems such as the Strategic Petroleum Reserve. JAS3D runs on UNIX-based systems, and can be run using either the iterative, indirect solution methods of dynamic relaxation, from SANTOS, or the conjugate gradient scheme, from JAC, to obtain a quasi-static solution. Because these are explicit solution methods, no stiffness matrix is formed during the calculation, which reduces the amount of data storage necessary for the code to execute.

The halite mined at both German sites and especially at Asse contains less argillaceous material than the salt at the WIPP site. However, because WIPP salt has had extensive mechanical property evaluation specifically for use in numerical modelling, it is used as the basis for the models. The creep behavior of rock salt is assumed to be described well by the multi-mechanism deformation (MD) model proposed by Munson and Dawson, and later extended by Munson et al. [1].

The spatial extend of the DRZ is determined as a post-processed parameter not directly calculated by JAS3D. The criterion defining the area of the disturbed zone is based on stress invariants, which was one of the criteria used for studies characterizing the potential for damage or healing of WIPP salt [2], as determined by dilation experiments on WIPP salt samples. The two stress invariants comprising the damage criterion are I_1, which is the first invariant of the stress tensor, and the square root of J_2, the second invariant of the deviatoric stress tensor. The mean stress is represented by I_1 while the shear stress is represented by J_2. The ratio of the two invariants provides a stress relationship indicating the potential for dilatancy and fracture, as follows:

$$\frac{\sqrt{J_2}}{I_1} = \begin{cases} \geq 0.27, & \text{damage occurs} \\ < 0.27, & \text{remains intact} \end{cases}$$

where:

$$\sqrt{J_2} = \sqrt{\frac{1}{6}\left[(\sigma_1 - \sigma_2)^2 + (\sigma_2 - \sigma_3)^2 + (\sigma_3 - \sigma_1)^2\right]}$$

$$I_1 = \sigma_1 + \sigma_2 + \sigma_3$$

σ_1, σ_2, σ_3 = principal stresses .

Both Morsleben and Asse have open vertical shafts within the salt selections, which from a modelling point of view are axisymmetric and therefore amenable to analysis using what are often referred to as "pineapple slice" models. Two-dimensional pineapple slice models include radian and axial components where the axis of rotation is the centerline of the shaft. In the three-dimensional model, two coordinate axes form the horizontal plane in which a sector, in this case a ninety-degree angular wedge of uniform thickness normal to the plane, resides. The axial centerline of the shaft is normal to the plane. Because the shaft dimensions are not known precisely, a range of sizes are modelled encompassing the possible diameters at both sites. Depth of the model along the shaft is varies as well, by adjusting the lithostatic pressure boundary conditions. Starting with undisturbed salt and a lithostatic stress state, the shaft is excavated at time zero, and then the creep deformation and changing stress state of the salt is calculated for a long time period, arbitrarily chosen to be 100 years. This is the end point of the calculations for simulating Asse, but not for Morsleben, for which the shaft is modelled as being backfilled with clay and the calculations continued for another 200 years.

These scoping calculations are preliminary in nature. They are not intended to be definitive models of the actual German repository sites, and do not represent the actual mechanical behavior of the mine openings. Initial evaluations are instead being done to provide a basis for comparison to the analyses that have been and will continue to be performed for the WIPP facility. Conclusions from the study include a description of the relationships between depth, time since excavation, excavated diameter of the shaft, radial extend of the disturbed rock zone, and, for the Morsleben model, the effect of backfilling the shaft.

REFERENCES

1. Munson, D.E., A.F. Fossum, and P.E. Senseny, "Advances in Resolution of Discrepancies Between Predicted and Measured In Situ WIPP Room Closures," SAND88-2948, Sandia National Laboratories, Albuquerque, New Mexico, 1989.

2. Ratigan, J.L., L.L. Van Sambeek, K.L. DeVries, and J.D. Nieland, "The Influence of Seal Design on the Development of the Disturbed Rock Zone in the WIPP Alcove Seal Tests, " RSI-0400, RE/SPEC Inc., Rapid City, South Dakota, 1991.

Basic and Applied Salt Mechanics, Cristescu, Hardy, Jr & Simionescu (eds)
© 2002 Swets & Zeitlinger, Lisse, ISBN 90 5809 383 2

Thermomechanical analyses for the TSS-experiment and comparison with in-situ measurements

A.Pudewills
Forschungszentrum Karlsruhe GmbH, Institut für Nukleare Entsorgungstechnik
Karlsruhe, Germany

T.Rothfuchs
Gesellschaft für Anlagen- und Reaktorsicherheit (GRS) mbH, Fachbereich
Endlagersicherheitsforschung, Braunschweig, Germany

ABSTRACT

The numerical simulation of the thermal and thermomechanical response of the large-scale in situ experiment "Thermal Simulation of Drift Storage" will be presented. The objective of the investigations is to assess the current capability of the thermomechanical models by comparison of numerical results with in situ measurements over a 10-years period. The temperatures measured in the test field and predicted by 3D model calculations compare very favorably. Analyses of thermomechanical effects were conduted using 2D, plane-strain conditions and two different constitutive models for the backfill material. For both models the quantitative agreements between the calculated drift closure rates and backfill compaction, respectively, are somewhat less satisfactory.

INTRODUCTION

During the last decade a repository concept for direct disposal of spent nuclear fuel in rock salt has been developed (Bechthold et al, 1993). The spent fuel elements are incapsulated in large self-shielding POLLUX casks which after an interim storage will be emplaced on the floor of long drifts at a depth of about 850m below the surface in a rock salt formation. Immediately after emplacement, the drifts are backfilled with dry crushed salt. To support this repository concept, the in situ experiment "Thermal Simulation of Drift Storage" (TSS) was started in 1989 in the Asse salt mine in Germany and is still running unter the current European project "Backfill and Material Behaviour in Underground Salt Repositories" (BAMBUS) (Bechthold et al, 1989, 1999). The objectives of the test are to demonstrate the emplacement technology and to study the thermal and thermomechanical consequences of the direct disposal of spent fuel in drifts on the rock salt and on technical barriers. A further aim is to evaluate the capability of the available codes to simulate the thermomechanical behaviour of the backfill and rock salt under representative repository conditions.

Within the framework of this project, the rise of temperature in the test field, which is induced by electrical heaters, the closure of the drifts followed by the compaction of the backfill. and the resulting stresses in the surrounding salt were studied numerically. Furthermore, this paper presents the results of comparing thermal and thermomechanical models with test data for a selected instrumentation during the heating phase and part of the cooling phase.

The numerical analyses have been carried out using two different finite element codes. The time-dependent temperature distributions in the test field were previously calculated using the three-dimensional code FAST-BEST (Ploumen. 1980) based on the coarse mesh method which was specifically developed for the investigation of waste repository structures. The thermomechanical calculations were performed with the two-dimensional, finite element code MAUS (Albers, 1991).

The current analysis extends the results of earlier model calculations (Bechthold et al 1999 Padewills, 1998) which revealed that the predicted drift closure and backfill compaction in the heated area differ from the in situ measurements. Sensitivity studies investigate how the assumed material models of crushed salt affect the numerical results. Attention is focused on the thermomechanical behaviour of the backfill material in the hot area of the experiment.

IN SITU EXPERIMENT

The TSS test field is located in a Staßfurt halite formation at 800m depth below the ground surface. Two parallel drifts were excavated with a length of about 75 m each. The drifts are 4.5m wide and 3.5m high. separated by a 10m thick pillar. In each test drift three electrically heated casks with a thermal power of about 6.4 kW were emplaced horizontally on the floor. Corresponding to the real disposal casks, the heaters have a length of 5.5 m, a diameter of 1.5 m, and a weight of 65 ton. Figure 1 shows a plan view of the experiment on the 800m level in the Asse salt mine and the positions of measuring sections are also indicated.

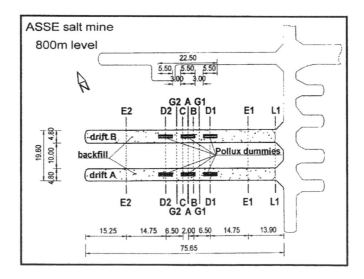

Figure 1   Plan View of the TSS Test Field at 800m Level and the Measuring Sections

After installation of the heaters and measuring equipment, the test drifts were backfilled with dry crushed salt. The excavation of the drifts began in spring 1989 and the heating started in September 1990. After more than eight years of operation, the heaters were switched off in February 1999. Data retrieval will be continued for another year during the cooling phase.

NUMERICAL MODEL

According to the real geometry of the experiment, a three-dimensional finite element model for temperature calculation was used. The model considers the finite length of the drifts and the 3m distance between the heaters. The symmetry planes are cut through the centre of the pillar width in the middle of the drift length. A detailed description of this numerical model with the assumed boundary and initial conditions is given in (Pudewills, 1997).

To perform the thermomechanical analyses with a reasonable numerical effort, a two-dimensional finite element model was used, assuming generalised plane strain conditions. The model represents a vertical cross-section perpendicular to the drifts at three measuring cross-sections; Sections A and G located in the zone of maximum temperatures and section E2 situated in the cooler region at 22m distance from the centre of the experiment (Figure 1). This model includes one drift and the half-pillar. The element discretisation and the boundary conditions are shown in Figure 2.

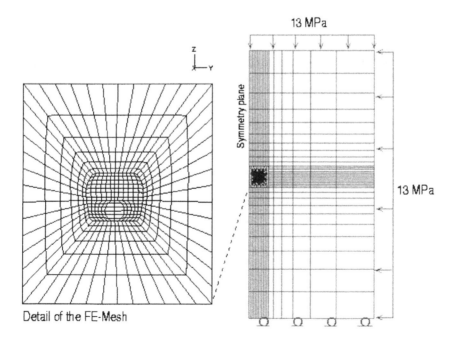

Figure 2   Finite Element Mesh and Boundary Conditions Used for the
Thermomechanical Analyses

The computational domain has a horizontal extention of 100m and a height of 200m. Symmetry conditions were assumed on the left border and a lithostatic pressure of 13 MPa at the top and right boundaries of the model, corresponding to the results of in situ stress measurements (Rothfuchs et al 1998 . The initial temperature of the entire model before heating was taken to be 37 C. The temperature predictions from thermal calculation were used as input data for thermomechanical calculation with the MAUS code.

MATERIAL MODEL AND PARAMETERS

The thermomechanical behaviour of the rock salt was described as a thermoelastic material model with temperature-dependent steady state creep given by Wallner et al, (1979) . The thermoelastic properties and the constitutive equation of the steady state creep strain rate are taken as follows:

Thermoelastic properties:

Young's modulus: 24000 MPa
Poisson's ratio: 0.25
Coefficient of thermal expansion: $4.2 \ 10^{-5} K$

Steady state creep strain rate : $\dot{\varepsilon}_c = A \cdot \sigma_{eff}^5 \cdot \exp(-Q/RT)$

with $A = 2.06 \ 10^{-6} \ MPa^{-5} s^{-1}$; $Q = 54.21$ kJ/mol; $R = 8.314$ J/(mol K) and σ_{eff} representing the effective stress. T is the temperature.

With respect to the material model of crushed salt, two different constitutive relations have been used. For instance, the constitutive relation of crushed salt proposed by Hein (1991) was adopted. It is based on a viscoplastic formulation and considers both volumetric and deviatoric strain rates under hydrostatic and deviatoric stress conditions. The material parameter values and the functions h_1 and h_2 are derived from extensive triaxial tests performed by Korthaus (1996).

For the earlier predictive calculations it was assumed that the backfill behaves as a hydrostatic material with a temperature-dependent creep consolidation (Korthaus, 1991). This formulation was adapted to oedometer tests at temperatures up to 200 °C (Diekmann and Stührenberg, 1990). Table 1 summarises both constitutive equations of crushed salt used in thermomechanical simulations together with the material parameters and the thermo-elastic properties.

COMPARISON OF NUMERICAL RESULTS WITH IN SITU MEASUREMENTS

The numerical simulations start with the calculation of the isothermal drift closure rates and stress distributions in rock salt around the drifts before the emplacement of the electrical heaters. After about 1.4 years, the heaters were switched on, the backfill material and the temperature development in the test field were taken into account. The thermomechanical calculations were continued for an experimental period of another eight years. The results of the finite element calculations performed for

Table I

Table I
Constitutive models and parameters for crushed salt

Deviatoric material model	Hydrostatic material model
Thermoelastic Properties	*Thermoelastic Properties*
Young's modulus: 2000 MPa	Young's modulus: 2000 MPa
Poisson's ratio: 0.25	Poisson's ratio: 0.25
Coefficient of thermal expansion: $4.2 \ 10^{-5}$K	Coefficient of thermal expansion: $4.2 \ 10^{-5}$K

Creep Behaviour	*Hydrostatic Consolidation Behaviour*
$\dot{\varepsilon} = A \cdot e^{-Q/R/T} \cdot (h_1 \cdot p^2 + h_2 \cdot q^2)^n \cdot$ $(h_1 \cdot p / 3 \cdot \mathbf{1} + h_2 \cdot \mathbf{S})$ $h_1(\eta) = a / (((\eta_0/\eta)^c - 1) / \eta_0^c + d)^m$ $h_2(\eta) = b \cdot h_1(\eta) + 1$ $\dot{\varepsilon}$: tensor of the strain rate T : absolute temperature p : mean normal stress **1** : unit tensor q : deviatoric stress invariant **S** : tensor of the deviatoric stress η : porosity η_0 : initial porosity (36%) Q : activation energy R : universal gas constant $A = 1.09 \cdot 10^{-6}$/s/MPa^{-5}; Q/R = 6520/K; n = 2; a = 2.469E-3; b = 0.9; c = 0.1 ; m = 2.25; d = 0.0003	$\dot{K} = c_1 \ (\eta_0 - K)^6 (P - c_3 \ K \ c_4)^2$ $c_1 = c_1 \exp(c_2 (1 / T_{ref} - 1 / T))$ $c_3 = c_3 (c_5 + c_6 (T - To) + c_7 (T - To)^2)$ $\dot{K}$: rate of compaction (1/d) η_0 : initial porosity (36%) K : total compaction (ΔV/ Vo) P : pressure (MPa) T : temperature (K) To = 273 K; T_{ref} = 392 K; c_1 = 170 (1/d); c_2= 4000 K; c_3 = 130; c_4= 1.8; c_5 = 2.6; c_6= -0.021 1/ K; c_7 = $5 \ 10^{-5}$K^{-2}

both material models of crushed salt will be compared to experimental data measured during the heating phase of the test only.

Thermal Results

The three-dimensional temperature calculation was performed previously and a detailed discussion of those results is reported in an earlier paper (Pudewills, 1997). A brief presentation of some temperature histories at different positions throughout the measuring cross-sections A, G and E is given in Fig. 3 together with the measured data. From this figure it can be seen that the agreement of the calculated with the measured results is rather good at all positions and the maximum difference is less than 10%.

Drift closure rates

The development of the measured and calculated horizontal and vertical drift closure rates for both material models and for the three cross-sections considered is illustrated in Figure 4. In the heated region (i.e. cross-section A and G) the calculation results for both material models of crushed salt differ significantly from the test data.

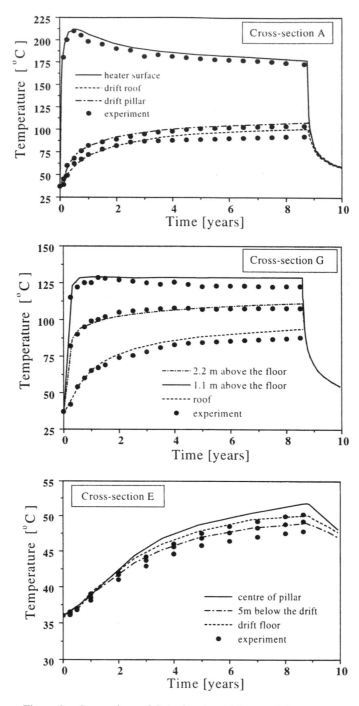

Figure 3   Comparison of Calculated and Measured Temperatures
at Different Positions in the Test Field

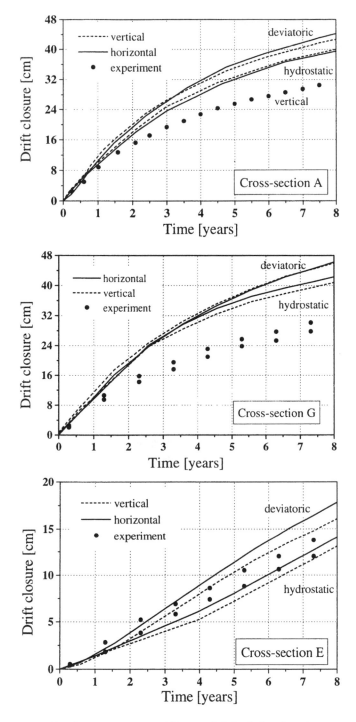

Figure 4　Drift Closures; Measured Data and Calculation Results of Two
Backfill Material Models for Three Vertical Cross-Sections

The numerical results suggest that the present deviatoric constitutive law of crushed salt, already validated by laboratory benchmark (Korthaus, 1998) behaves much softer than the crushed salt used in the TSS-experiment and also than the hydrostatic model used in earlier studies. The observed drift closure rates during the isothermal period of the experiment, prior to the start of heating and in the regions far away from the heaters, (i.e., cross section E) compare closely with the calculated drift closure.

Backfill porosity and compaction pressure

The development of the backfill porosity was not directly measured in the experiment. However, it was determined from the measured drift closure taking into account an initial backfill porosity of 35%. For this reason, the comparison between calculation results and the estimated values for in situ porosity now is of an orientative character only. Nevertheless, the development of the porosity averaged over the drift cross-section in the heated zone and the calculated porosity at two positions in the drift are shown in Figure 5. Generally, the same tendency as for the drift closure presented above can be seen (i.e. the porosity reduction is overestimated by both material models used). Due to the temperature-dependent compaction of the backfill and large temperature gradients in the drift, a non-uniform distribution of the backfill porosity over the drift cross section is expected. Contour plots of calculated porosity distribution in the drift should support this presumption. Figure 6 shows the spatial distribution of the porosity in the drift after 6 years of heating. In the planned post-test project an extensive sampling programme with laboratory investigations will be peformed to elucidate this presumption.

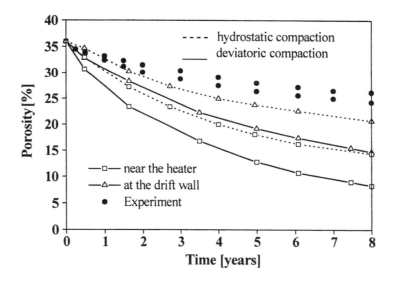

Figure 5   Comparison of Calculated and Measured Backfill Porosity in
the Heated Area for Two Material Models of Crushed Salt

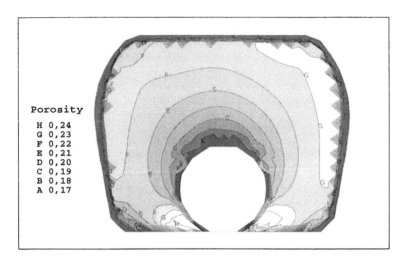

Figure 6   Distribution of Porosity in the Drift at Cross-Section A

The development of backfill pressure was measured by hydraulic pressure cells mounted in several sections at the roof, the floor and the walls. Figure 7 presents a comparison between measured and calculated backfill pressure at the drift roof and pillar in section G. The curves indicate that the compaction pressures obtained for the devitoric material model of crushed salt matches the measured pressure, while the results for the hydrostatic model again show an overestimation. Therefore, the results from earlier predictive calculations using the hydrostatic material model exhibit a reasonable agreement between calculated and measured responses in the sense of general data trends.

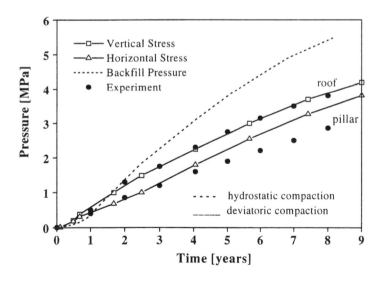

Figure 7   Comparison of Calculated and Measured Compaction Pressure
in the Backfill for Cross-Section A

Further numerical results of thermomechanical sensitivity analyses (see also the paper presented by Korthaus, (1999)) suggest that the present deviatoric constitutive model of crushed salt behaves softer than the crushed salt used in the TSS-experiment. The possible reason for this effect could be a separation of loose salt grains with different sizes during the backfilling process, causing an inhomogeneous distribution over the drift cross section.

CONCLUSIONS

Finite element analyses of the TSS-experiment were performed to investigate the thermomechanical behaviour of the rock salt and backfill material under repository relevant conditions. The results of the calculations were compared to experimental data for three vertical cross sections through the test field. For the moment the following conclusions can be drawn:

- The calculated temperatures agree quite well with the experimental data at all locations, indicating that three-dimensional modelling is necessary to obtain adequate thermal fields.

- The drift closures are overestimated by two-dimensional analyses in the warmer sections, but agree with the measurements in the cooler sections, thus indicating that the temperature dependence of the constitutive models for crushed salt and rock salt, respectively, are not yet well understood. As a consequence of the computed higher closure rates, the models predict lower porosities.

- Finally, the thermomechanical results based on a new constitutive equation for crushed salt describing both volumetric and deviatoric deformations lead to larger differences between the calculation results and the measurements. The discrepancy could be attributed to uncertainties in the material model of backfill. Therefore, material parameters obtained from laboratory tests have to be adjusted in a parametric study, where the measured and calculated drift closure rates and backfill compaction pressure are matched.

Acknowledgment: This work has been supported partly by the Commission of the European Communities through the Management and Storage of Radioactive Waste project.

REFERENCES

Albers, G., 1991. "MAUS - A Computer Code for Modelling Thermomechanical Stresses in Rock Salt", in Computer Modelling of Stresses in Rock, Proc. of a Technical Session, Brussels, 6-7 Dec. 1983, EUR 9355 EN, Brussels

Bechthold, W., Heusermann, S., Schrimpf, C., Gommlich, G., [1989], "Large scale test on in-situ backfill properties and behavior under reference repository conditions", NEA/CEC Workshop on Sealing of Radioactive Waste Repositories, May 22-25 1989, Braunschweig, Germany.

Bechthold, W.; Closs, K.D.; Knapp, U.; Papp, R., [1993], "System analysis for a dual-purposed repository", Final report, EUR 14595 EN, Luxembourg

Bechthold, W.; Rothduchs, T.; Poley, A.; Gheoreychi, M.; Heusermann, S.; Gens, A .; Olivella, S., [1999], "Backfill and Material Behaviour in Underground Salt Repositories (BAMBUS)", Final Report, Brussels (in print)

Diekmann, N. and Stührenberg, D., [1990], "Untersuchungen zum Kompaktionsverhalten von Salzgrus mit der Oedometerzelle TRE3002", KWA 58019, Archiv-Nr.: 107586, Bundesanstalt für Geowissenschaften und Rohstoffe (BGR).

Droste, J.; Feddersen, K. H.; Rothfuchs, T.; Zimmer, U., [1996], "The TSS-Project: Simulation of Drift Emplacement", Final Report Phase 2, 1993-1995, GRS-127, Gesellschaft für Anlagen- und Reaktorsicherheit mbH, Brunschweig.

Hein, H. J., [1991], "Ein Stoffgesetz zur Beschreibung des thermomechanischen Verhaltens von Salzgranulat" Dissertation, RWTH Aachen, 1991.

Korthaus, E., [1991], "Thermische und thermomechanische Prognoserechnungen zum TSS- Versuch (Thermische Simulation der Streckenlagerung)", Primärbericht Nr. 19.03.03.P.04 A, Kernforschungszentrum Karlsruhe GmbH, Karlsruhe.

Korthaus, E., [1996], "Consolidation and Deviatoric Deformation Behaviour of Dry Crushed Salt", Proc. of 4th Conf. on the Mechanical Behavior of Salt, Montreal, Canada, pp. 365-377

Korthaus, E., [1998], "Experiments on Crushed Salt Consolidation with True Triaxial Testing Device as a Contribution to an EC Benchmark Exercise", FZKA 6118

Korthaus, E., [1999], "Consolidation behaviour of dry crushed salt: Triaxial tests, benchmark exercise,and in situ validation", Proc. of 5th Conf. on the Mechanical Behavior of Salt, Bucharest, Roumania, (in print)

Ploumen, P., [1980], "Numerische Langzeitrechnung dreidimensionaler Temperaturfelder mit Hilfe eines speziellen Finite Element-Verfahrens am Beispiel der Endlagerung hochradioaktiver Abfälle im Salzgestein", Dissertation, RWTH-Aachen.

Pudewills, A., [1997], "Thermal Simulation of Drift Emplacement: Temperature Analyses", Topical Report, FZKA 5955, Forschungszentrum Karlsruhe, GmbH, Karlsruhe.

Pudewills, A., [1998], "Thermomechanical analysis of the TSS experiment", Proc. Internat. Congr. on Underground Construction in Modern Infrastructure, Stockholm, S, June 7-9, 1998, Rotterdam: Balkema, pp. 317-323.

Rothfuchs, R.; Droste, J.; Feddersen, K. H.; Heusermann, S.; Schneefuss, J.; Pudewills, A., [1998], "Special safelty aspects of drift disposal – the Thermal Simulation of the Drift Storage Experiment", Nucl. Technology, vol. 121, No. 2, pp.189-198.

Schneefuß, J.; Feddersen, N.; Jockwer, J; Droste, J., [1996], "The TSS Project: Research on compaction of and gas release in saliferous backfill used in drift emplacement of spent fuel", Final report, EUR 16730, Brussels.

Wallner, M.; Caninenberg, C.; Gonther, H., [1979], "Ermittlung zeit- und temperaturabhängiger mechanischer Kennwerte von Steinsalz", Proc. 4. Int. Cong. Rock Mech., Montreux, Switzerland, Sept.2-8, 1979, Vol. 1, pp. 313-318, Rotterdam, Balkema.

Part VIII

Mining application

Basic and Applied Salt Mechanics, Cristescu, Hardy, Jr & Simionescu (eds)
© *2002 Swets & Zeitlinger, Lisse, ISBN 90 5809 383 2*

Risk analysis for a well brine using DKR control method with application to slanic salt brine

Virgil I.Breaban
"Ovidius" University, Civil Engineering Dept., Constanta, Romania

Radu A.Canarache
Inicad Soft, Str. Popa Tatu 20, Sector 1, Bucharest, Romania

Gyorgy Deak & Stefania Deak
Minesa S.A, Str. T. Vladimirescu, 15-17, Cluj-Napoca, Romania

Liviu Draganescu
Salina Slanic Prahova, Str. Cuza Voda 21, Jud. Prahova, Romania

SUMMARY

Studies have been carried out on the stability of a well brine, using the DKR control method, based on the numerical computational methods. The DKR Control method was elaborated following some experiments and tests, both in the laboratory and in-situ, for some salt exploitations situated in Romania. The general form of the method consists of three mains steps:
- Analyzing of the available geologic, mechanical and topometric information:
  - the dynamics of the exploitation (the evolution during the time) and the exact geometry of the cavities today;
  - physical and mechanical properties of the material (the creep law), following laboratory and in-situ tests;
  - cavernometric measurements, made during the exploitation, containing the level of subsidence at several time steps;
  - the evolution of the exploitation in time.

It should be noted that the information available for the study was not complete following the analysis. The degree of knowledge available was estimated at 79.24%, which reduced the confidence of the numerical simulation.

- Creating the numerical computational model:
  - several two-dimensional model studies have been performed, representing characteristic sections of the well;
  - a diagram of the evolution in time for every section, according to the available data, corresponding to the total time of exploitation, from the initial opening of the well.

- Calibrating the numerical computational model:
  - the main control parameter was the measured subsidence; unfortunately, data was available for only a few points; and this subsidence level was accepted for the whole section;
  - the results have been compared to the results of the numerical simulation, considering both the dynamics of the exploitation and the creep law obtained in the laboratory. The creep law parameters have been corrected until an error of 5% has been

achieved. At this point it was considered that the calibration of the numerical model was accurate to within 95%.

- Exploitation using the numerical model:
  - With the calibrated model, the analysis with time was continued, considering that the actual geometry changes were only due to the collapse of some parts, where the admissible stresses have been overpassed. The analysis was carried on for 12,978 days, computing the possible subsidence and stresses.
  - Due to the fact that the well is situated in a seismic area, some seismic analysis was performed, computing the supplementary stresses induced by earthquakes, at several moments of the exploitation evolution. By this superposition, the collapse risk increases because the geometry changes faster than that predicted by the creep analysis, and the well behavior moves toward an unstable condition. Estimates the effect on stability of the Slanic brine, with or without seismic effects, and evaluates the risk for its collapse as a function of time.

Basic and Applied Salt Mechanics, Cristescu, Hardy, Jr & Simionescu (eds)
© *2002 Swets & Zeitlinger, Lisse, ISBN 90 5809 383 2*

Some new developments about safety analysis and mechanical behaviour of salt cavities

K.-H.Lux & Z.Hou
Professorship for Disposal Technology and Geomechanics, Clausthal University of Technology, Clausthal-Zellerfeld, Germany

Abstract

A non-linear strength function in dependence of the minimum principle stress and the stress geometry, and the on continuum damage mechanics based creep model *How/Lux* recently developed by *Hou (1997)* and *Hou & Lux (1998)*, waren applied to proof and analysis of safety for caverns in saliniferous formations. Because the material model *How/Lux* takes into account the influences from minimum stress, the coefficient of utilization, the stress geometry and dilatancy, the results regarding the ratio between time-to-rupture, strain intensity as well as coefficient of utilization and minimum stress are qualitatively plausible. An example is given for the bearing behavior of a salt cavern.

Strength function under consideration of the stress geometry

Equations (1) to (3) show the strength function in the invariant plane taking into consideration the stress geometry, developed by *Hou (1997)* and *Hou et al. (1998)*:

$$\beta^D(\theta,\sigma_3) = k(\theta,\sigma_3) \cdot (\beta^D)^{TC} \left(a9 \geq \frac{\sqrt{3}}{3}, a9 = \frac{\sqrt{3}}{3} \Rightarrow (\beta^D)^{TC} = (\beta^D)^{TE} \right) \tag{1}$$

$$k(\theta,\sigma_3) = \left[\frac{1}{\cos\left(\theta + \frac{\pi}{6}\right) + a9 \cdot \sin\left(\theta + \frac{\pi}{6}\right)} \right]^{\exp(-a10 \cdot \sigma_3)} \tag{2}$$

$$(\beta^D)^{TC} = \max\beta^D - \left(\max\beta^D - \frac{\sigma_{DTC}}{\sqrt{1,5}} \right) \cdot \exp(-cc \cdot \sigma_3) \tag{3}$$

With

$\beta^D(\theta, \sigma_3)$ deviatory shearing strength in the invariant plane, in MPa

$(\beta^D)^{TC}$ deviatory shearing strength under TC conditions, in MPa

$\max\beta^D$ maximum deviatory shearing strength, in MPa

σ_3 minimum stress, in MPa

σ_D unconfined compression strength, in MPa

$k(\theta, \sigma_3)$ a function to describe the influences of the stress geometry on the strength, in -

θ *Lode*-angle, in degrees

a8, a9, a10 parameter, in 1/MPa, - and 1/MPa

The fracture strength can alternatively be shown in the σ_3/σ_1-plane and in the Mohr's diagram. Equation (4) shows the fracture strength function in the σ_3/σ_1-plane, equations (5) and (6) show the fracture strength function in the Mohr's diagram:

$$\sigma_{1B}(\theta,\sigma_3)=\sqrt{2}\cdot\cos\theta\cdot k(\theta,\sigma_3)\cdot\left(\beta^D\right)^{TC}+\sigma_3 \qquad (4)$$

$$\tau_{nt}=g(\sigma_n)=\tan\phi\cdot\sigma_n+c=\frac{(\sqrt{1,5}\cdot\beta^D)^{'}}{2\cdot\sqrt{(\sqrt{1,5}\cdot\beta^D)^{'}+1}}\cdot\sigma_n+\frac{\sqrt{1,5}\cdot\beta^D-\sigma_3\cdot(\sqrt{1,5}\cdot\beta^D)^{'}}{2\cdot\sqrt{(\sqrt{1,5}\cdot\beta^D)^{'}+1}} \qquad (5)$$

$$\left(\beta^D\right)^{'}=\frac{d\beta^D(\theta,\sigma_3)}{d\sigma_3}=cc\cdot\left(\max\beta^D-\frac{\sigma_{DTC}}{\sqrt{1,5}}\right)\cdot\exp(-a8\cdot\sigma_3)\cdot k$$

$$-a10\cdot\beta^D\cdot k\cdot\exp(-a10\cdot\sigma_3)\cdot\ln\left[\frac{1}{\cos\left(\theta+\frac{\pi}{6}\right)+a9\cdot\sin\left(\theta+\frac{\pi}{6}\right)}\right] \qquad (6)$$

With

ϕ angle of internal friction, in degrees

c cohesion in dependence of θ and σ_3, in MPa

σ_{1B} maximum stress in dependence of θ and σ_3, in MPa

τ_{nt}, σ_n shearing stress and normal stress in the fracture, in Mpa

$(\beta^D)^{'}$ deviation of $\beta^D(\theta, \sigma_3)$ to σ_3, in -

Equations (4) and (6) show that the angle of internal friction ϕ and the cohesion c are shown in dependence of the minimum stress σ_3 and of the stress geometry θ. With an increase of the minim stress (here for a certain stress geometry), the angle of internal friction ϕ decreases and the cohesion c increases. If the minimum stress σ_3 is sufficiently high, the deviation $(\beta^D)'$ approaches zero according to equation (6). This result indicates that the *Mohr* parameters (c, ϕ) are not necessarily constant.

Material model *Hou/Lux* with creep rupture criterion

A new material model called the *Hou/Lux* material model has been developed as a differentiation to existing material models - based on the material model *Lubby2* according to *Lux (1984)* and based on the Continuum-Damage-Mechanics

The details about the material model *Hou/Lux* are written in *Hou & Lux (1999)* in this proceedings.

Case study: storage cavern

In the following, the bearing behavior of a storage cavern in rock salt with a height of h = 200 m and a radius of r = 50 m is examined as an example. The cavern is located inside a salt deposit (depth of salt surface z = 630 m) with varying roof depths of z = 700, 900 and 1100 m. This study mainly focuses on the introduction of the cavern located at a position of depth of z = 900 - 1100 m, and on its bearing behavior.

An internal pressure of p_i = 0 MPa represents the extreme case of strain in a gas storage cavern. Following the momentarily idealized excavation, the calculations continue according to the known *Lubby2* material model and according to the new *Hou/Lux* material mode until t = 500 d.

Figure 1 shows the extensive distribution of the effective stress in the area of observation according to the *Lubby2* material model and the *Hou/Lux* material model for the caverns with a roof depth of z = 900 m. The comparison shows that at a first glance, the effective stresses and thus the strain on the rock formation look similar. Closer examination, however, shows a serious difference: with the *Hou/Lux* material model, the extreme stress is not at the cavern contour but deeper inside the rock formation. At the cavern contour, the stress on the rock formation has dropped due to stress rearrangements - a result of loosening/destrengthening of the structure

The damage during the observed time span in a contour range of 2.5 to 7.5 m is limited and its intensity decreases with an increasing distance from the cavern. For the introduced cavern configuration, due to the bearing behavior, the maximum damages at a time t = 500 d are not located in the roof or bottom area, but in the wall areas.

Consequences for the design of salt caverns

The material models used for cavern design so far neither considered the dilatancy and the damage nor the resulting destrengthening and the additional creep deformations and stress rearrangements. Neither did they include a creep rupture criterion. Therefore, alternative evaluation criteria must be introduced, which allow an evaluation of the calculated state variables with sufficiently conservative distances to failure states, based on stresses as well as on deformations.

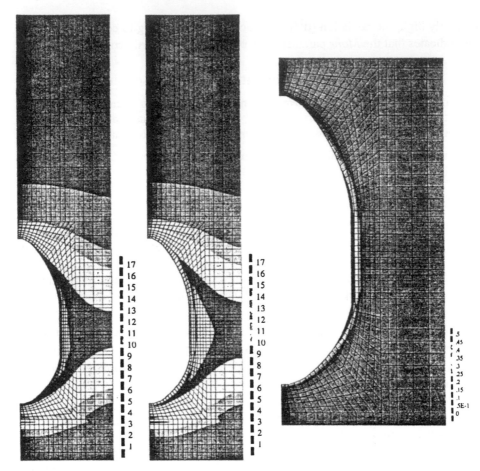

effective stress (*Lubby2*) effective stress (*Hou/Lux*)   damage (*Hou/Lux*)

Figure 1 Effective stress distribution (left) according to Lubby2 material model and effective stress distribution (middle) as well as damage distribution (right) according to Hou/Lux material model at a time t = 500 d (position of roof depth of z = 900 m)

The *Hou/Lux* material model at least in principle eliminates these disadvantages and thus offers new possibilities for the determination of the general behavior and for the cavern design:

- The spatial and temporal development of the loosening, destrengthening and dilatancy zones can be determined.
- Intensities of the loosening zone and resulting spalling in the course of time can be calculated.
- Design parameters such as optimum cavern form, cavern height, cavern diameter, width of the salt pillars, thickness of the residual salt roof , distance to saliniferous boundaries, minimum and maximum internal pressure and the corresponding maximum life as well as changes to the primary permeability resulting from dilatancy, can be more realistically deviated than before.

- It must be assumed that the conservativities of the cavern design can be reduced by the more realistic determination of state variables by means of the *Hou/Lux* material model. Consistent location-related material parameters are a prerequisite.
- The current version of the *Hou/Lux* material model allows the simulation of the operational phases of internal pressure reduction and constant internal pressure including the special case of atmospheric pressure (blow out).
- If the healing of damages is also considered, the operational phase of the internal pressure increase can be more accurately registered. According to today's view, the internal pressure increase and the operational phase at maximum pressure should contribute to the healing of damages.

References

Hou, Z. (1997): Untersuchungen zum Nachweis der Standsicherheit für Untertagedeponien im Salzgebirge, Dissertation TU Clausthal.

Hou, Z. & Lux, K.-H. (1998): Ein neues Stoffmodell für duktile Salzgesteine mit Einbeziehung von Gefügeschädigung und tertiärem Kriechen auf der Grundlage der Continuum-Damage-Mechanik. Geotechnik 21 (1998) Nr. 3, pp. 259 - 263.

Hou, Z. & Lux, K.-H. (1999): A material model for rock salt including structural damages as well as practice-oriented applications. 5th Conference on Mechanical Behavior of Salt, August 1999 in Bucharest.

Hou, Z.; Lux, K.-H. & Düsterloh, U. (1998): Bruchkriterium und Fließmodell für duktile Salzgesteine bei kurzzeitiger Beanspruchung. Glückauf - Forschungshefte, 59 (1998) Nr. 2, pp. 59 - 67.

Hou, Z.; Lux, K.-H. & Stampler, J. (1999): The constitutive model Hou/Lux for salt rock based on continuum damage mechanics and it's numerical implementation in the FEM Program. 7th Int. Symposium on Numerical Models in Geomechanics, Sept. 1999 in Graz.

Lux, K.-H. (1984): Gebirgsmechanischer Entwurf und Felderfahrungen im Salzkavernenbau. Ferdinand Enke Verlag Stuutgart.

Lux, K.-H.; Hou, Z.; & Düsterloh, U. (1998): Some new aspects for modelling of cavern behavior and safety analysis. SMRI Fall Meeting, October 1998 in Rome, pp. 359 - 390.

Lux, K.-H.; Düsterloh, U.; Bertram, J. & Hou, Z. (1997): Abschlußbericht zum BMBF-Forschungsvorhaben 02 C 0092 2. Professur für Deponietechnik und Geomechanik der TU Clausthal.

Lux, K.-H.; Rokahr, R.-B. & Kiersten, P. (1985): Gebirgsmechanische Anforderungen an die untertägige Deponierung von Sonderabfällen im Salzgebirge. STUVA-Tagung, Hannover, 1985.

Basic and Applied Salt Mechanics, Cristescu, Hardy, Jr & Simionescu (eds)
© 2002 Swets & Zeitlinger, Lisse, ISBN 90 5809 383 2

Elastic-viscoplastic instantaneous response around circular and non-circular cavities performed in rock salt

O.Simionescu
Department of Mechanics, Bucharest University, Bucharest, Romania

1. INTRODUCTION

Underground cavities of cross sectional shapes circular, or non-circular are often used in mining and civil engineering. The elasto-viscoplastic constitutive equations became in last decades important tools of phenomenological description of the main physical phenomena encountered at geomaterials, like yield, failure, dilatancy, or compressibility of the volume. In the study of stress behaviour around underground cavities we could distinguish two time periods : the first one, in which the cavity is excavated, followed by the time interval in which the cavity is exploited. The first time period is usually much shorter than the second one. Thus, in the constitutive model we could emphasize two tipes of mechanical behaviour : one related to *instantaneous elastic response* of the rock mass, corresponding to the first period of time, resp. a *creep deformation* of the material that lasts over a long period of time.

In the present paper we study the elastic-viscoplastic instantaneous response around one isolated square-like cavity, followed by the problem of the interaction between two circular nearby cavities. Within the elastic domain an interesting alternative to numerical approaches, like finite elements, is the combined use of complex potentials and conformal mappings, leading to explicit or semi-explicit solutions. Using the complete analysis of elastic stress distribution, we study the instantaneous viscoplastic behaviour around such cavities. Among others, the location of yield and failure domains, as well as dilatant and contractant characteristics are studied, based on recent researches by Prof. Cristescu. The examples given in our paper has been computed using the code MATHEMATICA for a rock salt, the numerical results showing quantitatively the effect of geometric and mechanical parameters on stress concentration and on location of instantaneous viscoplastic surfaces.

2. ELASTIC-VISCOPLASTIC CONSTITUTIVE EQUATIONS

We are using the well-known elastic-viscoplastic constitutive equations due to Cristescu [1]:

$$\dot{\varepsilon} = \left(\frac{1}{3K} - \frac{1}{2G} \right) \dot{\sigma} 1 + \frac{1}{2G} \dot{\sigma} + k(\sigma, d) \left\langle 1 - \frac{W^{I}(t)}{H(\overline{\sigma}, \sigma)} \right\rangle \frac{\partial F}{\partial \sigma}$$

where G and K are the elastic parameters, F is the viscoplastic potential and H the yield function. Here:

$$W^{I}(t) = \int_{0}^{T} \sigma(t) \cdot \dot{\varepsilon}^{I}(t) dt$$

is the irreversible stress work used as hardening parameter or internal state variable. Further, k is a viscosity parameter, while σ and τ are the mean stress, resp. octahedral shear stress.

The *compressibility-dilatancy boundary* is defined by the equality $X(\sigma,\tau)=0$, the volume being compressible for $X(\sigma,\tau)>0$, resp. dilatant for $X(\sigma,\tau)<0$. The *short-time failure surface* is done by $Y(\sigma,\tau)=0$, while the *yield surface* is taken as $H(\sigma,\tau)=W^I(t)$.

The examples of viscoplastic surfaces given in this article has been computed for a rock salt, the expressions of these surfaces being determined from triaxial tests (see Cristescu [2], Cristescu and Paraschiv [3]).

2. INSTANTANEOUS BEHAVIOUR AROUND AN ISOLATED SQUARE-LIKE CAVITY

We study the stress distribution around an isolated square-like cylindrical cavity performed in an elastic, homogeneous and isotropic material. At infinity we prescribe two far-field stresses, generally distinct, while the boundary of the cavity is supposed to be free of stresses. Using the method of complex potentials, we find the stress distribution around the cavity, as shown in Simionescu [6]. In order to describe the *instantaneous* response around the cavity, we inject the obtained elastic solution into the expressions of previous viscoplastic surfaces.

We analyse the evolution of the domain of dilatancy with the ratio k of far-field stresses: for k=0.3 the entire contour of the cavity is dilatant, while in the case k=1 only the domain near the corners remains dilatant. A rapid transition to a compressibility zone have place as we enter inside the massif. The failure due to compressive stresses occurs around the corners, at a certain (small) distance inside the massif. A possible failure due to tensile stresses can appear in the roof of the cavity, mainly for small values of k. The material yields, mainly around the corners, for an important zone.

3. INSTANTANEOUS BEHAVIOUR FOR THE INTERACTION OF TWO CIRCULAR CAVITIES

We analyse the elastic response, just after the excavation, of two circular cylindrical cavities having the diameters a, providing that the distance between the centres of the cavities is 2L. As boundary conditions we assume the cavities to be free of stresses, while at infinity we prescribe two far-field stresses, generally distinct.

Using the asymptotic analysis, developed in Simionescu [7], we find the stress distribution around the cavities. To obtain the instantaneous response in the problem of interaction of two circular cavities, we inject the elastic solution into the expressions of viscoplastic surfaces. The main aspect observed is the asymmetry of viscoplastic surfaces, even when the ratio k equals 1, which is due to the interaction effect. The cavity wall is dilatant for k<1, becoming compressive for k>1. The effect of the interaction decreases with L increasing, for L=1.5a this effect being weak.

4. REFERENCES

1. N. Cristescu, *Rock rheology*, Kluwer, Dordrecht, 1989.
2. N. Cristescu, *Viscoplasticity of geomaterials*, in *Time-dependent behaviour of geomaterials*, eds. N. Cristescu and G. Gioda, Springer, 1994.
3. N. Cristescu, I. Paraschiv, *Optimum design of large caverns*, Proc. 8th Int. Congr. Rock Mech., Tokyo, Balkema, 1995.
4. O. Simionescu, *Asymptotic analysis for stress concentration around square holes with rounded corners*, Rev. Roum. Sci. Techn.-Mec. Appl., **40**, 349-372, 1995.

5. O. Simionescu, S. Roatesi, *Stress concentration of the interaction of two square holes with rounded corners-a comparative study*, ZAMM, 76, Proc. ICIAM `95, vol. 5, 479-480, 1996.

6. O. Simionescu, *Stress concentration around one isolated square-like cavity* (in preparation).

7. O. Simionescu, *Asymptotic stress analysis for the interaction of two circular cavities` problem* (in preparation).

8. O. Simionescu, *Error analysis in the problem of stress concentration around cavities*, (in preparation).

9. O. Simionescu, *Mathematical methods in termomechanics of underground cavities*, Ed. Academiei, Bucharest (in preparation).

Basic and Applied Salt Mechanics, Cristescu, Hardy, Jr & Simionescu (eds)
© 2002 Swets & Zeitlinger, Lisse, ISBN 90 5809 383 2

Geolog – integrated system for the mining industry

Daniel Stanescu & Lucia Diaconu
Mining Computing Center Cluj, Romania

Ioan Macoviciuc
ROMTELECOM Cluj, Romania

Adrian Dadu
S.C. COMINEX S.A., Cluj, Romania

Liviu Draganes
SALROM, Salt Mine of Slanic Prahova, Romania

Mircea David
SALROM, Salt Mine of Ramnicu Valces, Romania

Francisc Laszloffi & Vasile Georgiu
SALROM, Salt Mine of Ocna Dej, Romania

ABSTRACT

The GEOLOG program package realizes a common base of geological and topographical data and the calculations related to these data, enabling the automatic drawing of maps and plans. So, it meets the needs of determining the models of the terrain, of the mineral deposits or of the workings, drawing maps for quarries or mines, tracing topographical plans, calculating the geological reserves and the change of the reserves. These needs appear in the usual activities in the mining industry.

The GEOLOG has modules for data acquisition and archiving and update. It performs analytical, graphical and complex functions (modeling the surface and the deposits).

The modules of this application can be combined in a user specific integrated system. This application is used in several salt mines from Romania, such as those from Slănic-Prahova, Ocna-Dej, Râmnicu-Vâlcea, Târgu-Ocna. The package can be tailored upon the specific needs of the users.

INTRODUCTION

In the traditional practice, keeping up to day the maps and the plans in the mines is a job done by operating the supervened changes on a copy of the original. The new chart is to be used instead of the original. This means a big amount of manual work and the alteration of the exactness of the chart. Another inconvenience of the traditional methods consists in the difficulty of elaborating a drawing board of a non-standard format at a different scale than the scale of the original. Manual methods encounter problems in dealing with elements delivered by different sources. Automatic drawing of the plans and maps eliminated these inconveniences.

We considered the necessity of simultaneously using geological and topographical information and of correlating this information for drawing plans and making geological calculus. This necessity imposed the creation of some unitary data structures for developing a common base of geological and topographical data, combined with the integration of all the modules that concur in realizing the functions for obtaining the drawings and geological situations into a unitary technology of work. The result was the GEOLOG program package.

THE MAIN FUNCTIONS OF THE GEOLOG PROGRAM PACKAGE

The GEOLOG package is meant as an instrument for achieving the various types of processing related to geological and topographical activities in the mining industry. Its functions can be classified upon several criteria:
a) Classification according to the stage of the processing:
- functions for acquisition and archiving of the data existing at a specified moment of time (topographical measurements, planimetric details, description of the geological and mining workings;
- functions for updating of the archived data.
b) Classification according to the type of the information being manipulated:
Analytical functions:
- input and listing of the geological research workings, samples, formations, sorts, prices;
- calculation of the contents, volumes and reserves.
Graphical functions:
- 3D representations of plans, solids;
- interactive update of graphical elements;
- determining and representation of horizontal and vertical sections through spatial objects;
- obtainment of different symbols or annotations along any type of curves or of symbols filling closed polygons.
Complex functions:
- modeling of the surface of the terrain;
- modeling of the deposits (numerical model generation for: galleries, rise headings, dissolving voids, roof, floor, safety pillars, deposits);
- obtainment of new points, curves or surfaces by correlating different existing such objects (points, curves, surfaces);

– generating of plans meant for geological interpretation.

THE MODULES OF THE GEOLOG PROGRAM PACKAGE

The GEOLOG program package is composed by several modules, which can be combined in different ways, according to
the necessities of the users. The modules are independent
(using however a common base of data) and they complete
each other. For example, the elements obtained by the data
acquisition module through the analytical module are combined in the modeling module to produce modeling objects.
These objects would be represented onscreen by the graphic
module. Finally, they would produce plans using the drawing
module.
The modules are not "pure" and usually consist of several types of functions (as described above).

Analytical and topographical data acquisition

The topographical data acquisition can be made either
by scanning the existing maps and subsequently digitizing
them onscreen, or by collecting the topographical measurements from the primary documents. For the points resulted
from measurements, a calculation of the rectangular coordinates through theodolite traverses and tacheometric surveys
is made.

The analytical data acquisition is realized interactively or in batches, using as starting point the description
of research workings, the primary documents, the existing
geological profiles, the standards defining the sorts for
different processing industries, etc.

Graphical modules

Graphical representation

The initial contours or the objects resulting through
modeling can be represented by the graphical module. All
these information have identical structure from the logical
point of view.
The representation is made by parallel orthogonal projection on an interactively chosen plane. It is possible to
represent any number of objects (solids, contours).

The graphical output program represent spatial objects
obtained from the topographical input or using the modeling
routines or by making horizontal or vertical section
through the spatial models.

Through the representation procedure, it is possible to make some other operations. The user is enabled to make several enlargements of the represented a zone or calculate distances between two interactively chosen points. It is also possible to represent solids with hidden lines, eliminating from the drawing the edges of the unseen faces (faces covered by other objects, situated nearer to the viewer).

The new version of the program package, now in work, will present major enhancements regarding graphical representation.

Figures 3-5 present sample images obtained using the new version of the representation program. The represented objects were creating using several methods that will be briefly described.

Updating of contours

The spatial objects can be modified using a graphical editor within the application package. This module is a useful tool during data validation process.

The editor operates on a collection of files containing contours or points. Any of these elements (contours, points) can be inserted, moved from one container object into another, multiplied, modified or erased. The editor also offers the possibility to interactively introduce data from the terrain notebooks.

In horizontal plane, the editor offers the possibility to determine distances between two points and to delimit and calculate the areas of portions of contours.

The editor also makes possible to edit vertical profiles. The modified lines from the profiles can subsquently be added to enrich the information used for spatial modeling.

Determining sections through solids

The graphical module can produce one or more horizontal and vertical sections through the selected objects. The vertical sections can be obtained following one or more parallel, interactively chosen directions, while the horizontal sections can be obtained at any chosen height (or at different distances from a chosen height).

Figures 1 and 2 represent fragment from automatically determined vertical section through the models of the deposit, workings and terrain at Ocna Dej and at Slănic Prahova, respectively.

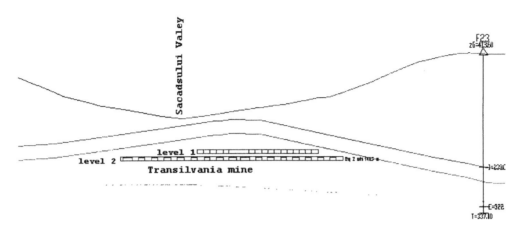

Figure 1 - Fragment from an Automatically Generated Vertical Profile
(Ocna Dej)

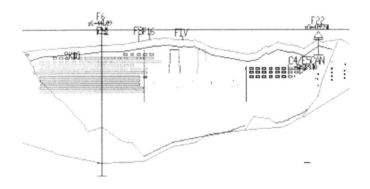

Figure 2 - Fragment from an Automatically Generated Vertical Profile
(Slănic Prahova)

The profiles, like any other drawing generated with the program package, can be completed with text and symbols from the library of the package. The output can be exported to AUTOCAD or to any GIS supporting DXF files.

Obtaining symbols or annotations for preparing maps

After obtaining all the objects to be represented on a plan, a final step before plotting it is to add symbols and texts that would complete the map.
The application permits to automatically generate graphical objects that contain:

371

- text that represent:
  - the value of the height on level curves;
  - information related to different objects (e.g. their names);
  - heights on a curve at specified distances along the curve;
  - heights of points belonging to a specified curve;
  - the names, surfaces and, optionally, point numbers (along with their coordinates) of the selected contours ;
- symbols that explain the type of object a line is representing on the map (either filling symbols or symbols on the frontier of a curve).

The texts and symbols constitute objects themselves. They can be consequently edited using the graphic editor. They are generated using libraries of characters and symbols and follow the national standards established for representing maps in the mining domain.

Map generation

The package gives the possibility to draw its results on a plotter. The maps are generated at a chosen scale, and the tablet is automatically filled through a dialogue. The legend is generated using a text file describing its elements. The program is compatible with a wide range of applications by generating DXF files.

The modeling modules

Modeling the deposits

The GEOLOG package has different modules for modeling the deposit, for different types of deposits: layered or massive deposits or lodes. For every type of deposit, we use as starting point the description of the research workings (drillings, galleries, etc).

For the massive deposits, two sets of points are obtained. These sets will lead to the model of the roof and of the floor of the deposit. For the layered deposits, each layer of the deposit is modeled using such two sets of points. The model consists of an optimal net of triangles, describing in every case the given surface. Beside the points obtained from the description of mining workings by the modeling procedure, limiting contours (obtained through digitizing or automatic procedures) are used.

In the case of the lodes, we use the information from the terrain notebooks and the topographical description of the axes of the lodes, taken from the existing maps. From these, the thickness of the lode is determined in any of its points, for every level taken into consideration.

Modeling the surface of the terrain

Using the points resulted from the topographical measurements and the information from the terrain sketches, a model of the surface is generated as a net of triangles.

This model is satisfactory for obtaining horizontal and vertical sections and can be used by the module for determining and drawing the level curves. Figure 3 represents the result of the surface modeling for a quarry at Aghireş. The two drawings reflect to positions of the light illuminating the quarry.

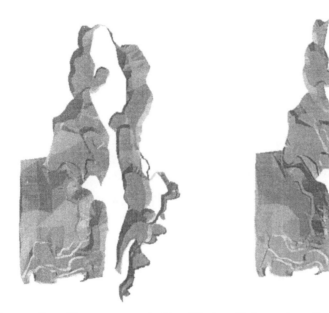

Figure 3 - The quarry at the Mining Enterprise from Aghireş

Modeling through elevation

Numerical models for the mining workings are obtained as polyhedrons starting from their bases' contours and determining their elevation at a given level or with a given distance on vertical direction. The facets of these solids are triangles or convex polygons. Parts of the galleries in a mine are constructed using this method.

Figure 4 represents the Cantacuzino Mine from Slănic Prahova and the old mines nearby.

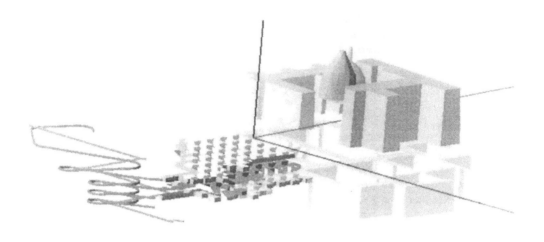

Figure 4 - The Cantacuzino Mine from Slănic Prahova

Underground modeling

Through specific algorithms, starting from the axes of
the galleries or of the heading rises, polyhedral solids
with the boundary surfaces formed by triangles or convex
polygons are built. In generating these solids, information
about the geometrical characteristics of the galleries and
of the rise headings (such as sides, height, cross section,
etc) is used.
Other methods of constructing spatial objects, such as
the ribbon modeling, are also used during the construction
of the models for a mine.

Figure 5 represents a mine from Ocna-Dej, modeled using
several methods: surface modeling, ribbon modeling.

Correlation modules

Different types of objects are obtained by applying
certain operators on other existing objects. For example,
we can obtain the limiting contour of the deposit at a
given time by intersecting the limiting contour at a preceding time with the line that defines the present exploitation line.

The modules do the following operations:
- determine the interior and/or the exterior points or
  contours compared to given contours;
- determine the close contours delimited by the intersection of two contours;

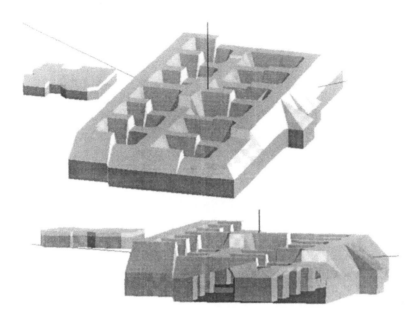

Figure 5 - The 23 August Mine from Ocna-Dej: aerial view and a section

- project points or contours on given surfaces (modeled using triangles);
- determine de intersection between two surfaces modeled with triangles.

Calculus

The GEOLOG program package calculates:
- medium concentration in the components from the deposit that are of interest;
- surfaces and volumes of computing units (such as
  panels);
- quantities of reserves from the deposit, differentiated by groups, categories, etc;
- The program also generates listings with the results
  of the calculuses.

CONCLUSIONS

The components of the program package being modularly
conceived, functionally independent from each other, the
program package can be easily configured for adapting to
different concrete situations from the mining plants or
from the mining research units.

Part IX

Case studies

Basic and Applied Salt Mechanics, Cristescu, Hardy, Jr & Simionescu (eds)
© 2002 Swets & Zeitlinger, Lisse, ISBN 90 5809 383 2

Case studies on the application of practical rock mechanics programs for selection of support for mine openings and shafts

Hamid Maleki
Maleki Technologies, Inc., Spokane, Washington, USA

Steve Finley, Collin Stewart & Michael Patton
Green River, Wyoming, USA

ABSTRACT

Two case studies are presented on the application of practical rock mechanics programs for selecting support systems for mine openings and shafts in trona mines. Case study 1 presents (1) the results of rock mechanics evaluations for characterizing geologic conditions near the position of a future mine shaft and (2) a ground support interaction analysis for addressing liner requirements. Case study 2 consists of detailed measurements of roof bolt strains, roof deformation, and gas pressure in a mine roof. These measurements provide a better understanding of lateral forces imposed on roof bolts as a result of creep deformation and differences in mechanical properties near shale-trona interfaces. Results are used to enhance numerical modeling procedures used routinely in evaluations of mining plans.

GEOTECHNICAL CHARACTERIZATION AND LINER STRESS ANALYSES

Rock mechanics data are available from both a continuous borehole drilled from the surface near the location of a future shaft and from underground holes drilled in the roof and floor within an instrumented panel. To provide site-specific data for construction of a shaft, a geotechnical program was implemented to obtain continuous core while characterizing hydrogeologic conditions. The core was used for lithological and structural studies and to determine the mechanical properties of the rock. In addition, both regional and local measurements were used in this study to infer the far-field stress regime near the location of the proposed shaft. These measurements indicated an isotopic stress field with a stress gradient of 0.03 MPa/m.

Lithology, rock quality designation (RQD), and mechanical properties for the entire tested section of the exploratory borehole were examined. The data reflect a large contrast in strength and deformation values for mudstones, near-surface sandstones, the Tower Sandstone, oil shales, and trona. Mudstones and near-surface sandstones are generally the weakest and least stiff material. The Tower Sandstone, limestone, and calcareous shales (marlstone) are generally both stiff and strong and have uniaxial compressive strengths reaching 220 MPa and a Young's modulus as high as 51 GPa. Oil shales and trona beds are stronger than mudstones, with average uniaxial compressive strengths of 41 MPa.

Four distinct zones (A, B, C, and D) were identified in the borehole core. Table 1 shows the calculated rock mass rating (RMR). Rock mass quality was calculated for the four zones using a rock classification technique (geomechanics classification or RMR) proposed by Bienawski (1981). These zones include both competent and weak sections and thus were used to analyze rock-support interactions. Rock mass quality is good to fair near the location of the proposed shaft (table 1).

Table 1. Rock mass quality for four analyzed rock zones.

Zone	Minimum unconfined compressive strength, MPA	Minimum YoungÁs modulus, GPA	PoissonÁs ratio	RQD, pct	Maximum groundwater flow, gpm	Rock mass rating	Rock mass quality	Cohesion, MPA	Angle of internal friction, deg
A-Near surface sandstones ..	9.6	2	0.29	85-100	12	56	Fair	NA	NA
B- Tower sandstone	85	19.3	0.26	90-100	1	74	Good	NA	38
C- Mudstones ..	11.7	3.8	0.24	70-100	1	57	Fair	2	33+
D- Trona, oil shale	34	7.5	0.26	100	1	66	Good	2.8	33+

+ Estimated based on regional measurements

The timing of installation of secondary support (liner) is important for controlling wall rock deformation and the extent of the inelastic zone around the shaft. Analytical solutions developed by Panet (1979) for elastic rocks in a hydrostatic stress field indicate that most wall deformation occurs within two diameters of the shaft from the face. At a distance of 1.2 radii (approximately 5 m) from the face, 85 percent of the wall deformation has taken place. Thus, a liner installed at such a distance will experience 15 percent wall deformation.

Stress concentration around the mine shaft and load transfer to the liner have been estimated using numerical modeling methods. The analysis excludes hydrostatic loads from water pressure because a careful dewatering system is planned. A 0.5-m-thick, unreinforced concrete liner with a 28-day strength of 27.5 MPa is assumed in these calculations. The liner will be installed 5 m from the face after 85 percent of the expected deformation has taken place. The focus of the stress analyses is on the calculations of tangential (hoop) stresses within the liner in a horizontal plane. These stresses are the most significant source of loading in a compressive, premining stress state.

The finite-difference code FLAC was also used to analyze the extent of the failure zones around an unlined shaft and to reveal the role of a liner in controlling inelastic deformation in the shaft walls. These analyses were completed for zones C and D using a hydrostatic stress field and Mohr-Coulomb plasticity models. Table 2 shows that maximum tangential stresses are within the allowable working stress of concrete (12.5 MPa) (American Concrete Institute, 1983a, b).

Table 2. Liner maximum tangential (axial) stresses and bending moments, finite-difference analy

Zone	Axial force, kN	Moment, cm-kN	Total axial stress, Mp
C	111	326	11.6
D	100	158	9.2

DESIGN OF MINE OPENINGS AND SUPPORT SYSTEMS

Case study 2 involved field measurements of roof deformation and support loading, and development of numerical modeling procedures for routine evaluation of mine layouts. The field program provided a better understanding of roof loading mechanisms and thus assisted in determination of the type of analyses and selection of what in situ properties would be appropriate for the numerical models.

A field program was implemented in a five-entry panel using 4.2- to 6.1-m-wide rooms and 20-m-wide pillars. The mine uses borer miners and flexible conveyor trains to extract a 3-m-thick trona bed, located 460 m below the surface. Borer miners mine oval entries 4.2 (single-cut entries) to 6.1

m wide (middle entry [# 3], one-and-one-half cut). The roof consists of 15 to 60 cm of trona overlain by a sequence of shales and siltstones, and trona. There is a 15-cm-layer of trona (Jewel Seam) approximately 1.5 m into the roof.

The field program consisted of installation of 32 strain-gaged roof bolts, 12 roof extensometeres, one mechanical packer to measure gas pressure in the roof and corehole drilling and pull tests. The bolts used in this study were standard-grade 60, No. 6 rebar. Two slots were milled on the opposite sides of each instrumented bolt, and strain gages were attached at 0.15- and 0.35-m spacings for the 1- and 2.1-m-long bolts, respectively. Using these strain gage configurations, both axial loads and bending moments can be calculated (Maleki and others 1985; Signer and others 1997). The bolts were calibrated in the SRL laboratory to establish a relationship between bolt load, voltage, and bolt strain for the entire deformation range of the bolts (0 to 12 percent strain).

Roof stability was excellent in the test section, as indicated by lack of significant loading in the roof bolts and a very small amount of roof deformation. Data from 12 extensometers showed total movement ranging from 0 to 8 mm. Analyses of data obtained from all 32 instrumented roof bolts indicate three mechanisms that influence stress distribution along roof bolts. (1) Gas pressure and draining processes (2) Gravity forces associated with extracting the ore and widening the rooms to final operational requirements and (3) Time-dependent material behavior.

Gas pressure increases roof loading if not drained. Using a packer installed 2.1 m above the mine roof, the authors measured a gas pressure of 0.14 MPa shortly after mining at one gassy location. After draining the gas reservoir through drain holes, this additional stress was relieved. Gas drainage thus caused the roof to rebound upward, which compressed the upper part of the roof bolts. Figure 1 presents average load distribution along two 2.1-m-long instrumented bolts and the location of trona-shale interface planes.

The immediate roof rocks deform downward because of gravity right after the entries are mined, putting the bottom of the bolts in tension (figure 2). The neutral plane appears to be about 1.5 m above the mine roof. Under the combined effects of gravity and drainage, the bolts were partially in tension and partially in compression. These combined effects were most pronounced in all 2.1-m-long bolts in room 5 (single-cut), but were not noted as consistently in the wider room (# 3). Mining and gravity effects overcome upward relaxation for wider entries (6.1 m or more), and thus compressive forces on the bolts are not significant in wide entries.

Widening the bolted entries to their final dimensions (6.1 m) imposed large strains on bolts adjacent to the new cut. Some bolts yielded locally, while the new bolts installed in the new cut showed little increase in loading (figure 2). Localized yielding in roof bolts can be reduced by (1) delaying installation of bolts until after the entry is widened, (2) using a milder, more ductile steel in roof bolts, and 3) using a resin point-anchored bolt system. The latter system is also useful for reducing lateral bolt loading at interfaces and for enhancing gas drainage. This system has been used in a new instrumented panel with the same results.

Time-dependent deformation along the shale-trona interface contributes to localized horizontal loading of the bolts at these interfaces, which in turn contributes to bolt bending and inelastic bolt strains that exceed 2 percent strain. Figure 3 present snapshots of maximum measured strain along instrumented bolts and shows two horizons where the concentration of strain is higher. These horizons, which are approximately 0.6 and 1.5 m above the mine roof, correspond to the location of bed 17-shale and Jewel Seam-shale interfaces, respectively. Note that all measured strains are significantly smaller than the critical strains (6 to 7 percent) that precede bolt failure. These measurements clearly demonstrate the importance of horizontal loading on roof bolt loads in trona with time-dependent properties and provide useful guidelines for including significant geologic features (interfaces, etc.) in the models.

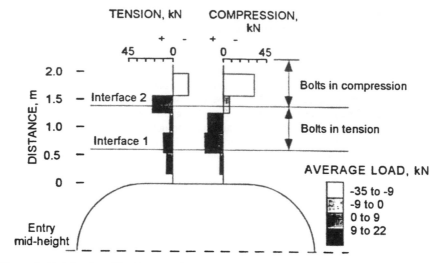

Figure 1. Bolt load profile along two instrumented roof bolts.

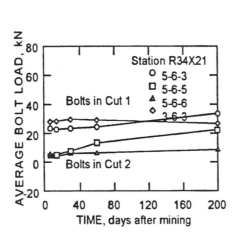

Figure 2. Average bolt load history for bolts installed in two steps as the entry was widened.

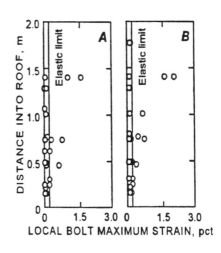

Figure 3. Maximum bolt strain profile at (A) one-weak and (B) six-month after mining.

References

American Concrete Institute, 1983a. Building Code Requirements for Structural Plain Concrete and Commentary. ACI-318-1-83.

American Concrete Institute, 1983b. Building Code Requirements for Reinforced Concrete. ACI-318-83, MI.

Bieniawski, Z. T., 1981. Rock Mechanics Design in Mining and Tunneling. Balkema.

Itasca 1995. Fast Lagrangian Analyses of Continua.

Maleki H., S. Signer, and R. King, 1994. Evaluation of Support Performance in a Highly Stressed Mine. Proc., 13th Intern. Conf. on Ground Control in Mining, Morgantown, WV.

Maleki, H., and H. Brest Van Kempen, 1985. Evaluation of Rock Bolt Tension Measuring Techniques. 26th U.S. Symp. on Rock Mechanics, June 1985.

Panet, M., 1979. Time-Dependent Deformations in Underground Works. Proc. of 4th. ISRM conf., V. 3, pp. 279-289. Balkema.

Signer and others, 1997. A Method for the Selection of Rock Support Based on Bolt Loading Measurements. Intern Support, Oslo, Norway.

Basic and Applied Salt Mechanics, Cristescu, Hardy, Jr & Simionescu (eds)
© 2002 Swets & Zeitlinger, Lisse, ISBN 90 5809 383 2

Mining subsidence as effected by the interaction between contemporary mining works and those from the 17th-18th century at ocna dej saline

Christian Marunteanu, Victor Niculescu & Sorin Mogos
University of Bucharest, Bucharest, Romania

Ervin Medves
S.C. IPROMIN S.A.Bucharest, Romania

Petre Raisz
SALROM, Bucharest, Romania

Francise Lukacs
SALROM, Ocna Dej Salt mine, Romania

1. OCNA DEJ SALT DEPOSIT

The salt deposit from Ocna-Dej has a tabular shape with some variation in thickness lensing formation as well as narrowing and pinching out. The maximum thickness of the salt body is 156 m. The salt formation lays on the complex of the Tuff of Dej (lower Badenian) and overburden consists of marls, tuffs and breccia (upper Badenian).

2. HISTORY OF THE MINING WORKS

The first mines in the geological region of Ocna-Dej date from the Daco-Roman period (100-200 a.c.), their traces being still visible nowadays. The salt extraction continued during the 13th and 14th centuries, conducted by different owners of these lands, Ocna-Dej becoming one of the most representative salt mines from Transylvania, with pure and high quality salt.

A more organised and industrial scale mining of salt started in the 17th century, when new mining methods were adopted, with large, bell-shape rooms, reaching in height 100 m and in diameter 45-50 m. Due to the geometry of the rooms and in accordance with the geomechanical parameters of the salt and the geomechanical conditions in the location area, most of these works were preserved in the initial form until the 20th century. Some of the rooms were flooded by the intrusion of the water from the floor and the dissolution phenomena generated in some case the collapse of the room vaults.

Beginning with the 19th century the mining was the performed by the square rooms and pillars method. The new mining works often intersected those operated in earlier times.

The industrial development from the 20th century determined the increase of salt mining in the area, new mines being opened, among them those named Ferdinand (23 August) and 1 Mai.

3. INTERACTION BETWEEN CONTEMPORARY AND OLD MINING WORKS

The contemporary mining works induced in the environment a modification of the stress state, that affected the equilibrium state of the old works. Many times the precise position and shape of the old works was not well known and they influenced at their turn the stability of the contemporary works, especially by the incorrect sizing of the safety pillars.

This geomechanical interaction between the old and contemporary works usually induced negative influences on the stability of the geological massif, with spectacular subsidence effects due to the sudden falling of the roofs. Large sinkholes were formed at the surface, determining afterwards their natural or artificial backfilling or generating salt lakes (Fig. 1).

4. GROUND COLLAPSE IN THE AREA OF THE SALT MINES
CICIRI – 23 AUGUST

A good example of interference between old and new mining works could be considered the subsidence sinkhole produces at the contact between the contemporary salt mine 23 August, closed in 1959, and the bell mine Ciciri, closed in 1754.

As one can see on the map of the ground instability phenomena (Fig. 1), some of the rooms of the mine 23 August are very near to the mine Ciciri. The inaccurate location of the mine Ciciri on the old maps determined the design of the mining works of the mine 23 August too near to the opening of the mine Ciciri. The safety pillar between the two mines became too thin after the exploitation of the adjacent rooms. This determined the overburden of the pillar and the change of the state of stress from the triaxial stress to biaxial or even uniaxial. Taking into account the correlation between the resistance and the behaviour of the salt and the state of stress as resulted from the laboratory tests - triaxial compressive strength more than 60.0 MPa (with radial component 3.0 MPa), uniaxial compressive strength 22.5 MPa and long term creep strength 11.7 MPa - the breaking in time of the safety pillar became inevitable. The breaking of the pillar between the two mines was probably the main cause of the sudden producing of a large sinkhole in January 1998. The cone had initially about 200 m diameter and 150 m depth, but its extension is continuing.

5. INVESTIGATIONS CARRIED OUT IN THE HAZARDOUS AREA

To forecast the subsidence phenomena and to avoid the negative effects on the surface buildings, a number of investigations in the hazardous area were initiated.

In the first stage geophysical investigations - electrometric and seismic measurements - were carried out. The result of these measurements allowed the identification and location of the underground cavities. The periodical recurrence of the measurements led to the determination of the dynamics of the displacements in the investigated area, the quantifying of the displacements at the ground surface being realised by means of topographic monitoring on alignments of topographic markers.

In the second stage a series of laboratory tests had in view the determination of the geomechanical properties of the salt and of the covering formations, with a special accent on the rheological behaviour.

The geophysical and topographic monitoring are going on and a permanent ESG microseimic monitoring (Hyperion configuration with 16 channels) is proposed.

6. FORECASTING OF SUBSIDENCE PHENOMENA

The processing and the interpretation of the investigation data allowed the identification and the prediction of subsidence phenomena within the hazardous area. In the Figure 2 are represented the outline of the actual sinkhole, the forecasted outline corresponding to the stable slope and the safety zone around the final outline.

The outlines were drawn on the basis of more radial vertical sections where forecasting depends on lithologic composition, geomechanic and rheologic parameters of salt and rocks, tectonics and microtectonics of the massif, hydrogeology, spatial location of the mining cavities etc (Fig. 3).

The evolution of the expansion of the sinkhole was estimated at 4 - 8.5 m per month towards SW and NE, 1 m per month towards W and 2 m per month towards SE. Thus, the safety limit could be reached in 3 years in S-W direction, 5-6 months in N-E direction and 5-6 years in the South direction. To prevent any undesirable evolution and effect of the phenomenon a permanent monitoring of the area by geophysical and topographical methods is recommended.

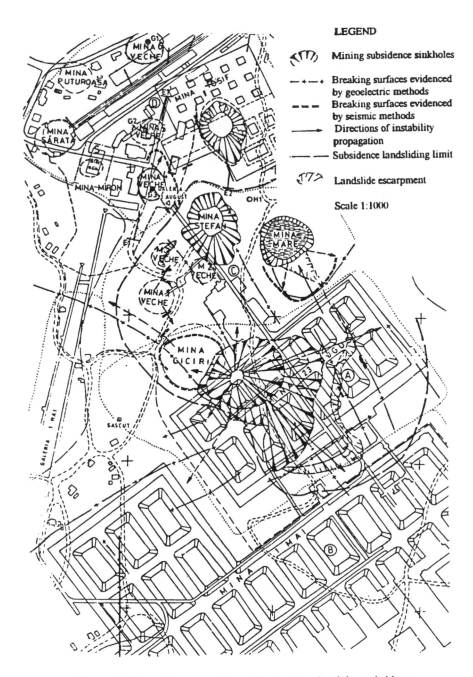

LEGEND

⟨᛭⟩   Mining subsidence sinkholes

—·—·→   Breaking surfaces evidenced by geoelectric methods

– – –   Breaking surfaces evidenced by seismic methods

———→   Directions of instability propagation

———   Subsidence landsliding limit

⟨᛭⟩   Landslide escarpment

Scale 1:1000

Figure 1. Map of the ground instability due to salt mining subsidence

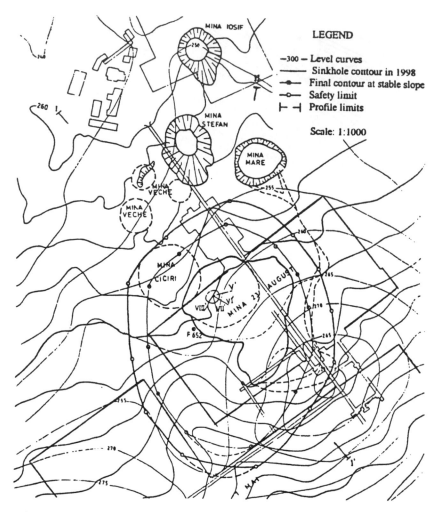

Figure 2. Forecasting map of the Ciciri - 23 August subsidence sinkhole evolution

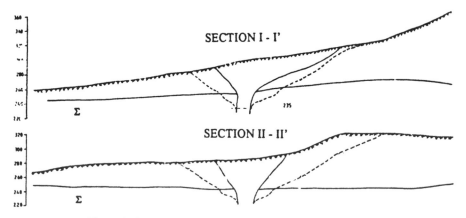

Figure 3. Forecasting sections through the subsidence sinkhole

Basic and Applied Salt Mechanics, Cristescu, Hardy, Jr & Simionescu (eds)
© 2002 Swets & Zeitlinger, Lisse, ISBN 90 5809 383 2

A methodology for rock mechanics design of brine fields based on case histories of sinkhole formation in Windsor-Detroit area

Leo Rothenburg & Maurice B.Dusseault
University of Waterloo, Waterloo, Canada

Dennis Z.Mraz
Mraz Project Consultants, Saskatoon, Canada

Introduction

The paper examines three instances of sinkhole formation in Windsor-Detroit area (Canada, USA) with the objective to formulate a rational rock mechanics-based design methodology to prevent sinkhole formation as a result of solution mining operations.

Review of published data for the Sandwich brine field and two fields at Grosse Ile suggests that conditions at all three fields were such that numerous cavities essentially coalesced and wide, practically unsupported spans were created by the time sinkholes were formed, R. Terzaghi, 1970, Landes and Piper, 1972. In all three cases sinkholes were formed in conditions when minimum dimensions of respective brine fields were 1150-1550 ft while estimated pre-sinkhole subsidence ranged 2-2.5 ft.

Analyses presented in the paper for the Sandwich field show that a combination of brine filed dimensions, subsidence measurements and data on mechanical properties of rocks can be used to explain the type of overburden deterioration that permitted formation of sinkholes. Analyses are used to illustrate a methodology of designing brine fields based on limiting the minimum span to avoid sinkhole formation.

Sinkholes due to solution mining in Windsor-Detroit area

There were three instances of sinkhole formation in the Windsor-Detroit area. In 1954, a 400-500 ft wide and 25 ft deep sinkhole developed at the Sandwich brine field owned jointly by Canadian Industries Ltd. and by Canadian Salt Co. Two sinkholes developed in 1971 at the Wyandotte Chemicals brine field. Geometric outlines of the three fields are illustrated in Figure 1. Table 1 summarizes other pertinent data.

Geology of the Windsor-Detroit area

Geologic sequences in Figure 1 presented by R. Terzaghi, 1970, for the Sandwich field are typical of general conditions in the Windsor-Detroit area. The most important stratum responsible for bridging large unsupported spans over cavities is Bass Island Dolomite described by K. Terzaghi as follows: "In that region, the uppermost two hundred feet of the formation is reported to be sound, strong and competent, and capable of bridging cavities with a span of several hundred feet without any indication of stopping."

Salt is usually mined in the Windsor-Detroit area from Unit B of the Salina formation. At the Sandwich field the Unit B is about 220 ft thick and starts at a depth of 1380 ft.

Development history and subsidence at the Sandwich field

Brine production at the Sandwich field started in 1902 when first wells were drilled to the base of the Salina formation. The amount of salt extracted from the Sandwich field is rather modest, about 9.75 millions short tones. If salt would have been extracted uniformly across the field, the average height of the cavity would have been 44 ft, out of the total salt thickness of 220 ft. It is estimated that the minimum unsupported span may have been 1700 ft at the time the sinkhole was formed. Ground distress at the

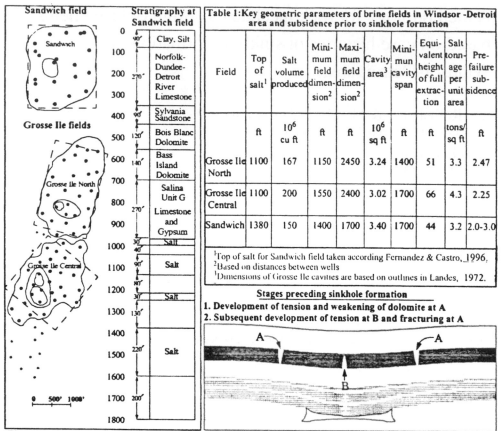

Figure 1: Sandwich and Grosse Ile brine fields and stratigraphy at Sandwich field.

Table 1: Key geometric parameters of brine fields in Windsor -Detroit area and subsidence prior to sinkhole formation

Field	Top of salt[1]	Salt volume produced	Mini-mum field dimen-sion[2]	Maxi-mum field dimen-sion[2]	Cavity area[3]	Mini-mun cavity span	Equi-valent height of full extrac-tion	Salt tonn-age per unit area	Pre-failure sub-sidence
	ft	10^6 cu ft	ft	ft	10^6 sq ft	ft	ft	tons/ sq ft	ft
Grosse Ile North	1100	167	1150	2450	3.24	1400	51	3.3	2.47
Grosse Ile Central	1100	200	1550	2400	3.02	1700	66	4.3	2.25
Sandwich	1380	150	1400	1700	3.40	1700	44	3.2	2.0-3.0

[1]Top of salt for Sandwich field taken according Fernandez & Castro, 1996,
[2]Based on distances between wells
[3]Dimensions of Grosse Ile cavities are based on outlines in Landes, 1972.

Stages preceding sinkhole formation
1. Development of tension and weakening of dolomite at A
2. Subsequent development of tension at B and fracturing at A

Figure 2: Stages of overburden deterioration preceding sinkhole formation at the Sandwich brine field

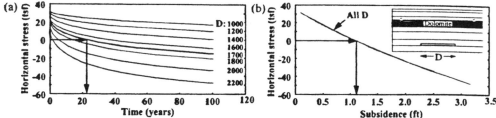

Figure 3. Simulated history of minimum horizontal stress at the top of dolomite above abutments.
(a) Horizontal stress versus time. (b) Horizontal stress versus subsidence for a range of cavity diameters.

Sandwich field became evident in 1948. It prompted regular surveys that indicated progressive increase in subsidence rates up to February 1954 when the area about 400 ft wide and 500 ft long rapidly settled about 25 ft. Accumulated subsidence prior to 1948 was estimated at about 1 ft based on tilt of buildings.

Ground response to solution mining

Numerical analyses presented in the paper demonstrate that a combination of brine filed dimensions, subsidence measurements and data on mechanical properties of rocks are compatible with the following mechanistic sequence leading to sinkhole formation. Initially, at a subsidence of about 1 ft, horizontal stress at the top of the stiff dolomite stratum is reduced to zero in areas above abutments of the brine field, Figure 2, locations A. Possible opening of joints may have impaired the ability of dolomite to span across the field with only minimum support from remnant pillars. This stage is likely to correspond to the observed acceleration of subsidence due to reloading of the field. With further subsidence, tension in the dolomite developed at abutments while horizontal stress above the brine field continued to drop. It is demonstrated that at the accumulated subsidence of 2.5 ft corresponding to sinkhole formation the computed horizontal stresses in the dolomite above the center of the field, Figure 2, location B was close to empirical tensile strength of dolomite.

Mechanical properties of rocks in the Windsor-Detroit area

Selection of parameters was based on available generic description of rocks in the area and assigning typical properties to such materials based on data summarized by Deere and Miller, 1966. Creep parameters of salt were chosen on the basis of mining experience in salts of the Salina formation. The selected parameters comfortably explain the post-sinkhole subsidence rate of 1 in/year at the Sandwich field.

Numerical analyses of the Sandwich field

In order to test the hypothesis that the overburden at the Sandwich field was damaged due to excessive subsidence caused by wide unsupported spans, a series of FEM analyses were carried out. Computations simulated instantaneous dissolution of 220 ft high cylindrical cavities with diameters between 1,000 ft and 2,200 ft to cover a range of possible unsupported spans at the Sandwich field.

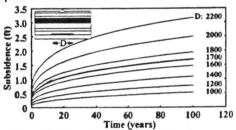

Figure 4 illustrates simulated peak subsidence history at centers of eight instantaneously excavated cylindrical cavities. The corresponding history of horizontal stress reduction in areas above abutments is illustrated in Figure 3a. In order to demonstrate that this stress reduction is subsidence-related, the time history of horizontal stress is replotted against subsidence on Figure 3b. Curves corresponding to different cavities collapse onto one line, irrespective of the cavity diameter.

Figure 4: Time history of peak subsidence for FEM-simulated cavities of various spans.

At a subsidence of about 1 ft the horizontal stress at the top of the dolomite stratum becomes zero. For subsidence in excess of 1 ft there is, therefore, a potential for opening joints in the stratum that largely responsible for the structural capacity of the overburden. This is consistent with an observation that subsidence above the Sandwich field started to accelerate in 1948 after about 1 ft of accumulated subsidence.

The sinkhole formation in 1954 was not possible without degradation of the overburden above the field. Figure 5a illustrates a simulated time history of horizontal stress reduction at the bottom of the Bass Island Dolomite stratum above the cavity center. As subsidence accumulates, tension progressively increases. Attention here should be paid to a cavity of 1,700 ft span corresponding to the minimum cavity span determined based on post-failure analysis. Tension for this cavity span is reached at a subsidence of

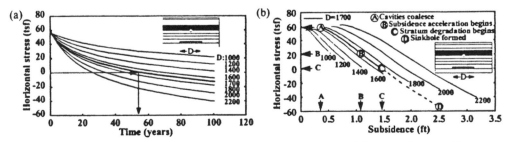

Figure 5. Simulated history of minimum horizontal stress at the bottom of dolomite above cavities.
(a) Horizontal stress versus time. (b) Horizontal stress versus subsidence

about 1.5 ft and it progressively increases with further subsidence. If the horizontal stress versus subsidence graph in Figure 5b is extrapolated to 2.5 ft corresponding to subsidence at which the sinkhole was formed, the corresponding horizontal tensile stress becomes 40-50 tsf. Tensile stress of 40-50 tsf corresponds to tensile strength of competent dolomite, according to data from Deere and Miller, 1966. The observation-computations loop becomes closed.

Design methodology

The analyses that were carried for the Sandwich field conditions suggest a simple and conservative methodology for determining dimensions of brine fields to avoid sinkhole formation.

To determine a safe span in specific conditions, a graph similar to the one illustrated in Figure 6 should be constructed through a series of FEM analyses similar to those described above. The graph relates the time required to develop tension in the overburden for the assumed cavity span. It can be seen that for practically all spans and reasonable times, tension first appears above abutments. It develops very quickly, in a matter of a few years for spans around 2000 ft and it takes about 50 years to develop tension for spans about 1550 ft. Spans below 1500 ft. are practically safe.

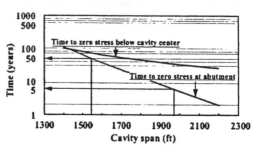

Figure 6: Time to onset of tension in dolomite versus spans of unsupported cavity.

References

Deere, D. U. and Miller R. P., 1966. Engineering classification and index properties of intact rock. Technical Report AFWL-TR-65-116, Air Force Weapons Laboratory, Kirkland Air Force Base, New Mexico.

Landes, K. K. and Piper, T. B., 1972. Effect upon environment of brine cavity subsidence at Grosse Ile, Michigan, 1971. Solution Mining Research Institute, April 1972.

Terzaghi, R. D., 1970. Brinefield subsidence at Windsor, Ontario. Third Symposium on Salt, pp. 298-307.

Fernandez, G. and Castro, A., 1996. Structural stability evaluations of expansion alternatives for solution mining operations at Windsor, Ontario. Solution Mining Research Institute. Fall 1996 Meeting, Cleveland, Ohio, Oct. 20-23, 1996.

Basic and Applied Salt Mechanics, Cristescu, Hardy, Jr & Simionescu (eds)
© 2002 Swets & Zeitlinger, Lisse, ISBN 90 5809 383 2

Research regarding the possible solutions related to the stability of drillings which overcross massive salt deposits

Professor Adrian Todorescu, Prof. Cornell Hirian,
Senior Lecturer

Vasile Gaiducov
Senior Lecturer

Victor Arad
Lecturer

Mihaela Toderas
University of Petrosani, Petrosani, Jud. Hunedoara, Romania

SUMMARY

In Romania the emphasizing of oil-field structures in deep geological conditions is related usually to the overcrossing of thick salt deposits (500-3000 m) located over the oil potential structures. To overcross by drilling such salt deposits induce major risks, which can compromise their initial development (drilling and cementation) and also their further behavior - the stability of the drill hole. During the drilling process, it has been observed from a practical point of view, certain difficulties, such as: excavations, hole diameter diminishment and technical injuries, which lead in some cases to the abandonment of drilling operations.

The above mentioned considerations and others, not detailed here, have required an approach involving the research of salt behavior with respect to deformation. These studies are directed to establishing the interaction mechanism between the salt massif and the drilled hole. The main goal being to provide hole stability until their intubation (mainly directed to the drilling mud properties) and also after this operation (cementing materials employed, type and quality of ducts, their sizes, etc.) in fact everything being related to drilling hole stability.

Part X

Salt pillars and cavities

Basic and Applied Salt Mechanics, Cristescu, Hardy, Jr & Simionescu (eds)
© 2002 Swets & Zeitlinger, Lisse, ISBN 90 5809 383 2

Computer aided echographic control system for underground salt dissolution process monitoring

Liviu Barbu
The Mining Computing Center, Cluj, Romania

Ilie Banuta
SALROM (Autonomous Adm. of Salt), Salt Mine of Targu Ocna, Romania

Maria Marcu
SALROM, Salt Mine of Ocna Mares, Romania

Ioan Salomia
SALROM, Mining Company of Ramnicu Valcea, Romania

SUMMARY

The Echographic control system (ECS) was designed for computer aided measuring and representation of the cavities resulting during the underground salt dissolution process. Cavity measurement was performed with a Polish USMS or with an EDO WESTERN sonar. The ECS system contains three modules, namely:

1. **The measuring module** contains the software dedicated to the data acquisition from the measuring process. The data acquisition function is done by a serial interface, between sonar and an IBM PC compatible computer, which sends data to the computer with a speed of 115.200 bauds. Actually, there are three different interfaces (two for the polish sonar and another for the EDO Western one). For each of them, a program performs the task of processing the signals, plotting them on the laptop's display in an intuitive form and saving them into a date base. The right upper corner of the display shows the measured signals, as they are seen during the measuring process. The scale is maximal, at the resolution of the measurement, the resolution being about 6 cm. This data, saved in a temporary InterBase database (Inprise), are later imported, from the laptop to the main database on a more powerful PC, which has better capabilities to hold the entire database and to plot the required images. Beside this task, the measuring module contains the program that is responsible to provide deviation measurement results in a tabular form, for insertion in the main database.

2. **The interpretation module** performs two tasks: optionally, it performs the filtration of the signal and, respectively, the calculation of the most probable contour, at each depth of measurement. In the case of measurements with the Polish USMS sonar, it was necessary to develop sophisticated methods to discriminate depth of measurement signal from the adjacent depth signal. It is noted that this type of sonar has inappropriate characteristics for this type of measurement. Finally, this module is useful for the operator to do his own interpretation.

3. **The representation module** makes it possible to obtain various geometric representations of a single cavity or of a set of cavities, for example sections, 2D and 3D representation of 3D surfaces. In use it is possible to see the same cavity, at two different moments, in horizontal/vertical sections and to plot the images of two cavities, at the same time (in 2D or 3D). This module also permits the plotting of the results of deviation measuring. This is very important for the correct positioning in the space of the plotted cavities. This module is also responsible for performing other tasks, for example: loading *text* concerning cavities, maintaining the database in a proper state, and printing various reports concerning the cavities (distances, areas, volumes, etc.).

The modules described reside in two programs, one for the first module, another for the next two. They run under Windows 3.1, respectively Windows95/98. The measuring data and all the other data concerning the cavities (including other types of measurements) are located in a single Interbase database file. As for hardware, the system needs a laptop (running Windows 3.1), another IBM PC compatible computer (running Windows95/98) in standard configuration, and an A3/A4 printer or a plotter (optional).

Basic and Applied Salt Mechanics, Cristescu, Hardy, Jr & Simionescu (eds)
© 2002 Swets & Zeitlinger, Lisse, ISBN 90 5809 383 2

Static and dynamic analysis of the 5th, 6th and 7th levels in the cantacuzino mine from Prahova

Adrian Dadu, Ioan Macoviciuc & Daniel Stanescu
Mining Computing Center Cluj, Romania

Liviu Draganescu
SALRON, Salt Mine of Slanic Prahova, Romania

1. GENERAL CONSIDERATIONS

In the last decades, the salt massifs had begun to be studied for knowing their behavior in time, during the exploitation of the massif and for excavation (or usage of abandoned exploitation works) of big oil silos for depositing radioactive residues or other chemicals and toxic products. But for these purposes we must be able to predict with certitude the stresses and the strains in underground mining works.

In this paper we suggested a complex numerical analysis of these mining works in salt, in static conditions (using Hooke's law) and rheological conditions (using visco-elastic model) with the finite elements method, in the domain of small displacements.

The salt rock is considered isotropic and homogeneous.

2. THEORETICAL FORMULATION

The literature on applications of the finite element techniques to mining mechanical equilibrium is not extensive. The approach in finite elements from this paper follows generally ideas from (Rao [1981, Oden 1967, Olariu 1986, Girbea 1990, Dadu 1985).

2.1. Static analysis

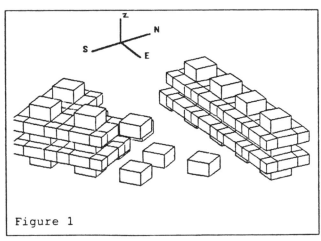

Figure 1

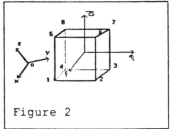

Figure 2

The structure (like the one in Fig.1) is divided in three-dimensional elements with 8 nodes (Fig.2). The geometry and the small displacements within a finite element 'e', are approximated with the shape functions:

$$N_i(\xi,\eta,\tau) = (1+\xi_i\xi)(1+\eta_i\eta)(1+\tau_i\tau)/8 \qquad i=1,8; \quad \xi,\eta,\tau \in [-1,1] \tag{1}$$

and with the following formulae:

$$x = \sum_{i=1}^{8} N_i x_i; \quad y = \sum_{i=1}^{8} N_i y_i; \quad z = \sum_{i=1}^{8} N_i z_i$$

$$u = \sum_{i=1}^{8} N_i u_i; \quad v = \sum_{i=1}^{8} N_i v_i; \quad w = \sum_{i=1}^{8} N_i w_i \tag{2}$$

or the matrix form:

$$[x,y,z]^T = [N][X]; [u,v,w]^T = [N][d]$$

where x, y, z are the coordinates of a point from finite element and

u, v, w are the small displacements on the axes Ox, Oy, Oz from a point of the finite element. The matrices [X], [d] contain the coordinates of the nodes of the finite element and the nodal displacements, respectively. The stiffness matrix and load vector are to be derived from the principle of minimum potential energy.

The vector of strains can be expressed in terms of nodal displacements using the the matrix form:

$$[\varepsilon] = [B][d]$$

Also the column vector of the stresses can be expressed using Hooke's law for isotropic material, like this:

$$[\sigma] = [E][\varepsilon]$$

where [E] is the inverse of the elastic coefficient matrix. So the potential energy on the finite element becomes:

$$\pi_P^{(e)} = \frac{1}{2} \int_{V^{(e)}} [d]^T [B]^T [E][B][d]dV - \int_{V^{(e)}} [d]^T [B]^T [E][\varepsilon]_0 dV - \int_{V^{(e)}} [d]^T [N]^T [p]dV - \int_{S^{(e)}} [d]^T [N]^T [q]dS$$

where

$$[\varepsilon]_0, [p], [q], [d]$$

are, respectively, the column vectors of the initial strains, of the body forces, of the surface forces and the nodal displacements on the finite element.

The static equilibrium configuration of the structure can be found by solving the following linear system

$$\frac{\delta \pi_P}{\delta Q_1} = \frac{\delta \pi_P}{\delta Q_2} = ... = \frac{\delta \pi_P}{\delta Q_M} = 0$$

where [Q] is the column vector of nodal displacements on the whole body.

This system can be written in matrix form:

$$[K][Q] = [F] \tag{3}$$

where:

$$[K] = \sum_{e=1}^{E} [K^e] \qquad \text{is the assembled stiffness matrix for the whole body,}$$

$$[F] = \sum_{e=1}^{E} [F_i^{(e)}] + \sum_{e=1}^{E} [F_s^{(e)}] + \sum_{e=1}^{E} [F_b^{(e)}] \quad \text{is the global vector of nodal loadings and,}$$

$$[K^e] = \int_{V^{(e)}} [B]^T [E][B] dV \qquad \text{is the stiffness matrix on finite element,}$$

$$[F_i^{(e)}] = \int_{V^{(e)}} [B]^T [E][\varepsilon]_0 dV \qquad \text{is the vector of initial strain loadings on finite}$$

element,

$$\{F_s^{(e)}\} = \int_{S^{(e)}} [N]^T [q] dS \qquad \text{is the column vector of nodal loadings of surface}$$

forces on finite element

$$[F_b^{(e)}] = \int_{V^{(e)}} [N]^T [p] dV \qquad \text{is the column vector of nodal loadings of body forces}$$

on finite element.

The solution in displacements of static equilibrium problem is obtained solving the linear equations system (3), with the unknown vector [Q]. The system's matrix [K] is sparse, positive defined and symmetric of very big size. The solving is done after the incorporation of some supplementary conditions to eliminate the rigid body displacements of structure (because without these supplementary conditions, matrix [K], in many cases, is singular). With the solutions of the system (3) (the displacements in every node of the structure) we can express, using (2), the small displacements in every point of the structure.

2.2 Rheological analysis

For the study of the rheological behavior of the salt we considered its comportment at flow when

$$f = F(\sigma_{ij}; \varepsilon_{ij}; \sigma'_{ij}; \varepsilon'_{ij})$$

the tensor or tensions is constant in time. In these conditions, the rheological status function has the form:
The models used to describe the behavior of salt are:
- Poyting – Thomson, when $|\sigma| < |\sigma_{cr}|$
- Bürgers, when $|\sigma| > |\sigma_{cr}|$

and their equations are described in the full paper.
For flow ($\sigma_{ij}(t) = \sigma_{ij}(t_0) = \sigma_{ij}^0 = $ constant) the Bürgers model becomes solvable and leads to the equations (4) and (5):

$$\varepsilon_{ij}(t) = \sigma_{ij}^0 \left[\frac{1}{\eta} t - \frac{1}{E} \left(3 - e^{\frac{-E}{\eta} t} \right) \right] \qquad (4)$$

$$E(t) = \frac{\eta E}{Et - \eta \left(3 - e^{\frac{-E}{\eta} t} \right)} \qquad (5)$$

where:
σ'_{ij} – the components of the stresses deviator (in time);
ε'_{ij}^0 – the components of the strains deviator at the initial moment;
E – the elasticity modulus of Young;
η – the coefficient of viscosity, measured in poise;
E(t) – the time dependant elasticity module.

For obtaining the computer programs for studying the salt flow at the Slanic Prahova salt mine through the finite element method we used the model described by the equations (4) and (5).

This permits the use of the already existing programs for the static equilibrium of the mining works in this salt mine. Every mathematical formula from the static case remains, there is only one modification when expressing the rigidity matrix where, instead of the elasticity module of Young, we use one of the expressions from (4) or (5).

During the exploitation, for every moment in time (measured in years) the rigidity matrix will be calculated. The global equations system will be solved and its results will be the displacements at the chosen moment, in every node of the system. The graphical representations of the distorted structure, or vertical and horizontal sections, at different moments of time, can be obtained, too. Through these it is possible to observe the evolution of the distortions in a given period of time.

3. THE SOFTWARE APPLICATION

We elaborated a computer program for IBM_PC for static and rheological analysis with finite elements for underground mining works, in the extraction industry of salt, where the mining methods use supporting pillar and roomwork on several levels. We considered the elastic behavior in static regime and the viscoelastic behavior in time.

The functions of the computer program are:
- the automatic discretization in finite elements (Fig.2) of structures like those from Fig.1, using the geometric description of the mining works;
- the solving of the equations system which describe the static equilibrium of the structure, whose solutions are the small displacements in nodes according to the three coordinate axes of the global system and the solving of the differential system (9);
- the calculus of the six components of the stress tensor in the nodes of the structure;
- the graphic representation on display and plotter of the stress curves (Fig.3 and Fig.4) and of the possible breaking curves (breaking criterions Nadai, Mohr, Murrel) (Fig.5 and Fig.6) in horizontal plane at any altitude and in vertical plane upon any sectioning direction;
- the 3D representation (orthogonal projection) of the deformed and not deformed structure and the strain evolutions in time;
- the representation of the horizontal and vertical sections through the deformed and not deformed structure.

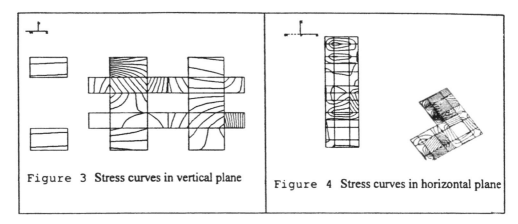

Figure 3 Stress curves in vertical plane   Figure 4 Stress curves in horizontal plane

The static and dynamic analysis, as it was described, has been used for studying the behavior of the 5th, 6th and 7th levels of the Cantacuzino salt mine from Slanic Prahova. During the exploitation, the models have been modified to reflect the reality. The programs successfully described the stresses appearing in the mining works and the strains produced as a result of these stresses.

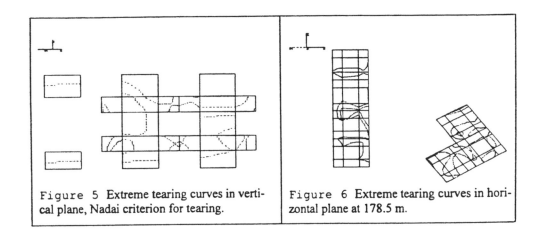

Figure 5 Extreme tearing curves in vertical plane, Nadai criterion for tearing.

Figure 6 Extreme tearing curves in horizontal plane at 178.5 m.

REFERENCES

Rao, S. S., 1981: The Finite Element Method in Engineering. Pergamon Press.
Olariu, V., 1986: Modelare numerica cu elemente finite. Ed. Tehnica, Romania.
Girbea, D., 1990: Analiza cu elemente finite. Ed. Tehnica, Romania.
Cristescu, N., 1990: Mecanica rocilor. Modele matematice reologice. Ed. St., Romania.
Oden, J.T. & Sato, T., 1967: Finite strains and displacements of elastic membranes by the finite element method. J. Solids Structures.3: 471-488
Petrila, T., 1987: Metode element finit si aplicatii. Ed. Acad., Romania.
Dadu, A. & Kopenetz, L., 1985: Numerical analysis of lightweight structures made of membrane, cable and beam elements. Romania, Timisoara.

Basic and Applied Salt Mechanics, Cristescu, Hardy, Jr & Simionescu (eds)
© 2002 Swets & Zeitlinger, Lisse, ISBN 90 5809 383 2

Three-dimensional verification of salt pillar design equation

Mark A.Frayne
RESPEC, Grand Bend, Canada

Leo L.Van Sambeek
RESPEC, Rapid City, USA

ABSTRACT

The time-dependent deformational behavior of salt makes the task of salt mine pillar design complex. Current methodologies for pillar design that involve numerical modeling require the commitment of significant computer and human resources to develop an acceptable solution. For this method, preliminary pillar characteristics (size and shape) are based on historical pillar designs, observed behavior, and engineering judgment. Then a worst case–best case approach is used in conjunction with a series of numerical simulations to define the appropriate pillar-design limits. However, as salt mines become deeper and mining economics drive pillar-design decisions, our observations database will become ever smaller. A need exists for simple pillar-design tools that help salt mining operations determine acceptable pillar characteristics and bounds the scope of analysis required to establish a formal pillar design.

Van Sambeek (1996) describes such a tool in the previous conference proceedings. The pillar-design equation developed in that paper is suitable for estimating stresses in square pillars, although assumptions were required to extend its use to rectangular salt pillars. Results from a series of two-dimensional finite element simulations using both axisymmetric and plane-strain analyses could not verify the assumptions used in developing the pillar-design equation for rectangular pillars.

Three-dimensional finite difference simulations of rectangular salt pillars were performed to examine the assumptions used in the extension of the pillar-design equation. The three-dimensional modeling results bolster use of the equation for estimating stress conditions in salt pillars of various shapes and sizes. In particular, this paper (1) reviews the initial pillar-design equation, (2) compares two- and three-dimensional numerical modeling results, and (3) presents equation modifications to improve the agreement with three-dimensional numerical modeling results.

INTRODUCTION

Simple pillar-design equations to calculate the average stresses in salt pillars were developed by Van Sambeek (1996) and presented in the previous conference. A series of two-dimensional axisymmetric and plane-strain numerical modeling analyses showed that the equations adequately reproduced the results of extensive modeling efforts. However, a comprehensive evaluation of the pillar equation was limited by the two-dimensional nature of the previous numerical modeling analyses. Axisymmetric analyses mirror the behavior of square salt pillars, and plane strain analyses simulate the behavior of long salt pillars. The capability to truly examine the behavior of rectangular pillars, which lie between the two extremes, is not available through two-dimensional numerical modeling.

The limitations presented by two-dimensional modeling were addressed for this paper by conducting a series of three-dimensional numerical modeling analyses to examine the behavior of rectangular pillars. The results of the three-dimensional modeling are used to verify the proposed pillar-design equations. The salt pillar-design equations are shown to be capable of reproducing the results of extensive three-dimensional modeling efforts. Use of these equations simplifies comparisons of different pillar sizes and shapes and provides a tool for estimating stress conditions in salt pillars.

TWO-DIMENSIONAL EQUATION REVIEW

Linear and nonlinear equations to predict the average horizontal stresses and the average effective stress for both the axisymmetric and plane-strain analyses are given by Van Sambeek (1996). The average vertical stress in the pillar is used as a normalizing parameter for the other stresses in the various shaped pillars. The average vertical stress ($\overline{\sigma}_v$) when the conditions are met for tributary loading is:

$$\overline{\sigma}_v = \gamma\, z /(1 - E) \tag{1}$$

where:

$\qquad$ E $\quad=\quad$ areal extraction ratio

$\qquad$ z $\quad=\quad$ depth

$\qquad$ γ $\quad=\quad$ average stress gradient for the overburden.

Proposed linear equations to calculate stresses in a rectangular-shaped salt pillar at a steady-state stress condition are:

$$\overline{\sigma}_W = \overline{\sigma}_v\left(0.1W : H\right) \tag{2}$$

$$\overline{\sigma}_L = \overline{\sigma}_v\left(0.1L : H\right) \tag{3}$$

$$\overline{\sigma}_{ef} = \overline{\sigma}_v\sqrt{(0.1W : H)^2\left(1 - \beta + \beta^2\right) - 0.1W : H(1 + \beta) + 1} \tag{4}$$

where:

$\qquad$ W, L $\quad=\quad$ pillar width and length (horizontal dimensions)

$\qquad$ H $\quad=\quad$ pillar height (vertical dimension)

$\qquad$ $\overline{\sigma}_W, \overline{\sigma}_L$ $\quad=\quad$ average horizontal stresses parallel to the width and length $\quad$ (5)

$\qquad$ $\overline{\sigma}_{ef}$ $\quad=\quad$ average effective stress.

$$\beta = \text{minimum} \left| \begin{matrix} L/W \\ \left(\dfrac{1 + 0.1\dfrac{W}{H}}{0.2\dfrac{W}{H}}\right) \end{matrix} \right.$$

Nonlinear, plane-strain versions of the equations (for infinitely long pillars) are as follows:

$$\bar{\sigma}_W / \bar{\sigma}_v = \exp(-2.931\,\text{H}:\text{W})$$ (6)

$$\bar{\sigma}_L / \bar{\sigma}_v = \frac{1}{2}(1 + \exp(-3.042\,\text{H}:\text{W}))$$ (7)

$$\bar{\sigma}_{ef} / \bar{\sigma}_v = \frac{\sqrt{3}}{2}(1 - \exp(-2.940\,\text{H}:\text{W}))$$ (8)

Nonlinear, axisymmetric versions of the equations (for square pillars) are as follows:

$$\bar{\sigma}_r / \bar{\sigma}_v = \exp(-3.418\,\text{H}:\text{W})$$ (9)

$$\bar{\sigma}_\theta / \bar{\sigma}_v = \exp(-3.934\,\text{H}:\text{W})$$ (10)

$$\bar{\sigma}_{ef} / \bar{\sigma}_v = \frac{\sqrt{3}}{2}(1 - \exp(-3.914\,\text{H}:\text{W}))$$ (11)

where:

$\bar{\sigma}_r$ = average radial stress

$\bar{\sigma}_\theta$ = average tangential stress.

THREE-DIMENSIONAL EQUATION DETERMINATION

The pillar-design equations from Van Sambeek (1996) were checked against average pillar stresses from the three-dimensional finite difference analyses. Forty three-dimensional numerical modeling simulations were performed to determine the pillar stresses for various pillar width to height (W:H) and length to width (L:W = β) ratios. In the analyses, the extraction ratio, pillar height, and saltback thickness were held constant. Average stresses were determined for each pillar and then an equation of the following form was fitted to the two average horizontal stresses and the effective stress:

$$\bar{\sigma} / \bar{\sigma}_v = A_1 + A_2 \exp[A_3(H:W)]\exp[A_4(H:L)]$$ (12)

where:

Equation Constant	$\bar{\sigma}_w / \bar{\sigma}_v$	$\bar{\sigma}_L / \bar{\sigma}_v$	$\bar{\sigma}_{ef}/ \bar{\sigma}_v$
A_1	0	0	1
A_2	0.94	0.86	−0.90
A_3	−2.84	−0.42	−1.862
A_4	−0.48	−2.46	−1.50

407

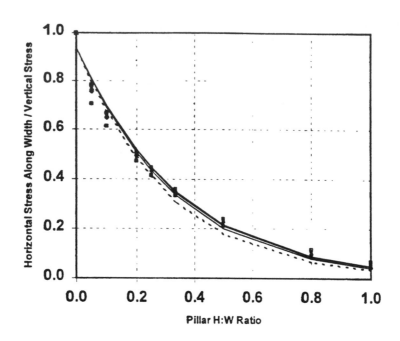

Figure 1. Average Horizontal Stress as a Function of Pillar Height to Width. Beta = 1 for Dashed Curve and Beta = 2 to 5 for Others.

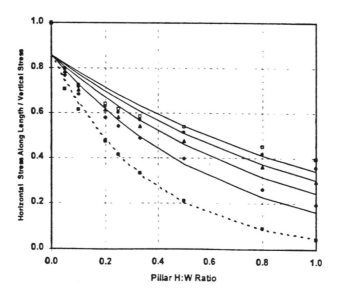

Figure 2. Average Horizontal Stress Along Length of Pillar as a Function of Pillar Height to Width. Dashed Curve for Beta = 1 and Other Curves Progressively Beta = 2 to 5.

Figures 1 through 3 illustrate the comparison of the normalized average stress values from the numerical modeling analyses and the pillar-design equation. The average vertical, horizontal, and effective stresses were determined at the midheight of the pillar. A comparison of average pillar

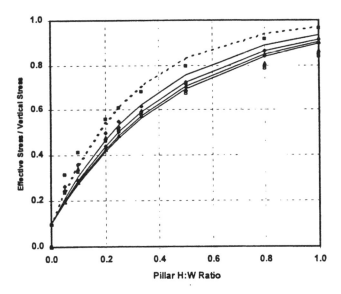

Figure 3. Average Effective Stress in Pillar as a Function of Pillar Height to Width. Dashed Curve for Beta = 1 and Other Curves Progressively Beta = 2 to 5.

stresses was made between the earlier version of the pillar-design equation and the equation developed using the three-dimensional numerical modeling results. The linear equations adequately reproduce the three-dimensional stresses for W:H ≤ 4. The nonlinear equations shown above are required for W:H > 4.

CONCLUSION

The previously published salt pillar-design equations were compared to stresses from a series of 40 three-dimensional numerical modeling simulations of rectangular-shaped pillars. The original version of the pillar-design equation had been compared to two-dimensional axisymmetric and plane-strain numerical modeling simulations only. The salt pillar-design equation proposed by Van Sambeek (1996) was found to adequately represent the three-dimensional stresses calculated for rectangular pillars. The same limitations noted in the original paper still apply; use of the salt pillar equation should be restricted to uniformly sized pillars in a regular pattern. Additional work is required to address the condition of yield and bearing pillar configurations or pillars in narrow workings. The equations can, however, be used to estimate the stress conditions in pillars when the pillar deformation rate and constitutive salt properties are known.

REFERENCE

Van Sambeek, L. L., 1996. "Salt Pillar Design Equation," *Proceedings, Fourth Conference of The Mechanical Behavior of Salt,* École Polytechnique de Montréal, Mineral Engineering Department, Québec, Canada, June 17 and 18, M. Aubertin and H. R. Hardy Jr. (eds.), Penn State University, Trans Tech Publications, Clausthal, Germany, 1998, pp. 495–508.

Basic and Applied Salt Mechanics, Cristescu, Hardy, Jr & Simionescu (eds)
© 2002 Swets & Zeitlinger, Lisse, ISBN 90 5809 383 2

Assessment of stress status in pillars based on laboratory trials

Cornel Hirian, Adrian Todorescu, Vasile Gaiducov, Victor Arad & Mihaela Toderas
University of Petrosani, Petrosani, Romania

1. INTRODUCTION

Being in the possession of several salt deposits (Ocna Mures, Praid, Cacica, Tg. Ocna, Ocnele Mari, etc.), Romania can be considered from this point of view as a fortunate country. The extraction rhythm is constantly increasing with the output growing from 368.000 tones in 1938 to 3.000.000 to tones in 1970 and 5.000.000 tones in 1985.

The first underground method of exploitations have employed conical and bell/shape rooms diameters of 60 m, and heights up to 100 m – Tg. Ocna, Slănic Prahova, Praid

Presently, since the extraction of salt is by dry-methods, two specific methods are generally in use:

1. Rectangular room – and – pillars method, with small room sizes.
2. Small rooms and square pillars method.

The first one can be applied for depths up to 200 meters, the second method being used for depths reaching 1000 meters. The first method is presently in use, at the Praid saline only, the small rooms and square pillars technique being generalized at Ocna Dej, Slănic Prahova, Feţele Târgului, Tg. Ocna, etc. The height of rooms is about 8 m, their width being of (14 – 16) m and the length of pillars ranging between 14 m and 17 m. When several levels are already excavated (Slănic Prahova, Tg. Ocna, etc.) between the levels is left a 8 m thick floor. For example, at Feţele Târgului, Tg. Ocna, the IXth excavation level corresponds to the depth of 250 m.

2. DETERMINATION OF ROCK SALT CHARACTERISTICS IN LABORATORY

In the laboratory, a set of mechanical characteristics were obtained from prismatic cylindrical samples and also using models of room, pillar and floor (see Figures 1, 2 and 3), the results obtained being synthesized in Table 1.

Interpreting the results included in **Table 1** it can be concluded that the dilatancy threshold is reached at 130 daN/cm^2 in simple compression, and at 204 daN/cm^2 in triaxial tests similar to the stress state existing underground.

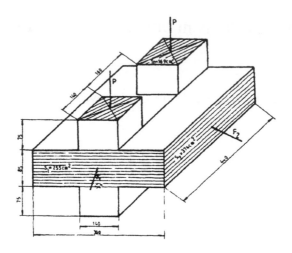

Fig. 1. Schematic representation of the tested model
Scale 1:100.

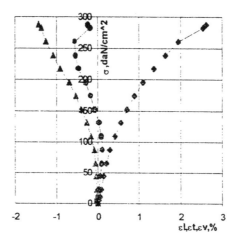

Fig.2 . Stress-strain curves obtained on
prismatic samples 240 x 155 x 80mm.

Table 1. Mean values obtained from specimens and models

Apparent specific weight γ_o $\cdot 10^4 [N/m^3]$	Compression strenght [MPa]						Tensile strength σ_{rt} [MPa]	Cohesion C [MPa]	Internal friction angle φ [°]	Double shearing strength τ_f [MPa]
	d=h = 42 mm	Prism		Lateral reaction model, σ_o						
		σ_{rc}	σ_{8d}	σ_{rc}	σ_{8d}	σ_o				
2,2	24,5	28,8	13,0	42,1	20,4	2,2	2,2	3,5	53	2,2

412

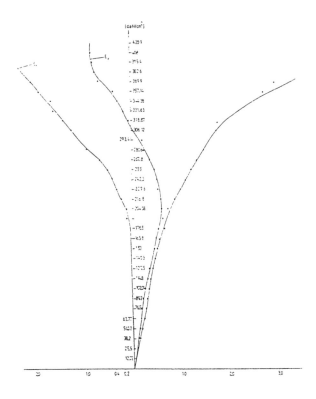

Fig.3   Dilatancy   obtained   from
model testing.

3. STRESS STATES IN PILLARS

Worldwide, the researchers who have considered the design of pillars in salt
mines, have agreed that on the pillars used in room, and pillars method of excavation,
only   static compression stresses are acting, induced by the weight of overburden. If
regarding the nature of stresses acting on pillars. there is a clear and generally accepted
opinion on concerning their magnitude, the approaches are highly different. Thus,
using the photo – elasticity procedures. G. Dorstewitz arrived at the conclusion that in
the pillars are stress concentrations – at their ribs-. and these concentrations is depending
on sev   l parameters, such as: the shape and dimensions of pillars; distance between
the pi.... s; nature of the rock existing in the pillar; type and characteristics of the rocks in
the roof and floor. From his experiment, the author arrives at the conclusion that the
compression stresses are not homogeneously distributed in thr pillar. i.e. at the ribs the
stresses may be three times bigger. L.D. Seviakov [4] assumes a uniform distribution of
the compression stresses in the horizontal cross section of the pillar. while K. Kegel [4]
assumes a non-homogeneous repartition of the compression stresses, with a concentration
located in the center of the pillar [2-4].

Our research team. following M.Stamatiu [4]. is accepting for the design of the
pillar, the hypothesis of the homogeneous repartition: a safety coefficient is introduced

to take care of the higher stresses towards the surface of the pillar cross-section. For square pillars their weight is neglected in the stability condition, and we arrived at the relationship:

$$\sigma_{mp} = (1 + L_c / L_p)^2 \, \sigma_z \, .$$

where σ_{mp} is average stress in the pillar $[tf/m^2]$; L_c is the room width, $[m]$; L_p – the pillar width, $[m]$; σ_z - is the vertical component of the stress state $[tf/m^2]$.

For: $L_c = 15$ m, $L_p = 15$ m, $\gamma_a = 2.2$ tf/m^3 and for various depths, the stresses in the pillar are comprised between 440 tf/m^2 and 3960 tf/m^2 at the depth 450 m. Since from the laboratory tests we have obtained a dilatancy threshold at $\sigma^* = \sigma_{lld} = 204 \, da \, N/cm^2$, if follows that the square pillars with ribs at 15 m, will be stable up to the depths :

$$H = \frac{\sigma_{kl}}{\left(1 + \dfrac{L_c}{L_p}\right)^2 \cdot \gamma_a} = \frac{2040}{4 \cdot 2.2} = 232 \text{ m.}$$

For greather depths the pillars of 15 m ribs are too small, and the ratio $L_c/L_p = x$ must be changed with a factor of 0.95 for $H = 250$ m, up to $x = 0.43$ for $H = 450$ m according to the equations given below:

$$\sigma_{mp} = (1 + x)^2 \, \sigma_z \quad ; \quad \frac{\sigma_{mp}}{\sigma_z} = (1 + x)^2 \quad ; \quad \sqrt{\frac{\sigma_{mp}}{\sigma_z}} - 1 = x \, .$$

The relationship relating the pillar width to the depth is shown in Fig.4

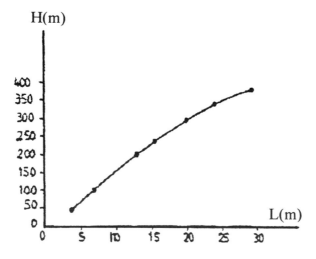

Fig.4.   Relationship between pillar width and depth.

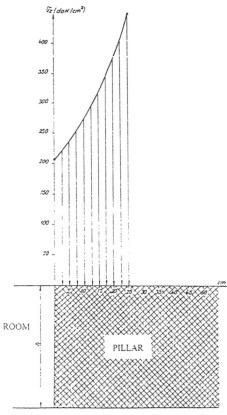

Fig. 5 Variation of stresses in the pillar.

The stress state in a pillar depends on the stability of the rock existing in the pillar, on its cohesion, on the angle of internal friction, and on the stresses induced at where we have: for cohesion $C = 35$ daN/cm^2 ; for internal friction angle $\varphi = 53^\circ$; for height of pillar $h = 800$ cm ; and x is the distance between the pillar surface up to the point where stress is estimated. For various values to x, we get the stress distribution shown in Figure 5.

REFERENCES

[1] **Stamatiu M.,** *Rock Mechanics* (in Romanian). 1962, Editura Didactică si Pedagogică, Bucuresti.
[2] **Hirian C.,** *Rock Mechanics* (in Romanian), 1981, Editura Didactica si Pedagogica, Bucuresti.
[3] **Popescu Al., Todorescu A.,** *Rock Mechanics in Mining* (in Romanian), 1982, Editura Tehnică, Bucuresti.
[4] **Stamatiu M.,** *The problem of designing the pillars in the salt mines from Romania* (in Romanian) , 1959, Editura Academiei Române. Bucuresti.
[5] * * * *Textbook of the mining engineer* (in Romanian), 1985, Editura Tehnica Bucuresti, vol.2.

Basic and Applied Salt Mechanics, Cristescu, Hardy, Jr & Simionescu (eds)
© 2002 Swets & Zeitlinger, Lisse, ISBN 90 5809 383 2

Characteristics of Salt-Bearing Rocks in the "Beyond-Limit" state, under various conditions of deformation

M.P.Nesterov & O.V.Salzseiler
R & D Institute of Hallurgy, Perm, Russia

SUMMARY

When forecasting the increase with time of the earth surface subsidence resulting from deforming of interchamber pillars left during mining rock salt and potash ores, one may be interested in such characteristics of these pillars as: modulus of reduction in their response capability; their minimum remaining strength; modulus of increase in the pillars' response after the start of the massif consolidation (this massif formed from these pillars by the time of reducing their remaining strength down to a minimum value) and other characteristics. Modulus of reduction may be used, moreover, for evaluating the capability of salt-bearing interchamber pillars, being in the "beyond-limit" state, to brittle failure. Under the term "beyond-limit" state in this case is defined as the state (or condition) of a rock after its loading until the ultimate strength or after its deformation reaches an ultimate value.

Laboratory investigations have been carried out to evaluate the relationship of the above named characteristics and conditions for deforming of interchamber pillars, using as an example various kinds of sylvinite ore and rock salt being mined at the Upperkama potash ore deposit. In the process of the investigation the following conditions were varied: height-to-diameter ratio of test samples; rate of sample deformation after deformations (strains) reach beyond-the-limit values; values of beyond-the-limit deformations after which the samples were deformed at various rates; dimensions of the space surrounding the samples under testing (into this space the samples could deform as pillars into the rooms). The results of these studies have led to conclusions that are of important practical interest.

Basic and Applied Salt Mechanics, Cristescu, Hardy, Jr & Simionescu (eds)
© *2002 Swets & Zeitlinger, Lisse, ISBN 90 5809 383 2*

Master bibliography

The Proceedings of the first four conferences on the Mechanical Behavior of Salt contain a comprehensive master bibliography including some 1350 entries. The following is a continuation of this earlier bibliography. It consists of a listing of papers, reports and texts associated with the subject of mechanical behavior of salt referenced in the papers presented at the fifth conference, as well as additional reference material made available by a number of the participants. Those references already presented in the earlier bibliographies are not repeated here. Furthermore, the bibliography does not include material relative to conventional or solution mining of salt except where such material relates directly to the conference theme. The editors appreciate the fact that the bibliography is not complete and it is planned to continue updating it in the proceedings of future conferences.

****

Anon., (1996a), Waste Isolation Pilot Plant Shaft Sealing System Compliance Submittal Design Report, Volume 1 of 2: Main Report. Appendices A, B, C, and D. SAND96-1326/1, Sandia National Laboratories, Albuquerque, NM.

Anon., (1996b), Crushed-Salt Constitutive Model Refinement, Calculation No. 325/04/05, RE/SPEC Inc., Rapid City, SD, Sandia WIPP Central Files [WPO#48071], Sandia National Laboratories, Albuquerque, NM.

Anon., (1998), Annual Ground Control Operating Plan for the Waste Isolation Pilot Plant, Westinghouse Geotechnical Engineering, Feb. 1998.

Albrecht, H., U. Hunsche, and O. Schulze, (1993), "Results from the Application of the Laboratory Test Program for Mapping Homogeneous Parts in the Gorleben Salt Dome," Geotechnik-Sonderheft, Glückauf, Essen, pp. 155-158.

Alden, T. H., (1963), "Latent Hardening in Rock Salt Structure Crystals," Jour. Metals, Vol. 15, p. 116.

Alheid, H.-J., M. Knecht & R. Lüdeling, (1998), "Investigation of the Long-Term Development of Damaged Zones Around Underground Openings in Rock Salt," Proceedings NARMS'98, Cancun 1998, Int. J. Rock Mech. & Min. Sci. Vol. 35, No. 4-5, Paper No. 27.

Aubertin, M., (1996), "Triaxial Stress Relaxation Tests on Saskatchewan Potash: Discussion," Canadian Geotechnical Journal, Vol. 33, pp. 375-377.

Aubertin, M., M. R. Julien, and L. Li, (1998), "The Semi-Brittle Behavior of Low Porosity Rocks," Proceedings NARMS'98, Cancun, Mexico, Vol. 2, pp. 65-90.

Aubertin, M., M. R. Julien, S. Servant, and D. E. Gill, (1999), "A Rate Dependent Model for the Ductile Behavior of Salt Rocks," Canadian Geotechnical Journal, Vol. 36, pp. 660-674.

Aubertin, M., J. Sgaoula, S. Servant, D. E., Gill, M. Julien, and B. Ladanyi, (1996), "A Recent Version of a Constitutive Model for Rocksalt," Proceedings 4th Conference on the Mechanical Behavior of Salt, Montreal 1996, Trans Tech Publications, Clausthal-Zellerfeld, Germany, pp. 129-134.

Aubertin, M., J. Sgaoulla, S. Servant, M. R. Julien, D. E. Gill, and B. Ladanyi, (1998), "An Up-To-Date Version of SUVIC-D for Modeling the Behavior of Salt," Aubertin, M., Hardy Jr., H. R. (Eds.), Proceedings 4th Conference on the Mechanical Behavior of Salt, (MECASALT IV). Montreal 1996, Trans Tech Publications, Clausthal-Zellerfeld, Germany, pp. 205-220.

Balthasar, K., M. Haupt, Ch. Lempp, and O. Natau, (1987), "Stress Relaxation Behavior of Rock Salt: Comparison of In-Situ Measurements and Laboratory Test Results," Proceedings 6th International Congress Rock Mechanics, Vol. 1, Montreal 1987, pp. 11-14.

Barber, D. J., (1962), "Etching of Dislocations in Sodium Chloride Crystals," Jour. Applied Physics, Vol. 33, p. 3141.

Berest, P., B. Brouard, and G. Durup, (1997), "Behavior of Sealed Solution-Mined Caverns," Proceedings 4th Conference on the Mechanical Behavior of Salt, Ecole Polytechnique, Montreal, June 1996, Trans Tech Publications, Clausthal-Zellerfeld, Germany, pp. 511-524.

Bertram, B. M., (1999), "Salt-Industrial Minerals 1998," Mining Engineering, Vol. 51, pp. 45-47.

Borm, G., and M. Haupt, (1988), "Constitutive Behavior of Rock Salt: Power Law or Hyperbolic Sine Creep?" Numerical Methods in Geomechanics, Balkema, Rotterdam, pp. 1883-1893.

Brodsky, N. S., (1993), Thermomechanical Damage Recovery Parameters for Rocksalt From the Waste Isolation Pilot Plant, SAND93-7111, (Prepared by RE/SPEC Inc., Rapid City, SD), Sandia National Laboratories, Albuquerque, NM.

Brodsky, N. S., F. D. Hansen, and T. W. Pfeifle, (1997), "Properties of Dynamically Compacted WIPP Salt," Proceedings 4th Conference on the Mechanical Behavior of Salt, Ecole Polytechnique, Montreal, June 1996, Trans Tech Publications, Clausthal-Zellerfeld, Germany, pp. 303-316.

Callahan, G. D., (1999), Crushed Salt Constitutive Model, Sandia Report SAND98-2680, Sandia National Laboratories, Albuquerque, New Mexico, 178 pp.

Callahan, G. D., M. C. Loken, L. D. Hurtado, and F. D. Hansen, (1997), "Evaluation of Constitutive Models for Crushed Salt," Proceedings 4th Conference on the Mechanical Behavior of Salt, Ecole Polytechnique, Montreal June 1996, Trans Tech Publications, Clausthal-Zellerfeld, Germany, pp. 317-330.

Callahan, G. D., K. D. Mellegard, and F. D. Hansen, (1998), "Constitutive Behavior of Reconsolidating Crushed Salt," Journal Rock Mechanics and Mining Sciences, Vol. 35, No. 4/5, pp. 422-423.

Campos de Orellana, A. J., (1997), "Non-Associated Pressure Solution Creep in Salt Rock Mines," Proceedings 4th Conference on the Mechanical Behavior of Salt, Ecole Polytechnique, Montreal, June 1996, Trans Tech Publications, Clausthal-Zellerfeld, Germany, pp. 429-444.

Carter, N. L., (1993), "Rheology of Salt Rock," Journal Structural Geology, Vol. 15, No. 10, pp. 1257-1272.

Chan, K. S., S. R. Bodner, A. F. Fossum, and D. E. Munson, (1992), "A Constitutive Model for Inelastic Flow and Damage Evolution in Solids Under Triaxial Compression," Mechanics of Materials, Vol. 14, No. 1, pp. 1-14.

Chan, K. S., S. R. Bodner, K. P. Walker, and U. S. Lindholm, "A Survey of Unified Constitutive Theories," Proceedings 2nd Symposium Nonlinear Constitutive Relations for High Temperature Application, NASA/Lewis, Cleveland, OH, pp. 1-24.

Chan, K. S., D. E. Munson, A. F. Fossum, and S. R. Bodner, (1997), "A Constitutive Model for Representing Coupled Creep, Fracture and Healing in Rock Salt," Proceedings 4th Conference on the Mechanical Behavior of Salt, Ecole Polytechnique, Montreal, June 1996, Trans Tech Publications, Clausthal-Zellerfeld, Germany, pp. 221-234.

Chumbe, D., A. Lloret, and E. Alonso, (1997), "Creep and Permeability Tests on Compacted Granular Salt," Proceedings 4th Conference on the Mechanical Behavior of Salt, Ecole Polytechnique, Montreal, June 1996, Trans Tech Publications, Clausthal-Zellerfeld, Germany, pp. 331-339.

Clabaugh, P. S., (1962), "Petrofabric Study of Deformed Salt," Science, Vol. 136, pp. 389-391.

Coble, R. L., (1963), "A Model for Boundary Diffusion Controlled Creep in Polycrystalline Materials," Journal Applied Physics, Vol. 34, pp. 1679-1682.

Corthesy, R., and D. E. Gill, (1990), "A Novel Approach to Stress Measurements in Rocksalt," Int. J. Rock Mech. Min. Sci., Vol. 27, pp. 95-107.

Cosenza, Ph., M. Ghoreychi, and B. Bazargan-Sabet, (1997), "In Situ Gas and Brine Permeability Measurements in Salt," Proceedings 4th Conference on the Mechanical Behavior of Salt, Ecole Polytechnique, Montreal, June 1996, Trans Tech Publications, Clausthal-Zellerfeld, Germany, pp. 3-16.

Cosenza, Ph., and M. Ghoreychi, (1998), "Effect of Added Fluids on Mechanical Behavior of Rock Salt," Euroconference on Pore Pressure, Scale Effects and the Deformation of Rocks, Aussois, France.

Cosenza Ph., M. Ghoreychi, B. Bazargan-Sabet, and de G. Marsily, (1999), "In Situ Rock Salt Permeability Measurement for Long Term Safety Assessment of Storage," International Journal Rock Mechanics and Mining Sciences, Vol. 36, pp. 509-526.

Cristescu, N., (1997), "Evaluation Damage in Rock Salt," Proceedings 4th Conference on the Mechanical Behavior of Salt, Ecole Polytechnique, Montreal, June 1996, Trans Tech Publications, Clausthal-Zellerfeld, Germany, pp. 131-141.

Cristescu, N., and U. Hunsche, (1998), "Time Effects in Rock Mechanics," Materials, Modelling and Computation, John Wiley and Sons, Chichester. 342 pp.

Dale, T., and L. D. Hurtado, (1997), "WIPP Air-Intake Shaft Disturbed-Rock Zone Study," Proceedings 4th Conference on the Mechanical Behavior of Salt, Ecole Polytechnique, Montreal, June 1996, Trans Tech Publications, Clausthal-Zellerfeld, Germany, pp. 525-535.

Davidge, R. W., and P. L. Pratt, (1963), "Formation of Color Centers in Undeformed and Deformed NaCl," Physica Status Solidi, Vol. 3, pp. 665-670.

DeSouza, E. M., (1988), "Ground Control, Brine Inflow Arrest, and Backfill Designs in the Search for New Mining Strategies for the Potash Industry," PhD Thesis, Queen's University, Department of Mining Engineering.

DeVries, K. L., G. D. Callahan, and D. E. Munson, (1997), "WIPP Panel Simulation with Gas Generation," Proceedings 4th Conference on the Mechanical Behavior of Salt, Ecole Polytechnique, Montreal, June 1996, Trans Tech Publications, Clausthal-Zellerfeld, Germany, pp. 537-550.

Diggs, T. N., J. L. Urai, and N. L. Cartyer, (1997), "Rates of Salt Flow in Salt Sheets, Gulf of Mexico: Quantifying the Risk of Casing Damage in Subsalt Plays," Terra Nova, Vol. 9, Abstract 11/4B29.

Drăgănescu, L., (1998), "Geological Study of the Salt Deposit of the Slanic Prahova and the Impact of the Exploitation on the Environmental," University of Bucharest, Bucharest, Romania.

Edgell, H. S., (1996), "Salt Tectonism in the Persian Gulf Basin," Geological Society of London Special Publication, Vol. 100, pp. 129-151.

Evans, H. E., and G. Knowles, (1978), "Dislocation Creep in Non-Metallic Materials," Acta Met., Vol. 26, pp. 141-145.

Fokker, P. A., (1997), "The Micro-Mechanics of Creep in Rocksalt," Proceedings 4th Conference on the Mechanical Behavior of Salt, Ecole Polytechnique, Montreal, June 1996, Trans Tech Publications, Clausthal-Zellerfeld, Germany, pp. 49-61.

Förster, S., et al., (1981), "Pneumatic and Hydraulic Fracturing Tests in Boreholes for Proof of Tightness and for the Stress Sounding in the Salt Rock Mass," (in German), Neue Bergbautechnik, Vol. 11, No. 5.

Fossum, A. F., D. E. Munson, K. S. Chen, and S. R. Bodner, (1997), "Constitutive Basis of the MDCF Model for Rock Salt," Proceedings 4th Conference on the Mechanical Behavior of Salt, Ecole Polytechnique, Montreal, June 1996, Trans Tech Publications, Clausthal-Zellerfeld, Germany, pp. 235-247.

Fossum, A. F., T. W. Pfeifle, K. D. Mellegard, and D. E. Munson, (1994), "Probability Distributions for Parameters of the Munson-Dawson Salt Creep Model," Rock Mechanics: Models and Measurements, Challenges From Industry, Proceedings 1st North American Rock Mechanics Symposium, University of Texas at Austin, June 1-3, 1994, Eds. P. P. Nelson and S. E. Laubach. Brookfield, VT: A. A. Balkema, pp. 715-722.

Fuenkajorn, K., and J. J. K. Daemen, (1989), "Borehole Closure in Salt," Proceeding 29th US Symposium Rock Mechanics, Balkema, Rotterdam, Netherlands, pp. 191-198.

Ghaboussi, J., R. E. Ranken, and A. J. Hendron, (1981), "Time-Dependent Behaviour of Solution Caverns in Salt," *Proc. ASCE, J. Geol. Eng. Div.,* Vol. 107, pp. 1379-1401.

Glabisch, U., (1997), "Stoffmodell für Grenzzustände im Salzgestein zur Berechnung von Gebirgshohlräumen," Dissertation an der TU Braunschweig, Braunschweig, Germany.

Gupta, I., and C. M. Li, (1970), "Stress Relaxation, Internal Stress and Work Hardening in LiF and NaCl Crystals." Mat. Science Engng., Vol. 6, pp. 20-26.

Hampel, A., and U. Hunsche, (1998), Die Beschreibung der rißfreien transienten und stationären Verformung von Steinsalz mit dem Verbundmodell, Geotechnik, Vol. 21, No. 3, pp. 264-267.

Hampel, A., U. Hunsche, P. Weidinger and W. Blum, (1998), "Description of the Creep of Rock Salt With the Composite Model – II Steady State Creep," Proceedings 4th Conference on the Mechanical Behavior of Salt, Ecole Polytechnique, Montreal, June 1996, Trans Tech Publications, Clausthal-Zellerfeld, Germany, pp. 277-285.

Hampel, A., and U. Hunsche, (2000), "Extrapolation of Creep of Rock Salt with the Composite Model," Proc. 5th Conf., (MEGASALT V), Bucharest 1999.

Hansen, F. D., (1997), "Reconsolidating Salt: Compaction, Constitutive Modeling and Physical Processes," International Journal of Rock Mechanics & Mining Sciences, SAND97-0484C, Vol. 34, No. 3-4, p. 492.

Hansen, F. D., (1979), Triaxial Quasi-Static Compression and Creep Behavior of Bedded Salt From Southeastern New Mexico, SAND79-7045, Sandia National Laboratories, Albuquerque, NM.

Hansen, F. D., and E. H. Ahrens, (1998), "Large-Scale Dynamic Compaction of Natural Salt,: Proceedings 4th Conference on the Mechanical Behavior of Salt, Ecole Polytechnique, Montreal, June 1996, Eds. M. Aubertin and H. R. Hardy, Jr., Trans Tech Publications, Clausthal-Zellerfeld, Germany, pp. 353-364.

Hansen, F. D., E. H. Ahrens, A. W. Dennis, L. D. Hurtado, M. K. Knowles, J. R. Tillerson, T. W. Thompson, and D. Galbraith, (1996), "A Shaft Seal System for the Waste Isolation Pilot Plant," Proceedings *SPECTRUM '96, Nuclear and Hazardous Waste Management*, International Topical Meeting, American Nuclear Society/Department of Energy Conference, Seattle, WA, (August 1996).

Hansen, F. D., G. D. Callahan, M. K. Loken, and K. D. Mellegard, (1998), Crushed Salt Constitutive Model Update, SAND97-2601, (RE/SPEC Inc., Rapid City, SD), Sandia National Laboratories, Albuquerque, NM.

Hansen, F. D., and K. D. Mellegard, (1979), Creep Behavior of Bedded Salt From Southeastern New Mexico at Elevated Temperature, SAND79-7030, (RE/SPEC Inc., Rapid City, SD), Sandia National Laboratories, Albuquerque, NM.

Hansen, F. D., and K. D. Mellegard, (1980), Further Creep Behavior of Bedded Salt From Southeastern New Mexico at Elevated Temperature, SAND80-7114, (RE/SPEC Inc., Rapid City, SD), Sandia National Laboratories, Albuquerque, NM.

Hardy, H. R., Jr., (1997), "Strength and Acoustic Emission in Salt Under Tensile Loading," Proceedings 4th Conference on the Mechanical Behavior of Salt, Ecole Polytechnique, Montreal, June 1996, Trans Tech Publications, Clausthal-Zellerfeld, Germany, pp. 143-162.

Heroy, W. B., (1968), "Thermicity of Salt as a Geologic Function," Mattox, R. B., et al. *Editors*, Saline deposits: Geol. Soc. America Special Paper 88, pp. 619-627.

Heusermann, S., S. Koss, and J. Sonnke, (1997), "Analysis of Stress Measurement in the TSDE Test at the Asse Salt Mine," Proceedings 4th Conference on the Mechanical Behavior of Salt, Ecole Polytechnique, Montreal, June 1996, Trans Tech Publications, Clausthal-Zellerfeld, Germany, pp. 17-29.

Hickman, S. H., and B. Evans, (1988), "Influence of Normal Stress and Contact Radius on Pressure Solution Along Halite Silica Contacts," EOS Transactions, American Geophysical Union, Vol. 69, No. 44, p. 1426.

Holcomb, D. J., and D. W. Hannum, (1982), Consolidation of Crush Salt Backfill Under Conditions Appropriate to the WIPP Facility, SAND82-0630, Sandia National Laboratories, Albuquerque, NM.

Hou, Z., (1997), "Untersuchungen zum Nachweis der Standsicherheit für Untertagedeponien im Salzgebirge," Dissertation an der TU Clausthal, Clausthal-Zellerfeld, Germany.

Hou, Z., and K-H. Lux, (1998), "Ein neues Stoffmodell für duktile Salzgesteine mit Einbeziehung von Gefügeschädigung und tertiärem Kriechen auf der Grundlage der Continuum-Damage-Mechanik," Geotechnik, Vol. 21, No. 3, pp. 259-263.

Hou, Z., K.-H. Lux, and U. Düsterloh, (1998), "Bruchkriterium und Fließmodell für duktile Salzgesteine bei kurzzeitiger Beanspruchung," Glückauf - Forschungshefte, Vol. 59, No. 2, pp. 59-67.

Hunsche, U., (1994), "Uniaxial and Triaxial Creep and Failure Tests on Rock: Experimental Technique and Interpretation," Visco-Plastic Behaviour of Geomaterials, Cristescu, N.D., and Gioda, G. (Eds.), CISM-Courses and Lectures No. 350, Springer, Vienna, pp. 1-54.

Hunsche, U., (1998), "Determination of the Dilatancy Boundary and Damage Up to Failure for Four Types of Rock Salt at Different Stress Geometries," Proceedings 4th Conference on the Mechanical Behavior of Salt, Ecole Polytechnique, Montreal, June 1996, Trans Tech Publications, Clausthal-Zellerfeld, Germany, pp. 163-174.

Hunsche, U., and A. Hampel, (1999), "Rock Salt - The Mechanical Properties of the Host Rock Material for a Radioactive Waste Repository," Engineering Geology, (Special Issue on Nuclear Waste Management in Earth Sciences, Eds.: Ch. Talbot & M. Langer, Elsevier, Amsterdam, Vol. 52, pp. 271-291.

Jackson, M. P. A., and C. J. Talbot, (1985), "External Shapes, Strain Rates and Dynamics of Salt Structures," Bulletin Geological Society America, Vol. 97, pp. 305-328.

Jin, J., W. D. Cristescu, and U. Hunsche, (1997), A New Elastic/Viscoplastic Model for Rock Salt," Proceedings 4th Conference on the Mechanical Behavior of Salt, Ecole Polytechnique, Montreal, June 1996, Trans Tech Publications, Clausthal-Zellerfeld, Germany, pp. 249-262.

Julien, M. R., (1999), <u>Une modélisation constitutive et numérique du comportement mécanique du sel gemme</u>. Ph.D. Thesis, Ecole Polytechnique de Montreal.

Julien, M. R., R. Foerch, M. Aubertin, and G. Cailletaud, (1997), "Some Aspects of the Numerical Implementation of SUVIC-D," <u>Proceedings 4th Conference on the Mechanical Behavior of Salt</u>, Ecole Polytechnique, Montreal, June 1996, Trans Tech Publications, Clausthal-Zellerfeld, Germany, pp. 389-403.

Jumikis, A. R., (1979), <u>Rock Mechanics</u>, Trans. Tech. Publ., Clausthal-Zellerfeld.

Kentir, C. J., S. J. Doig, H. P. Rogaar, P. A. Fokker, and D. R. Davies, (1990), "Diffusion of Brine Through Rock Salt Roof of Caverns," SMRI Fall meeting, Paris, France.

Kern, H., T. Popp, and T. Takeshita, (1992), "Characterization of the Thermal and Thermomechanical Behaviour of Polyphase Salt Rocks by Means of Electrical Conductivity and Gas Permeability Measurements," Tectonophysics, Vol. 213, pp. 285-302.

Kestin, J., H. E. Khalefa, and R. J. Correia, (1981), "Tables of the Dynamic and Kinematic Viscosity of Aqueous NaCl Solutions in the Temperature Range 20-150°C and the Pressure Range 0.1 - 35 MPa.", Journal Physical and Chemical Reference Data, Vol. 10, pp. 71-86.

Knowles, M. K., and C. L. Howard, (1996), "Field and Laboratory Testing of Seal Materials Proposed for the Waste Isolation Pilot Plant," <u>Proceedings Waste Management 1996 Symposium</u>, Tucson, AZ, February 1996, SAND95-2082C, Sandi National Laboratories, Albuquerque, NM.

Korthaus, E., (1997), "Consolidation and Deviatoric Deformation Behavior of Dry Crushed Salt at Temperatures up to 150°C," <u>Proceedings 4th Conference on the Mechanical Behavior of Salt</u>, Ecole Polytechnique, Montreal, June 1996, Trans Tech Publications, Clausthal-Zellerfeld, Germany, pp. 465-475.

Kwon, S., and H. Miller, (1997), "Using a Neural Network to Predict the Deformation of Underground Excavations in Rock Salt," <u>Proceedings 4th Conference on the Mechanical Behavior of Salt</u>, Ecole Polytechnique, Montreal, June 1996, Trans Tech Publications, Clausthal-Zellerfeld, Germany, pp. 405-416.

Langer, M., (1991), "General Report: Rheological Behavior of Rock Salt," <u>Proceedings 7th International Congress Rock Mechanics</u>, Aachen 1991, Vol. III, pp. 1811-1819.

Langer, H., and H. Offerman, (1982), "On the Solubility of Sodium Chloride in Water," Journal Crystal Growth, Vol. 60, pp. 389-392.

Lehner, F. K., (1990), "Thermodynamics of Rock Deformation by Pressure Solution," <u>Deformation Processes in Minerals, Ceramics and Rocks</u>, D. J. Barber, and P. G. Meredith, Editors, Unwin Hyman, London, pp. 296-333.

Lehner, F. K., and J. Bataille, (1984), "Non-Equilibrium Thermodynamics of Pressure Solution," Pure and Applied Geophysics, Vol. 122, pp. 53-85.

Leite, M. H., B. Ladanyi, and D. E. Gill, (1994), "Determination of Creep Parameters of Rock Salt by Means of an *in situ* Sharp Cone Test," Int. J. Rock Mech. Min. Sci., Vol. 30, pp. 219-232.

Loken, M. C., (1998), "Predicting the Consolidation of a Crushed-Salt Seal Using the Combined Mechanism Model," Calculation File 325/16/01, WIPP Central Files, WPO#52498, Sandia National Laboratories, Albuquerque, NM.

Lu, M., and E. Broch, (1997), "Study of Large Volume Gas Storage in Rock Salt Caverns by FEM," Proceedings 9th International Conference Computer Methods and Advances in Geomechanics, Wuhan, PRC, Yuan (ed.), Rotterdam: A.A. Balkema, Vol. 2, pp. 1489-1494.

Lux, K.-H., and Z. Hou, (1999), "Gefügeschädigungen als Grundlage zur Formulierung von neuartigen Stoffmodellen für viskoplatische Salinargesteine," Glückauf - Forschungshefte, Vol. 60, No. 1, pp. 23-24.

Maleki, H., and L. Chaturvedi, (1999), "Geotechnical Factors Influencing Stability in a Nuclear Repository in Salt," Proceedings 37th US Rock Mechanics Symposium, Vail, Colorado, pp. 1043-1049.

Martin, J. L., and A. S. Argon, (1986), "Low Energy Dislocation Structures Due to Recovery and Creep," Mat. Science Engng., Vol. 81, pp. 337-343.

Massier, D., (1997), "Mecanica Rocilor," University of Bucharest, Romania.

Matei, A., and N. D. Cristescu, (1999a), "The Effect of Volumetric Strain on Elastic Parameters for Rock Salt," Mechanics of Cohesive-Frictional Materials.

Matei, A., and N. D. Cristescu, (1999b), "Variation in Time of the Elastic Parameters of Rock Salt," Proceedings 1999 Congress International Society Rock Mechanics, Paris, August 1999, pp. 635-639.

Mellegard, K. D., G. D. Callahan, and P. E. Senseny, (1992), Multiaxial Creep of Natural Rock Salt, SAND91-7083, (RE/SPEC Inc., Rapid City, SD), Sandia National Laboratories, Albuquerque, NM.

Mellegard, K. D. and T. W. Pfeifle, (1993), Creep Tests on Clean and Argillaceous Salt From the Waste Isolation Pilot Plant, SAND92-7291. (RE/SPEC Inc., Rapid City, SD), Sandia National Laboratories, Albuquerque, NM.

Mellegard, K. D., T. W. Pfeifle, A. F. Fossum, and P. E. Senseny, (1993), "Pressure and Flexible Membrane Effects on Direct-Contact Extensometer Measurements in Axisymmetric Compression Tests," Journal Testing and Evaluation, Vol. 21, No. 6, pp. 530-538.

Mellegard, K. D., T. W. Pfeifle, and F. D. Hansen, (1999), Laboratory Characterization of Mechanical and Permeability Properties of Dynamically Compacted Crushed Salt, SAND98-2046, (RE/SPEC Inc., Rapid City, SD), Sandia National Laboratories, Albuquerque, NM.

Mellegard, K. D., and D. E. Munson, (1997), Laboratory Creep and Mechanical Tests on Salt Data Report (1975-1996), SAND96-2765. UC-721, 521 pp., Sandia National Laboratories, Albuquerque, NM.

Morgan, H. S., and R. D. Krieg, (1988), A Comparison of Unified Creep-Plasticity and Conventional Creep Models for Rock Salt Based on Predictions of Creep Behavior

Measured in Several in situ and Bench-Scale Experiments, SAND87-1867, 45 pp., Sandia National Laboratories, Albuquerque, NM.

Mraz, D., L. Rothenburg and J. Unrau, (1997), "Stress and Strain Around Openings in a Deep Potash Mine," Proceedings 4th Conference on the Mechanical Behavior of Salt, Ecole Polytechnique, Montreal, June 1996, Trans Tech Publications, Clausthal-Zellerfeld, Germany, pp. 459-467.

Munson, D. E., and P. R. Dawson, (1982), "A Workhardening/Recovery Model of Transient Creep of Salt During Stress Loading and Unloading," Proceedings 23rd U.S. Symposium on Rock Mechanics, University of California, Berkeley, CA, August 1982, Eds. R. E. Goodman and F. E. Heuze, American Institute of Mining, Metallurgical and Petroleum Engineers, New York, pp. 299-306.

Munson, D. E., K. L. DeVries, and G. D. Callahan, (1990), "Comparison of Calculations and *In Situ* Results for a Large Heated Test Room at the Waste Isolation Pilot Plant (WIPP)," Proceedings 31st U.S. Symp. on Rock Mechanics, (Edited by Hustrulid W. A. and G. A. Johnson) A. A. Balkema, Rotterdam, Netherlands, pp. 389-396.

Munson, D. E., R. L. Jones, and K. L. DeVries, (1991), "Analysis of Early Creep Closure in Geomechanically Connected Underground Rooms in Salt," Proceedings 32st U.S. Symp. on Rock Mechanics, A. A. Balkema, Rotterdam, Netherlands, pp. 881-888.

Munson, D. E., J. R. Weatherby, and K. L. DeVries, (1993), "Two- and Three-Dimensional Calculation of Scaled *in Situ* Tests Using the M-D Model of Salt Creep," International J. Rock Mech. Min. Sci. & Geomech. Abstr. Vol. 30, pp. 1345-1350.

Nguyen-Minh, D., and A. Pouya, (1992), "Une méthode d'étude des excavations souterraines en milieu viscoplastique - Prise en compte d'un état stationnaire des contraintes," Rev.Fr. Geotechnique, Vol. 59, pp. 5-14.

Olivella, S., J. Carrera, A. Gens, and E. E. Alonso, (1997), "Porosity Variations in Salt Aggregates Caused by Dissolution/Precipitation Processes Induced by Temperature Gradients," Proceedings 4th Conference on the Mechanical Behavior of Salt, Ecole Polytechnique, Montreal, June 1996, Trans Tech Publications, Clausthal-Zellerfeld, Germany, pp.77-86.

Olivella, S., A. Gens, E. E. Alonso, and J. Carrera, (1997), "Analysis of Creep Deformation of Galleries Blockfilled with Porous Salt Aggregates," Proceedings 4th Conference on the Mechanical Behavior of Salt, Ecole Polytechnique, Montreal, June 1996, Trans Tech Publications, Clausthal-Zellerfeld, Germany, pp. 379-386.

O'Neil, T., (1999), "Salt--Important Throughout History," Mining Engineering, Vol. 51, p. 45.

Ong, V., J. Unrau, J. Jones, A. Coode, and D. Mackintosh, (1997), "Triaxial Creep Testing of Saskatchewan Potash Samples," Proceedings 4th Conference on the Mechanical Behavior of Salt, Ecole Polytechnique, Montreal, June 1996, Trans Tech Publications, Clausthal-Zellerfeld, Germany, pp. 469-480.

Paraschiv, I., and N. Cristescu, (1993), "Deformability Response of Rock Salt Around Circular Mining Excavations," Rev. Roum. Sc. Tech. - Mécanique Appliquée, Vol. 38, No. 3, pp. 257-276.

Paraschiv-Munteanu, I., (1997), <u>Metode Numerice în Geomecanica</u>, Ph.D. Dissertation, University of Bucharest, Bucharest, Romania.

Paraschiv-Munteanu, I., (1998), "A Numerical Analysis of Behaviour of the Elastoviscoplastic Rock Around Vertical Boreholes," Analele Univ. Bucuresti. Mat., Vol. XLVII, pp. 83-92.

Peach, C. J., and C. J. Spiers, (1996), "Influence of Crystal Plastic Deformation on Dilatancy and Permeability Development in Synthetic Salt Rock," Tectkonophysics, Vol. 256, pp. 101-128.

Peach, C. J., R. H. Brzesowsky, P. M. T. M. Schutjens, J. L. Liezenberg, and H. J. Zwart, (1989), "Long-term Rheological and Transport Properties of Dry and Wet Salt Rocks," Nuclear Science and Technology, EUR 11848 EN. Official Publications, European Communities, Luxembourg.

Perami, R., C. Caleffi, M. Espagne, and W. Prince, (1993), "Fluage et microfissuration dans les stockages souterrains," <u>Geoconfine '93</u>, Arnould, Barrès et Côme (Eds.), Balkema, Rotterdam, pp. 99-104.

Pfeifle, T. W., (1995), "WIPP Crushed Salt Database for Constitutive Model Evaluations," Calculation No. 325/04/03, (RE/SPEC Inc., Rapid City, SD), WIPP Central Files, WPO#36823, Sandia National Laboratories, Albuquerque, NM.

Pfeifle, T. W., F. D. Hansen, and M. K. Knowles, (1996), "Salt-Saturated Concrete–Strength and Permeability," <u>Proceedings 4th Materials Engineering Conference</u>, ASCE Materials Engineering Division, Washington, DC, November 1996.

Pfeifle, T. W., K. D. Mellegard, and D. E. Munson, (1992), "Determination of Probability Density Functions for Parameters in the Munson-Dawson Model for Creep Behavior of Salt," Probabilistic Methods in Geomechanics, AMD, Vol. 134, (A. F. Fossum, ed.), ASME, New York, pp. 41-56.

Pfeifle, T. W., T. J. Vogt, and G. A. Brekken, (1997), "Correlation Analysis of Dome Salt Characteristics," <u>Proceedings 4th Conference on the Mechanical Behavior of Salt</u>, Ecole Polytechnique, Montreal, June 1996, Trans Tech Publications, Clausthal-Zellerfeld, Germany, pp. 87-100.

Pilvin, P., (1988), <u>Approches multiéchelles pour la prévision du comportement anélastique des métaux</u>, Ph.D. Thesis, Université Paris VI, Paris, France.

Popp, T., and H. Kern, (1988), "Ultrasonic Wave Velocities, Gas Permeability and Porosity in Natural and Granular Rock Salt," Phys. Chem. Earth, Vol. 23, No. 3, pp. 373-378.

Possum, A. F., (1977), "Visco-Plastic Behaviour During Excavation Phase of a Salt Cavity," Int. J. Num. Anal. Met. Geomech., Vol. 1, pp. 45-55.

Pouya, A., (1993), "Correlation Between Mechanical Behavior and Petrological Properties of Rock Salt," <u>Proceedings 32nd U.S. Symposium Rock Mechanics</u>, pp. 385-392.

Pudewills, A., (1997), "Influence of Anhydrite Strata on a Waste Disposal Drift," <u>Proceedings 4th Conference on the Mechanical Behavior of Salt</u>, Ecole Polytechnique, Montreal, June 1996, Trans Tech Publications, Clausthal-Zellerfeld, Germany, pp. 551-560.

Raj, R., (1982), "Creep in Polycrystalline Aggregates by Matter Transport Through a Liquid Phase," Journal Geophysical Research, Vol. 87, No. B6, pp. 4731-4739.

Ratigan, J. L., L. L. Van Sambeek, K. I. DeVries, and J. D. Nieland, (1991), The Influence of Seal Design on the Development of the Disturbed Rock Zone in the WIPP Alcove Seal Tests, RSI-0400, RE/SPEC Inc., Rapid City, South Dakota.

Roatesi, S., (1997), Numerical Methods in Mathematical Modelling of Mining Supports, Ph.D. Dissertation, University of Bucharest, Bucharest, Romania.

Roatesi, S., (1998), "A Finite Element Model for a Lined Tunnel in a Viscoplastic Rock Mass", Proceedings 3rd EUROMECH Solid Mech. Conference, Stockholm, pp. 170-171.

Roatesi, S., and A. H. C. Chan, (1998), "FEM for Underground Openings Problems in Viscoplastic Rock Mass", Rev. Roum. Sci. Techn.- Mec. Appl., Vol. 43, p. 2.

Robin, P.-Y. F., (1978), "Pressure-Solution at Grain-to-Grain Contacts," Geochimica et Cosmochimica Acta, Vol. 42, pp. 1383-1389.

Rutter, E. H., (1976), "The Kinetics of Rock Deformation by Pressure-Solution," Philosophical Transactions Royal Society London, Vol. 283, pp. 203-219.

Rutter, E. H., (1983), "Pressure Solution in Nature, Theory and Experiment," Journal Geological Society, London, Vol. 140, pp. 725-740.

Salzer, K., and W. Schreiner, (1977), "Long-Term Safety of Salt Mines in Flat Beddings," Proceedings Fourth Conference on the Mechanical Behavior of Salt, Ecole Polytechnique, Montreal, June 1995, Trans Tech Publications, Clausthal-Zellerfeld, Germany, pp. 481-494.

Saulnier, G. J., Jr., and J. D. Avis, (1988), Interpretation of Hydraulic Tests Conducted in the Waste-Handling Shaft at the Waste Isolation Pilot Plant (WIPP) Site, SAND88-7001, Sandia National Laboratories, Albuquerque, NM.

Senseny, P. E., (1986), Triaxial Compression Creep Tests on Salt From the Waste Isolation Pilot Plant, SAND85-7261, (RE/SPEC Inc., Rapid City, SD), Sandia National Laboratories, Albuquerque, NM.

Senseny, P. E., (1990), Creep of Salt From the ERDA-9 Borehole and the WIPP Workings, SAND89-7098, (RE/SPEC Inc., Rapid City, SD), Sandia National Laboratories, Albuquerque, NM.

Senseny, P. E., and A. F. Fossum, (1997), "Testing to Estimate the Munson-Dawson Parameters," Proceedings 4th Conference on the Mechanical Behavior of Salt, Ecole Polytechnic, Montreal, June 1996, Trans Tech Publications, Clausthal-Zellerfeld, Germany, pp. 263-276.

Senseny, P. E., T. W. Pfeifle, and K. D. Mellegard, (1985), Constitutive Parameters for Salt and Nonsalt Rocks From the Detten, G. Friemel, and Zeeck Wells in the Palo Duro Basin, Texas, BWI/ONWI-549, (RE/SPEC Inc.), Office of Nuclear Waste Isolation, Battelle Memorial Institute, Columbus, OH.

Senseny, P. E., T. W. Pfeifle, and K. D. Mellegard, (1986), Exponential Time Constitutive Law for Palo Duro Salt from J. Friemel No. 1 Well, BMI-ONWI-595, (RE-SPEC Inc.), Office of Nuclear Waste Isolation, Battelle Memorial Institute, Columbus, OH.

Skrotzki, W., H.-J. Dornbusch, K. Helming, R. Tamm and H.-G. Brokmeier, (1997), "Development of Microstructure and Texture in Pure Deformed Salt," Proceedings 4th Conference on the Mechanical Behavior of Salt, Ecole Polytechnic, Montreal, June 1996, Trans Tech Publications, Clausthal-Zellerfeld, Germany, pp. 101-114.

Spiers, C. J., and N. L. Carter, (1997), "Microphysics of Rocksalt Flow in Nature," Proceedings 4th Conference on the Mechanical Behavior of Salt, Ecole Polytechnic, Montreal, June 1996, Trans Tech Publications, Clausthal-Zellerfeld, Germany, pp. 115-128.

Sprackling, M. T., (1973), "Dislocation Mobility and Work Hardening in Sodium Chloride," Phil Mag, Vol. 27, pp 265-271.

Stamatiu, M., (1959), The Problem of Designing the Pillars in the Salt Mines From Romania (in Romanian), Editura Academiei Române, Bucharest, Romania.

Stead, D., Z. Szczepanik, and W. Gaskin, (1997), "Acoustic Characterization of Potash," Proceedings Fourth Conference on the Mechanical Behavior of Salt, Ecole Polytechnic, Montreal, June 1996, Trans Tech Publications, Clausthal-Zellerfeld, Germany, pp. 31-45.

Stührenberg, D., and C. L. Zhang, (1995), "Results of Experiments on the Compaction and Permeability Behavior of Crushed Salt," Proceedings 5th Int. Conf. on Radioactive Waste Management and Environmental Remediation, Berlin, 1995, (ICEM '95), ASME, New York, Vol. 1, pp. 797-801.

Tada, R., and R. Siever, (1986), "Experimental Knife-Edge Pressure-Solution of Halite," Geochimica et Cosmochimica Acta, Vol. 50, pp. 29-36.

Talbot, C. J., (1979), "Fold Trains in a Glacier of Salt in Southern Iran," Journal Structural Geology, Vol. 1, pp. 5-18.

Talbot, C. J., (1993), "Spreading of Salt Structures in the Gulf of Mexico," Tectonophysics, Vol. 228, pp. 151-166.

Talbot, C. J., (1998), "Extrusions of Hormuz Salt in Iran," Blumdell, D. J. and Scott, A. C., editors, Lyell: the Past is the Key to the Present, Geological Society London, Special Publications, Vol. 143, pp. 315-334.

Talbot, C. J., and A. Jackson, (1987), "Internal Kinematics of Salt Diapirs," Am. Assoc. Petr. Geol. Bull., Vol. 71, No. 9, pp. 1068-1093.

Talbot, C. J. and E. A. Rogers, (1980), "Seasonal Movements in a Salt Glacier in Iran," Science, Washington, Vol. 208, pp. 395-397.

Thorel, L., M. Ghoreychi, Ph. Cosemza, and S. Chanchole, (1997), "Rocksalt Damage and Failure Under Dry or Wet Conditions," Proceedings 4th Conference on the Mechanical Behavior of Salt, Ecole Polytechnic, Montreal, June 1996, Trans Tech Publications, Clausthal-Zellerfeld, Germany, pp. 189-202.

Van Sambeek, L. L., (1997), "Salt Pillar Design Equation," Proceedings 4th Conference on the Mechanical Behavior of Salt, Ecole Polytechnic, Montreal, June 1996, Trans Tech Publications, Clausthal-Zellerfeld, Germany, pp. 495-508.

Vogler, S., and W. Blum, (1990), "Micromechanical Modelling of Creep in Terms of the Composite Model," Wilshire, B. and R. W. Evans (Eds), Proceedings 4th Int. Conf. on Creep and Fracture of Engineering Materials and Structures, Institute of Materials, London, pp. 65-79.

Wawersik, W. R., and C. M. Stone, (1989), "A Characterization of Pressure Records in Inelastic Rock Demonstrated by Hydraulic Fracturing Measurements in Salt," Int. J. Rock Mech. Min. Sci. & Geomech. Abstr., Vol. 26, pp. 613-627.

Wawersik, W. R., and D. H. Zeuch, (1986), "Modeling and Mechanistic Interpretation of Creep of Rocksalt Below 200°C., Tectonophysics, Vol. 121, pp. 125-152.

Weidinger, P., (1998), Verformungsverhalten natürlicher Steinsalze: Experimentelle Ermittlung und mikrostrukturell begründete Modellierung, Dissertation, Universiät Erlangen-Nürnberg, 164 pp.

Weidinger, P., A. Hampel, W. Blum, and U. Hunsche, (1997), "Creep Behaviour of Natural Rock Salt and its Description With the Composite Model," Proceedings 11th Int. Conf. on the Strength of Materials (ICSMA-11), Prague, 1997, pp. 646-648.

Wu, S., A. W. Bally, and C. Cramez, (1990), "Alochthonous Salt, Structure and Stratigraphy of the North-Eastern Gulf of Mexico, Part II, Structures," Marine & Petroleum Geology, Vol. 7, pp. 334-370.

Zirngast, M., (1996), "The Development of the Gorleben Salt Dome (Northwest Germany) Based on Qualitative Analysis of Peripheral Sinks," Geological Society London Special Publication, Vol. 100, pp. 203-226.

Basic and Applied Salt Mechanics, Cristescu, Hardy, Jr & Simionescu (eds)
© 2002 Swets & Zeitlinger, Lisse, ISBN 90 5809 383 2

Conference participants

Victor Arad

Petrosani University
20 University Street
Petrosani HD 2675
ROMANIA

Ilie Banuta

SALROM
The Salt Mine of Târgu Ocna
Târgu Ocna
ROMANIA

Liviu Barbu

The Mining Computing Center,
Str. N. Titulescu, nr.4
Cluj-Napoca
ROMANIA

Pierre Berest

Laboratoire de Mécanique des Solides
Ecole Polytechnique - LMS
Palaiseau
FRANCE 91128

Sorin Berchimis

S.C. IPROMIN S.A.
Mendeleev St., no. 36-38
Bucharest
ROMANIA

P. A. Blum

Institut Physique du Globe de Paris
Paris
FRANCE Cedex 75272

Virgil Breaban

Ovidius University, Civil Eng. Dept.
Bd. Mamaia 124
Constanta
ROMANIA 8700

E. Broch

Norwegian University of Sci. & Technology
Trodheim
NORWAY

G. D. Callahan

RESPEC Inc., P.O. Box 725
Rapid City, SD
USA 57709

Galina Camenscki

University of Bucharest
Str. Academiei 14
Bucharest
ROMANIA RO 70109

Antonio J. Campos de Orellano

Consulting Mining Engineer
Av. de los Reyes Catolicos, 23
El Escorial, 28280
SPAIN

Radu A. Canarache

Inicad Soft
Str. Popa Tatu 20, Sect. 1
Bucharest,
ROMANIA 70772

J. P. Carpentier

Laboratoire de Mecanique des Solides
Ecole Polytechnique, LMS
Palaiseau
FRANCE 91128

Lokesh Chaturvedi

Environmental Evaluation Group
7007 Wyoming Blvd NE, #Suite F2
Albuquerque NM
USA 87109

Sanda Cleja-Tigoiu

Faculty of Mathematics
University of Bucharest
Str. Academiei 14
Bucharest
ROMANIA 70109

Philippe Cosenza

Ecole Polytechnique
Université Paris VI
UMR 7619 Sisyphe
Paris
FRANCE

Nicolae Cristescu

University of Florida
Mechanics & Engineering Science Dept.
Gainesville, FL
USA 32611-6250

Adrian Dadu

The Mining Computing Center
Cluj-Napoca
Str. Titulescu, nr. 4
ROMANIA 3400

Mircea David

SALROM
The Salt Mine of Râmnicu-Vâlcea
Râmnicu-Vâlcea
ROMANIA

Gyorgy Deak

Minesa S.A.
Str. T. Vladimirescu 15-17
Cluj-Napoca
ROMANIA 3400

Stefania Deak

Minesa S.A.,
Str. T. Vladimirescu 15-17
Cluj-Napoca
ROMANIA 3400

Justus M. de Jong

Salt 2000
7550 GC Hengelo OV
THE NETHERLANDS

Lucia Diaconu-Muresan

The Mining Computing Center
Str. N. Titulescu nr. 4,
Cluj-Napoca
ROMANIA 3400

Liviu Drăgănescu

SALROM
Salt Mine of Slanic-Prahova
Str. Cuza Voda 21
Slanic Prahoga
ROMANIA

Maurice B. Dusseault

Dept. of Civil Engineering & Earth Science
University of Waterloo
Waterloo, Ontario
CANADA N2L-3G1

Steve Finley	General Chemical Corporation P.O. Box 551 Green River, WY USA 82935
Peter A. Fokker	Nedmag Ind. Mining Manufacture PO Box 241, NI 9640 AE Veendau THE NETHERLANDS
Mark A. Frayne	RESPEC Inc. Grand Bend, Ontario CANADA
Vasile Gaiducov	University of Petrosani Str. Universitātii 20 Petrosani HD ROMANIA 2675
Vasile Georgiu	SALROM Salt Mine of Ocna-Dej Ocna-Dej ROMANIA
Mehdi Ghoreychi	Groupement pour l'étude des Structures Souterraines de Stokage Ecole Polytechnique Palaiseau FRANCE 91128
Ralf-Michael Günther	Institut für Gebrigsmechanik Friederikenstr. 60 Leipzig 04279 GERMANY
Andreas Hampel	Development & Application of Constutive Equations Kleiststrasse 32 Isernhagen GERMANY 30916
Francis D. Hansen	Sandia National Laboratories, MS 1395 1115 N. Main Carlsbad, NM USA 88220
H. Reginald Hardy, Jr.	Dept. of Energy & Geo-Environ. Eng. The Pennsylvania State University University Park, PA USA 16802

A. Hausdorf

Freiberg University of Mining & Tech.
Freiberg
GERMANY

F. Heinrich

Freiberg University of Mining & Tech.
Freiberg
GERMANY

Cornel Hirian

University of Petrosani
Str. Universitatii 20
Petrosani HD
ROMANIA 2675

Zheng Meng Hou

University of Technology
Erzstrasse 20
Clausthal-Zellerfeld
GERMANY D-38678

Udo Hunsche

Federal Institute for Geosciences
   & Natural Resources
Stilleweg 2
Hannover
GERMANY D-30655

Diane L. Hurtado

Sandia National Laboratories
Repository Technology
PO Box 5800 - MS 1322
Albuquerque NM
USA 87185

Mariana Ionita

S.C. IPROMIN S.A.
Mendeleev St., no.36-38
Bucharest
ROMANIA

Georgi Janev

GEOSOL
Provadia
BULGARIA

Vijen Javeri

Gesellschaft fur Reaktorsicherheit (GRS)
Koln
GERMANY D-50667

H. Kern

Institut für Geowissenschaften
Universität Kiel
Olshausenstr. 40
Kiel
GERMANY D-24098

D. Khrushchov

Institut of Geological Sciences
55-b Gonchara str.
Kiev
UKRAINE 252054

Kathy Knowles

Sandia National Laboraties
115 N. Main - MS 1395
Carlsbad NM
USA 88220

Ekhard Korthaus

Forschungszentrum Karlsruhe
Institut für Nukleare Entsorgungstechnik
Postfach 3640, Krlsruhe
GERMANY D-76021

Francisc Laszloffi

SALROM
Salt Mine of Ocna-Dej
Dej CLJ
ROMANIA

A. Lindert

Institut für Gebirgsmechanik
Friederikenstr. 60
Leipzig 04279
GERMANY

Ming Lu

SINTEF Civil & Environmental Eng.
Rock and Mineral Engineering
Trondheim
NORWAY N-7034

Francisc Lukas

SALROM
The Salt Mine of Ocna-Dej
Dej CLJ
ROMANIA

K.-H. Lux

University of Technology
Erzstrasse 20
Clausthal-Zellerfeld
GERMANY D-38678

Ioan Macovicius

The Mining Computing Center
Str. N. Titulescu nr. 4,
Cluj-Napoca
ROMANIA 3400

Hamid Maleki

Maleki Technologies, Inc.
5608 South Magnolia
Spokane WA
USA 99223

Maria Marcu

SALROM
Salt Mine of Ocna Mures
Ocna Mures
ROMANIA

James Martin

Morton Salt
Mortin International, Inc.
100 N. Riverside Plaza
Chicago, IL
USA 60606-1597

Cristian Marunteanu

University of Bucharest
Traian Vuia str. No.6
Bucharest
ROMANIA

Doina Massier

University of Bucharest
Str. Academiei 14
Bucharest
ROMANIA RO-70109

Ervin R. Medvés

S.C. IPROMIN S.A.
Mendeleev St.no. 36-38
Bucharest
ROMANIA RO-71104

Kirby D. Mellegard

RESPEC, Inc.
P.O. Box 725
Rapid City, SD
USA 57709-0725

Joel D. Miller

WIPP, Sandia National Laboratories
MTS, P.O. Box 5800
Albuquerque NM
USA 87185-0779

Sorin Mogos

University of Bucharest
Traian Vuia St.no. 6
Bucharest
ROMANIA

Dennis Z. Mraz

Mraz Project Consultants, Ltd.
Saskatoon
CANADA Saskatchewan

Mariana Nicolae

Ministry of Finance,
General Division for Forecasts
12 Libertatii Blvd.
Bucharest
ROMANIA

Victor Niculescu

University of Bucharest
Traian Vuia St., no.6
Bucharest
ROMANIA

Iuliana Paraschiv-Munteanu

University of Bucharest
Str. Academiei 14
Bucharest
ROMANIA RO-70109

Michael Patton

General Chemical Corporation
PO Box 551
Green River WY
USA 82935

W. M. Phillips

Dean of Graduate School
University of Florida
Gainesville, FL
USA 32611-6250

T. Popp

Institut für Geowissenshaften
Universitaet Kiel, Olshausenstr. 40
Kiel
GERMANY D-24098

Alexandra Pudewills

Forschungszentrum Karlsruhe
Inst. für Nucleare Entsorgungstechnik
Postfach 3640, Karlsruhe
GERMANY D-76021

Petru Reisz

SALROM Bucharest
Calea Victoriei, No. 220
Bucharest
ROMANIA RO 71104

Simona Roatesi

Department of Mechanics
University of Bucharest
Str. Academiei 14, Bucharest
ROMANIA RO 70109

Ioan Rosca

University of Bucharest
Str. Academiei 14
Bucharest
ROMANIA RO 70109

Leo Rothenburg

Dept. of Civil Engineering & Earth Sci.
University of Waterloo
Waterloo, Ontario
CANADA N2L 3G1

T. Rothfuches

Gesellschaften für Anlagen-und
Reaktorsicherheit
Postfach 2126
Braunschweig
GERMANY D-38011

Ioan Salomia

SALROM
Mining Company of Ramnicu Valcea
Ramnicu Valcea
ROMANIA

Klaus Salzer

IfG Institut für Gebirgsmechanik
Friederikenstr. 60, Leipzig
GERMANY 04279

Nicolae Sandru

University of Bucharest
Str. Academiei 14
Bucharest
ROMANIA RO 70109

Wolfgang Schreiner

Institut für Gebirgsmechanik
Friederikenstr. 60, Leipzig
GERMANY 04279

Otto Schulze

Federal Institute for Geosciences and
Natural Resources
Stilleweg 2
Hannover
GERMANY D-30655

S. Shehunova

Institute of Geological Sciences
55-b Gonchara str.
Kiev
UKRAINE 252054

Olivian Simionescu

University of Bucharest
str. Academiei 14
Bucharest
ROMANIA RO-70109

441

Daniel Stănescu

The Mining Computing Center
Str. N. Titulescu nr. 4,
Cluj-Napoca
ROMANIA 3400

Collin Stewart

General Chemical Corporation
PO Box 551
Green River WY
USA 82935

Peshna Stoeva

University of Mining and Geology
Sofia
BULGARIA 1700

Cristian Stoicescu

SALROM Bucharest
Calea Victoriei, No.220
Bucharest
ROMANIA RO-7114

Cristopher Talbot

Department of Earth Sciences
Uppsala University
36 Uppsala
SWEDEN SE-752

Mihaela Toderas

University of Petrosani
Str. Universitatii nr.20
Petrosani, HUN
ROMANIA 2675

Adrian Todorescu

University of Petrosani
Str. Universitatii nr.20
Petrosani, HUN
ROMANIA 2675

Georgi Valev

University of Architecture,
Civil Engineering and Geodesy
1 Chr. Smirnenski Blvd.
Sofia 1421
BULGARIA

Leo L. Van Sambeek

RESPEC Inc.
Rapid City SD
USA 57709-0725

Hartmut von Tryller

SOCON Sonar Control
Schachtstrasse 3b
Giessen
GERMANY D-31180

Alan Williams   Cleveland Potash Ltd.
                                 Boulby Mine, Loftus
                                 Saltbutn, Cleveland
                                 UNITED KINGDOM TS13 4UZ

    ****
              ****

CONFERENCE PARTICIPANTS

Fifth Conference on the Mechanical Behavior of Salt

University of Bucharest

August 9 -11, 1999

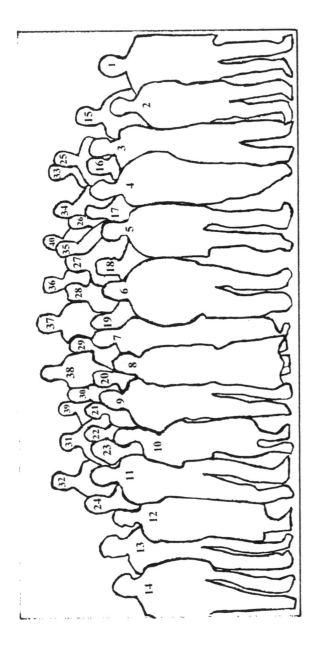

1. M. Cristescu, 2. Ming Lu, 3. O. Simionescu, 4. Otto Schulze, 5. P. Fokker, 6. H. R. Hardy,
7. Frank Hansen, 8. Kathy Knowles, 9. Udo Hunsche, 10. S. Cleja-Tigoiu,
11. Andreas Hampel, 12. Lucia Diaconu, 13. L. Drăgănescu, 14. Z. Hou, 15. C. Talbot,
16. Ervin Mendvés, 17. N. Sandru, 18. D. Khrushchov, 19. A. Pudewills, 20. D. Massier,
21. M. Nicolae, 22. I. Paraschiv-Munteanu, 23. S. Roatesi, 24. Daniel Stănescu,
25. A. Williams, 26. Mehdi Ghoreychi, 27. K. Salzer, 28. Von Tryller, 29. V. Javeri,
30. xxxxxxxx, 31. E. Korthaus, 32. T. Popp, 33. P. Berest, 34. J. P. Charpentier,
35. A. Reitze, 36. xxxxxxxx, 37. I. Rosca, 38. B. Vernescu, 39. G. Camenschki, 40. xxxxxxxx

International distribution of conference participants

A total of 94 participants from 12 countries contributed to the conference activities. The various countries and the associated participants are listed below, bracketed quantities (xx) indicate total from each country.

Bulgaria (3)
G. Janev
P. Stoeva
G. Valev
Canada (4)
M. B. Dusseault
M. A. Frayne
D. Z. Mraz
L. Rothenburg
France (5)
P. Berest
P. A. Blum
J. P. Carpentier
P. Cosenza
M. Ghoreychi
Germany(18)
R-M. Günther
A. Hampel
A. Hausdorf
F. Heinrich
Z. M. Hou
U. Hunsche
V. Javeri
H. Kern
E. Korthaus
A. Lindert
K.-H. Lux
T. Popp
A. Pudewills
T. Rothfuches
K. Salzer
W. Schreiner
O. Schulze
H. von Tryller
Netherlands (2)
J. M. de Jong
P. A. Folker

Norway (2)
E. Broch
M. Liu
Romania (39)
V. Arad
I. Banuta
L. Barbu
S. Berchimis
V. Breaban
G. Camenscki
R.A. Canarache
S. Cleja-Tigoiu
A. Dadu
M. David
G. Deak
S. Deak
L. Diaconu-Muresan
L. Drăgănescu
V. Gaiducov
V. Georgiu
C. Hirian
M. Ionita
F. Laszloffi
F. Lukas
I. Macovicius
M. Marcu
C. Marunteanu
D. Massier
E. R. Medvés
S. Mogos
M. Nicolae
V.Niculescu
I. Paraschiv-Munteanu
P. Reisz

Romania (continued)
S. Roatesi
I. Rosca
I. Salomia
N. Sandru
O. Simionescu
D. Stănescu
C. Stoicescu
M. Toderas
A. Todorescu
Spain (1)
A. J. Campos de Orellano
Sweden (1)
C. Talbot
U.S.A. (16)
G. D. Callahan
L. Chaturvedi
N. Cristescu
S. Finley
F. D. Hansen
H. R. Hardy
D. L. Hurtado
K. Knowles
H. Maleki
J. Martin
K. D. Mellegard
J. D. Miller
M. Patton
W. M. Phillips
C. Stewart
L. L. Van Sambeek
Ukraine (2)
D. Khrushchov
S. Shehunova
United Kingdom (1)
A. Williams

Author index

This index includes a listing of the authors of the papers in the current proceedings (underlined entries) as well as the authors of the literature referenced in these papers which were considered to be of direct interest to persons involved in the study of the mechanical behavior of salt. In some cases a particular author's name may be invisible on the indicated text page since it is associated with an "et al." listing. In such cases the author's name will be found in the reference section of the paper.

****

Kenter, C.J. 95, 96
Kern, H. 95, 98, 99, 109, 111, 145, 437
Khrushchov, D. 313, 438
Kiersten, P. 97, 112, 113
Kim, R.Y. 4, 6
Knecht, M. 83
Knowles, K. 438
Kopenetz, L. 399
Korthaus, E. 257, 258, 259, 263, 268,
   344, 348, 350, 438
Kranz, R.L. 63

L

Ladanyi, B. 172
Landes, K.K. 389
Langer, M. 73, 154
Laszloffi, F. 367, 438
Le Cleac'h, J.M. 73, 84
Lemaitre, J. 152, 159, 178
Lemp, Ch. 63
Levykin, A.I. 145
Lindert, A. 317, 438
Lister, G.S. 68, 127
Loken, M.C. 240, 242
Lu, M. 317, 318, 322, 327, 438
Lüdeling, R. 83
Lukacs, F. 385
Lukas, F. 438
Lux, K.-H. 151, 154, 156, 159, 161,
   163, 165-167, 357, 359, 438

M

Macoviciuc, I. 367, 399, 438
Maleki, H. 329, 330, 332-334, 379, 381,
   438
Marcu, M. 397, 439
Marsily, B. 59
Martin, J. 439
Marunteanu, C. 89, 385, 439
Massier, D. 209, 212, 273, 439
Matei, A. 146
Mc Tigue, D.F. 66
Medvedev, S. 37
Medvés, E.R. 23, 29, 89, 223, 385, 439
Mellegard, K.D. 172, 173, 242, 245,
   253, 439
Menzel, W. 178, 187
Meredith, P.G. 100, 101
Miller, J.D. 337, 439
Mingerzahn, G. 194

Minkley, W. 152, 187
Mogos, S. 89, 385, 439
Mönig, J. 113
Mraz, D.Z. 171, 175, 389, 439
Mrugala, M. 172
Munson, D.E. 73, 80, 127, 156, 161,
   244, 245, 330, 332, 338
Murrel, S.A.F. 100, 101

N

Natau, D. 63
Nawrocki, H. 146
Nesterov, M.P. 417
Neuber, H. 209, 212
Nguyen, Q.S. 232
Niandou, H. 146
Nicolae, M. 281, 440
Niculescu, V. 89, 385, 440
Nieland, J.D. 240

O

Obert, L. 273
Oden, J.T. 399
Olariu, V. 399

P

Panet, M. 380
Paraschiv-Munteanu, I. 273, 279, 440
Passaris, E.K.S. 73
Patton, M. 379, 440
Peach, C.J. 66, 73, 84, 95, 98
Pérami, R. 69
Pfeifle, T.W. 172, 173, 240, 242
Pharr, G.M. 73, 80, 83
Phillips, W.M. 440
Piper, T.B. 389
Plischke, I. 83
Ploumen, P. 342
Popp, T. 95, 98, 99, 109, 111, 145, 440
Pouya, A. 135, 140, 175
Pratt, P.L. 65
Press, W.H. 198
Prince, W. 69
Pudewills, A. 341, 342, 343, 440

R

Raisz, P. 385
Rao, S.S. 399
Ratigan, J. 84
Reisz, P. 440

Basic and Applied Salt Mechanics, Cristescu, Hardy, Jr & Simionescu (eds)
© 2002 Swets & Zeitlinger, Lisse, ISBN 90 5809 383 2

Subject index

This index includes a listing of the major key words appearing in the Conference proceedings. In general the material presented has not been cross indexed.

****

representation, 209, 215, 216
Maxwell's displacement representation, 221
Mean
    failure strength, 114-116, 118
    glide velocity, 195
    stress, 64, 103, 114, 274
Mechanical
    behavior, 209, 357
        of salt, 3, 57
    parameters, 29
    stability, 29, 71, 110, 118, 122
Meteorological factors, 46
Micro
    fissures, 152
    fracturing, 68
Micro-acoustic emissions, 74, 102
Microcrack
    propagation, 63
    structure, 98
Microcracking, 144, 290
Microcracks, 66, 69, 73, 74, 83, 85, 104, 109, 118, 120, 143-145, 187, 189
Microfracture
    generation, 3
    propagation, 3
Microscopic activation energy, 195, 200, 207
Microstructure, 198
    evolution, 198, 199
Middle Miocene, 35
Milky salt, 85
Mines de Potasse d'Alsace, 71
Minimum
    principal stress, 112
    stress, 164, 167
        distribution, 185
Mining
    applications, 353
    conditions, 171, 175
    experience, 152
    subsidence, 385
    temperature conditions, 180
Miocene formation, 89
Mirovo salt deposit, 41-42

Mises stress, 323, 324
Model
    calculations, 84
    parameters, 199, 203
    predictions, 242, 284, 285
Model-predicted consolidation, 239
Modelling, 37
Modified composite model, 194-196, 198-199, 204, 206
Mohr-coulomb criteria, 64
Monoaxial compression tests, 31
Morera's representation, 222
Morsleben mine, 204
MTS programmable testing facility, 9
Multiple regression procedure, 332
Multi-step
    creep paths, 59
    jacketed relaxation tests, 67
    triaxial compressive creep tests, 78
    unjacketed creep tests, 68
Munson-Dawson
    creep law, 171
    model, 173, 174
    steady state model, 173

N

Natural
    brine, 80
    gas storage caverns, 311
    rock salt, 108, 198, 204
    salt structures, 35
Neuber's
    displacement representation, 220
    representation, 209, 212
Non-associated flow model, 157
Non-constant
    elastic parameters, 143
    stress, 220
Non-dilatant
    behavior, 98, 102
    deformation, 97, 109
Non-dislocation obstacles, 194
Non-elastic instantaneous response, 223
Non-inverse transient creep, 179
Nonlinear elastic, 209
Normal

rate, 333, 334
Room-and-pillar system, 183
Room-creation phase, 183
Rupture, 152

S

Safe containment, 240
Safety
  analysis, 357
  assessment of underground storage, 71
  pillar, 386
Safety-integrity criterion, 120
Salado formation, 253, 389
Saliniferous structures, 162
Dalnic saline, 89
Salt, 51
  creep, 54
  dissolution, 37
  dome, 110, 112, 313
  extrusion, 37, 38
  extrusion-rates, 35
  flow, 37
  formations, 57
    of Ukraine, 313
  kinematics, 36
  massifs, 23
  massive Slanic Prahova, 24
  matrix, 200
    impurities, 194
  mechanics, 35
  mine pillar design, 405
  minerals, 205
  myloniytes, 39
  pillar, 183, 360, 395
  pillar-design equations, 406, 410
  rock-mass, 129
  rocks, 23
  velocity, 37, 38
  viscosity, 38
  brine system, 63
  marl varieties, 51
  withdrawal basins, 35
Sample preparation, 95
Sandia National Laboratory, 330, 338
Sandwich brine field, 389

Schwartz-Cauchy inequality, 235
Secondary creep, 179, 180
  parameters, 178
  phase, 189
  support, 380
Seismic
  danger, 50
  measurements, 386
  occurrences, 42, 47
  travel times, 74
Selecting creep laws, 327
Shaft seal
  calculations, 250
  system, 239
  design, 253
Shear, 115
  consolidation, 247
    creep, 253
    tests, 240-242, 254
  modulus, 174, 196, 197
  stress, 158
Shearing stress criteria, 320
Short term failure, 103
Short-period incremental creep tests, 20
Short-term
  failure, 275
    surface, 186
  strength, 160, 162, 167, 168
Signal-to-noise ratio, 9
Sinkhole, 386
  formation, 389, 391
Site-specific constitutive models, 330
Situ data, 28
Skempton's coefficient, 66
Slanic Prahova
  deposit, 29, 31
  salt, 30, 32
    mines, 27, 30, 31, 223, 402
Slanic salt
  brine, 355
  deposit, 89
  massive, 30
Slanic syncline, 30
Slinger backfilling procedure, 268
Small-crystal salt, 51
Softening, 168

Young's modulus, 11, 66, 67, 181, 187, 344

Z

Zagros
  Halokinetics research programme, 36
  mountain range, 35
Zero-based incremental loading, 10
Zones of fracture, 293

9 789058 093837